Electrochemical Engineering
Hartmut Wendt and Gerhard Kreysa

Springer
*Berlin
Heidelberg
New York
Barcelona
Hong Kong
London
Milan
Paris
Singapore
Tokyo*

Hartmut Wendt and Gerhard Kreysa

Electrochemical Engineering

Science and Technology in Chemical and Other Industries

With 177 Figures and 45 Tables

Springer

Prof. Dr. Hartmut Wendt
Institut für Chemische Technologie
TU Darmstadt
Petersenstraße 20
D-64287 Darmstadt
Germany

Prof. Dr. Gerhard Kreysa
Karl Winnacker Institut
DECHEMA e. V.
Theodor-Heuss-Allee 25
D-60486 Frankfurt am Main
Germany

ISBN 3-540-64386-9 Springer-Verlag Berlin Heidelberg New York

Library of Congress Cataloging-in-Publication Data
Wendt, Hartmut, 1933–
Electrochemical engineering : science and technology in chemical and other industries / Hartmut Wendt, Gerhard Kreysa.
p. cm.
Includes bibliographical references.
ISBN 3-540-64386-9 (hardcover : alk. paper)
1. Electrochemistry, Industrial. I. Kreysa, Gerhard. II. Title.
TP255.W46 1999
660'.297--dc21

This work is subject to copyright. All rights are reserved, whether the whole part of the material is concerned, specifically the rights of translation, reprinting, reuse of illustrations, recitation, broadcasting, reproduction on microfilm or in any other way, and storage in data banks. Duplication of this publication or parts thereof is permitted only under the provisions of the German Copyright Law of September 9, 1965, in its current version, and permission for use must always be obtained from Springer-Verlag. Violations are liable for prosecution under the German Copyright Law.

© Springer-Verlag Berlin Heidelberg 1999
Printed in Germany

The use of general descriptive names, registered names, trademarks, etc. in this publication does not imply, even in the absence of a specific statement, that such names are exempt from the relevant protective laws and regulations and therefore free for general use.

Typesetting: MEDIO, Berlin
Coverdesign: Design & Production, Heidelberg

SPIN: 10675807   2/3020 - 5 4 3 2 1 0 – Printed on acid-free paper.

Preface

Electrochemical Engineering sounds very much like chemical engineering, but the chemists, electrochemists, material scientists and whoever else comes into touch with technical electrochemical systems very soon gets the feeling, that chemical engineering wisdom will not get them very far in enhancing their understanding and helping them to solve their problems with technical electrochemical devices. Indeed not only the appearance of but also the physics and physical chemistry in electrochemical reactors – electrolyzers, batteries or fuel cells and others – are quite different from that of normal chemical reactors. Next to interfacial charge transfer and current density distributions is the relatively high importance of mass transfer and its hindrance in liquid electrolytes which distinguishes electrolyzers from chemical reactors. Therefore electrochemical engineering science became a science branch which at first developed with little reference to chemical engineering treating the relevant topics on a high mathematical level. This has led to a certain perfection, which today – in principle – allows us to model almost any desired electrolyzer or cell configuration with numerical methods to a degree and precision which satisfies the highest demands. This is classical chemical engineering stuff, which, however, neglects the chemical side of electrochemical technology. Therefore the present authors decided to write a book, which adds to these fundamentals (presented in chapters 1 through 8 and which can be found in almost every book on electrochemical engineering published during the last fifteen years) three more chapters which also cover the more chemical-oriented and technology oriented side of electrochemical engineering science and technology and an additional twelvth chapter on fuel cells. This was also the reason to include in the chapter on electrochemical kinetics a paragraph on electrocatalysis, in the chapter on electrochemical cell and plant engineering some paragraphs on materials, corrosion and materials engineering and why they wrote in chapter 7 a paragraph on utilization on porous electrocatalysts which refers to the same problem in chemical reaction engineering. They also gave examples in chapter 6 – Electrochemical Reaction Engineering – about the coupling of electrolyzers to chemical reactors for the case of chlorate formation.

In this context they believe that chapters 10 and 11 – Industrial Electrodes and Industrial Processes – are very important parts of the book and they also cover in chapter 11 – Industrial Processes – the non-electrochemical steps of the processes with their respective chemical engineering implications. The authors want to stress that electrochemical engineering is not confined to the relatively few electrochemical processes in the chemical process industries, but that electrochemical engineers are increasingly earning much more money in the metallurgical, galvanic and surface treatment business and also the electronic industries than in CPI. Therefore they extended their description of processes to these topics and also to environmental pollution control.

Last but not least they chose to write chapter 12 on the emerging fuel cell technology, a technology, where chemical and electrochemical engineering are more closely associated with each other than in any other electrochemical process.

H. Wendt, who composed the text in the joint manuscript of the two authors, is very much obliged to his secretary, Mrs. Antje Pappenhagen, who so patiently, competently and cleverly mastered the computer and to Mr. Böttiger, who with great skill made all the drawings which appear in this book and whose quality is therefore not impaired by the limited abilities of computer programs.

Darmstadt, December 1998							H. Wendt

Contents

Chapter 1
The Scope and History of Electrochemical Engineering

1.1 Carl Wagner and the Beginning of Electrochemical Engineering Science 1
1.2 Electrochemistry and Electrochemical Engineering Science 2
1.3 Electrochemical Engineering Science and Technology Since the Mid-1960s 3
1.4 What Means Electrochemical Engineering Science and Technology Today? 5
References 7
Further Reading 7

Chapter 2
Basic Principles and Laws in Electrochemistry

2.1 Stoichiometry of Electrochemical Reactions 8
2.2 Faraday's Law 10
2.3 Production Rates and Current Densities 11
2.4 Ohm's Law and Electrolyte Conductivities 12
2.5 Parallel Circuits and Cells with Electrolytic Bypass and Kirchhoff's Rules 14
Further Reading. 16

Chapter 3
Electrochemical Thermodynamics

3.1 Equilibrium Cell Potential and Gibbs Energy 17
3.2 Electrode Potentials, Reference Electrodes, Voltage Series, Redox Schemes 21
3.3 Reaction Enthalpy, Reaction Entropy, Thermoneutral Cell Voltage and Heat Balances of Electrochemical Reactions 28
3.4 Heat Balances of Electrochemical Processes 29

3.5	Retrieval of Thermodynamic Data and Activity Coefficients		31
3.6	Thermodynamics of Electrosorption.		35
References			37

Chapter 4
Electrode Kinetics and Electrocatalysis

4.1	The Electrochemical Double Layer		39
4.2	Kinetics of Interfacial Charge Transfer		41
4.3	Electrode Kinetics of Multielectron Charge Transfer Reactions.		45
4.4	Thermal Activation and Activation Energies of Electrochemical Reactions		49
4.5	Electrochemical Reaction Orders		49
4.6	Current Density/Potential Correlations for Different Limiting Conditions		51
	4.6.1	Micro- and Macrokinetics of Electrochemical Reactions.	51
	4.6.2	Mass Transfer Controlled Current Potential Curves	52
	4.6.2.1	Reaction Controlled Current Voltage Curves	54
	4.6.3	Charge Transfer Controlled Current Voltage Correlation.	55
	4.6.4	Combined Activation and Mass Transport Control	56
4.7	Reaction Controlled Current Voltage Curves		57
	4.7.1	Introductory Remarks	57
	4.7.2	Fast Preceding Reaction of an Electroactive Minority Species	58
	4.7.3	Fast Consecutive Reactions	60
4.8	Electrocatalysis		61
	4.8.1	Principles of Electrocatalysis	61
	4.8.2	Heterogeneous Electrocatalysis in Cathodic Evolution and Anodic Oxidation of Hydrogen	61
	4.8.2.1	The Volcano Curve	62
	4.8.3	Electrocatalysis in Anodic Oxygen Evolution and Cathodic Oxygen Reduction	64
	4.8.4	Redox Catalysis	66
4.9	Catalyst Morphology and Utilisation		68
	4.9.1	Structural Features and Catalyst Morphology of Electrocatalysts for Gas Evolving and Gas Consuming Electrodes	68
	4.9.2	Utilisation of Porous Electrocatalyst Particles.	69
4.10	Electrocatalysis in Electroorganic Synthesis.		71
	4.10.1	Introduction into the Field of Electroorganic Synthesis.	71
	4.10.1.1	Mediated Electrochemical Conversions of Organic Substrates	71

	4.10.1.2	Direct Anodic and Cathodic Electrochemical Conversions of Organic Substrates	72
	4.10.2	Electrocatalytic Oxidations by Oxides of Multiply-Valent Metals	72
	4.10.2.1	The Heterogeneously Catalysed Benzene Oxidation at Pb/PbO$_2$ Electrodes in Sulfuric Acid	74
	4.10.3	Electrocatalytic Hydrogenation and Electrocatalyzed Mediated Reduction	74
	4.10.4	The Electrode Surface as Medium Catalysing Chemical Reactions of Electrogenerated Reactive Organic Intermediates	75
	4.10.4.1	Electrocatalytic Action of Electrosorbed Non-Reactant Species –Electrocatalysis of the Second Kind	78
	4.10.5	Kinetics and Selectivity of Homogeneous Chemical Consecutive Reactions Following Charge Transfer	79
References			80
Further Reading			80

Chapter 5
Mass Transfer by Fluid Flow, Convective Diffusion and Ionic Electricity Transport in Electrolytes and Cells

5.1	Introduction.		81
5.2	Fluid Dynamics and Convective Diffusion		81
5.3	Fluid Dynamics of Viscous, Incompressible Media		84
	5.3.1	Laminar vs Turbulent Flow	86
	5.3.2	Velocity Distributions for Laminar Flow	87
	5.3.2.1	Singular Electrode: Unidirectional Laminar Flow Along a Plate	87
	5.3.2.2	Pair of Planar Electrodes	88
	5.3.2.3	Circular Capillary Gap Cell	89
5.4	Mass Transport by Convective Diffusion		90
	5.4.1	Fundamentals	90
	5.4.2	Dimensionless Numbers Defining Mass Transport Towards Electrodes by Convective Diffusion	92
	5.4.3	Hydrodynamic Boundary Layer and Nernst Diffusion Layer: Planar Electrodes	93
	5.4.4	Mass Transport Towards a Singular Planar Electrode Under Laminar Forced Flow	95
	5.4.5	Channel Flow and Mass Transfer to Electrodes of Parallel Plate Cells for Free and Forced Convection .	97
	5.4.5.1	Free Convection at Isolated Planar Electrodes and between Two Vertical Electrodes	97
	5.4.5.2	Convective Mass Transfer for Parallel Plate Cells with Forced Convection: Planar Plate Cells	98

	5.4.5.3	Mass Transfer in Circular Capillary Gap Cells	101
	5.4.6	Convective Mass Transfer Toward Rotating Electrodes	102
	5.4.6.1	Rotating Cylinder	102
	5.4.6.2	Rotating Disc Electrode	102
	5.4.7	Mass Transfer at Gas Evolving Electrodes	103
	5.4.7.1	Calculating $k_{m,\text{bubble}}$ According to the Penetration Model or Model of Periodic Boundary Layer Renewal	105
	5.4.7.2	Calculating Bubble-Enhanced Mass Transfer According to Flow Model	105
	5.4.8	Mass Transfer in Three-Dimensional Electrodes	106
	5.4.9	Summary	107
5.5	Heat Transport		107
	5.5.1	Chilton–Colburn Analogy of Mass and Heat Transfer	107
	5.5.2	General Description of Heat Generation and Heat Transfer in Electrolyzers and Fuel Cells	108
	5.5.2.1	Heat Balance and Steady State-Temperature of Cells	109
5.6	Ionic Charge and Mass Transport in Electrolytes		110
	5.6.1	Strong Electrolytes	110
5.7	Temperature Dependence of Electrolyte Conductivities		111
5.8	Molten Salt Electrolytes		113
5.9	Segregation in Stagnant Electrolytes of Binary Molten Carbonates in Fuel Cells		114
5.10	Current Density Distribution in Cells and Electrochemical Devices		117
5.11	Primary Current Density Distribution		119
5.12	Secondary Current Density Distribution		121
5.13	Secondary Current Density Distribution and "Throwing Power" in Electrodeposition and Electrocoating		122
5.14	The Wagner Number		124
5.15	Tertiary Current Distribution		125
References			127
Further Reading			127

Chapter 6
Electrochemical Reaction Engineering

6.1	Introductory Remarks		128
6.2	Microkinetic Models		128
6.3	Mode of Operation		129
6.4	Electrical Control of Cells		131
6.5	Macrokinetic Models		131
	6.5.1	Stirred-Batch Tank Reactor	131
	6.5.2	Continuously Stirred Tank Reactor	132

	6.5.3	Plug-Flow Reactor (PFR)	133
	6.5.3.1	Plug Flow Electrolyzer with Uniform Current Density	135
	6.5.3.2	PFR Operated at Mass Transfer Limited and Higher Current Density	135
	6.5.4	Cell Cascades	136
	6.5.5	Extended Modelling of Electrolyzers	138
	6.5.6	Residence-Time Distribution	139
	6.5.7	The Selectivity Problem of Consecutive Reactions in Batch Reactors	142
6.6		Coupling of Electrochemical and Chemical Reactors	146
6.7		Electrolyzer Design and Chemical Yield Losses Due To Parasitic Chemical Reactions	148
6.8		Performance Criteria of Electrochemical Reactors	149
	6.8.1	Fractional Conversion, X	150
	6.8.2	Relative Amount of Charge-Q_r	150
	6.8.3	Overall Conversion Related Yield Θ_p	150
	6.8.4	Current Efficiency Φ^e	151
	6.8.5	Parameters for Energy Considerations	152
References			152
Further Reading			152

Chapter 7
Electrochemical Engineering of Porous Electrodes and Disperse Multiphase Electrolyte Systems

7.1		Introduction	153
7.2		Three-Dimensional Electrodes	154
	7.2.1	General Considerations	154
	7.2.2	Fundamental Equations	155
	7.2.2.1	Nanoporous Electrode Particles	156
	7.2.2.2	Microporous Electrodes	156
	7.2.2.3	Packed and Fluidized Bed Electrodes	157
	7.2.3	Gas Consuming Nanoporous Electrodes for Fuel Cells and Nanoporous Catalyst Particles and Layers for Gas Evolving Electrodes	157
	7.2.3.1	Physical Structure of Particulate, Gas Consuming Nanoporous Gas Diffusion Electrodes	157
	7.2.3.2	Physical Structure of Raney Nickel Coatings for Hydrogen Evolving Cathodes	159
	7.2.3.3	Modelling Hydrogen Concentration Profiles and Catalyst Efficiencies for Hydrogen Consuming Fuel Cell Anodes or Other Gas Diffusion Electrodes	160
	7.2.3.4	Modelling of Hydrogen Concentration Profiles and Catalyst Efficiencies for Hydrogen Evolving Nanoporous Raney-Nickel Catalyst Coatings	165

	7.2.4	Porous Battery Electrodes	171
	7.2.5	Packed Bed and Fluidized Bed Electrodes Composed of Coarse Particles.	173
	7.2.5.1	Fluidized Bed Electrodes.	178
7.3	Ionic Conductivity of Electrolytes Containing Dispersed Gas Bubbles in Gas Evolving Electrolyzers.		179
7.4	Electrolyzers with Gaseous Reactants		183
7.5	Electrochemical Liquid/Liquid Systems		186
References			186
Further Reading			186

Chapter 8
Electrochemical Cell and Plant Engineering

8.1	Materials Choice and Corrosion Problems.		187
	8.1.1	Metals.	188
	8.1.2	Carbon	192
8.2	Electrode Materials		193
	8.2.1	Stainless Steel	194
	8.2.2	Nickel.	194
	8.2.3	Lead.	195
	8.2.4	Titanium	195
	8.2.5	Noble Metals.	195
	8.2.6	Massive Carbon	196
8.3	Electrode Design		196
	8.3.1	Gas Evolving Electrodes	196
	8.3.2	Gas Consuming Electrodes, Gas Diffusion Electrodes . .	197
8.4	Separators: Membranes and Diaphragms		199
	8.4.1	Membranes	201
	8.4.2	Diaphragms	203
8.5	Polymeric Materials for Cell Bodies and Electrolyte Loops		203
8.6	Gaskets.		205
8.7	Electrodes		206
	8.7.1	Horizontal Electrodes	206
	8.7.2	Membrane Electrolyzer.	207
8.8	Cell and Electrode Design		208
	8.8.1	Zero Gap Electrolysis Cells.	208
	8.8.2	Vertical/Horizontal Electrodes	209
	8.8.3	Divided/Undivided Monopolar/Bipolar Cells and Modes of Electrolyte Flow	209
	8.8.4	Special Cell Designs.	210
	8.8.5	Capillary Gap Cells	216
	8.8.6	Swiss Roll Cell	216
	8.8.7	Cells with Three-Dimensional Electrodes	217
8.9	Power Supply for Electrochemical Plants		218

	8.9.1	Rectifiers	218
	8.9.2	Transformer Wiring	218
	8.9.3	Further Equipment	219
Further Reading			220

Chapter 9
Process Development

9.1	Scope and Purpose of Laboratory and Pilot Plant Measurements		221
9.2	Laboratory Methods		222
	9.2.1	Steady-State Measurements of Current Density Potential Correlations	222
	9.2.1.1	General Remarks	222
	9.2.1.2	Measuring Devices	223
	9.2.1.3	Evaluation of Rotating Disc Measurements	223
	9.2.1.4	Current-Voltage Correlation for Competing Reactions by Non-Electrochemical Methods	225
	9.2.1.5	The Ring Disc Electrode	226
	9.2.2	Non-Steady State Methods	230
	9.2.2.1	General Remarks	230
	9.2.2.2	Potentiodynamic Polarisation Curves	230
	9.2.2.2.1	Cyclic Voltammetry and Linear Potential Sweep Method	231
	9.2.2.2.2	Initial Polarisation Curves	233
	9.2.2.3	Square-Wave Pulses	233
	9.2.2.4	Eliminating the IR Drop	235
	9.2.2.4.1	Galvanostatic Methods	236
	9.2.2.4.2	Potentiostatic Procedures	236
9.3	Pilot Plant Methods		236
	9.3.1	General Considerations	236
	9.3.2	Mass-Transfer Measurements	237
	9.3.3	Determination of Residence-Time Distributions	238
9.4	Mathematical Modelling and Optimisation by Factorial Design of Experiments		239
	9.4.1	Introduction	239
	9.4.2	General Procedure for Optimum Finding by Experiment	239
	9.4.3	Factorial Design of Experiments	240
9.5	Cost Analysis		243
	9.5.1	Composition of Productions Costs	243
	9.5.2	Total and Specific Investment Costs	244
	9.5.3	Cost Optimisation with Respect to Current Density	245
	9.5.4	Optimisation of Non-Selective Electrolysis Processes	248
	9.5.4.1	Current Density Against Current Efficiency	249

		9.5.4.2	Temperature vs Current Efficiency	250
	9.5.5		Examples Including Influences of Process Parameters on the Equipment for Non-Electrochemical Unit Operations and Corresponding Costs	250
Further Reading				251

Chapter 10
Industrial Electrodes

10.1	Catalytically Activated Electrodes.		252
10.2	Functioning, Longevity and Application of Electrocatalyst Coatings		253
10.3	Design of Industrial Electrodes		255
	10.3.1	Monopolar Electrodes and Current Density Distribution on Their Surface	255
	10.3.2	Electrodes for Bipolar Electrode Stacks	257
	10.3.3	Gas Evolving Electrodes	258
10.4	Structural Features of Electrocatalysts for Gas Evolving and Gas Consuming Electrodes		260
10.5	Electrocatalytically Activated Dimensionally Stable Chlorine-Evolving Electrodes		260
	10.5.1	Technological History	260
	10.5.2	Electrocatalysis and Selectivity of Anodic Chlorine Evolution at RuO_2-Anodes	261
	10.5.3	Preparation and Formulation of the Coatings.	261
	10.5.4	Improvement of Adhesion and Strength of the Coatings	261
	10.5.5	Design of Cells Using DSAs	262
	10.5.6	Lifetime of Dimensionally Stable Chlorine Evolving Anodes	263
	10.5.7	DSAs for Chlorate and Hypochlorite Production	264
10.6	Oxygen Evolving Anodes.		265
	10.6.1	Technical Processes	265
	10.6.2	Electrocatalysis of Oxygen Evolution in Advanced Alkaline Water Electrolysis.	265
	10.6.2.1	Coatings Containing Cobalt and Iron Oxides	265
	10.6.3	Electrocatalysis of the Anodic Oxygen Evolution by Raney-Nickel Coatings	266
	10.6.4	Catalyst-Coated Titanium Electrodes for Oxygen Evolution From Acid Solutions	266
10.7	Hydrogen Evolving Cathodes		268
	10.7.1	Technoeconomical Significance of Cathodic Hydrogen Evolution	268
	10.7.2	Electrocatalyst Coatings for Hydrogen Evolution from Alkaline Solution	268

	10.7.2.1	Technically Applied Coatings	268
	10.7.2.2	Nickel Sulfide Coatings	269
	10.7.2.3	Raney-Nickel Coatings	269
	10.7.2.3.1	Precursor Alloys and Fabrication of Coated Cathodes	269
	10.7.2.3.2	Utilisation of the Catalyst in Raney-Nickel Coatings	271
	10.7.2.3.3	Performance and Ageing of Raney-Nickel Coatings	272
	10.7.3	Coatings of Platinum Metal Oxides	273
	10.7.4	Active Coatings of Flame Sprayed, Doped Nickel Oxide	273
	10.7.5	Platinum and Platinum Metal Cathodes in Membrane Water Electrolyzers	273
10.8	Fuel-Cell Electrodes		274
	10.8.1	Low- and High-Temperature Fuel Cells	274
	10.8.2	Structural Design of Gas-Diffusion Electrodes in Low-Temperature Fuel Cells	275
	10.8.3	Oxygen Reduction Catalysts in Low-Temperature Cells	276
	10.8.4	Catalysts for Anodic Hydrogen Oxidation	276
	10.8.5	Properties, Preparation and Improvement of Electrocatalysts in Gas Diffusion Electrodes for Low Temperature Cells	277
	10.8.5.1	Pt-Activated Active Carbon	277
	10.8.5.2	Particle Size of Pt Nanocrystals on Active Carbon and Their Effective Catalytic Activity	278
	10.8.5.3	Pt-Alloy Catalysts	278
	10.8.6	Morphology and Structure of Complete PTFE-Bonded Active-Carbon Electrodes	279
	10.8.7	Ageing of Pt-Catalysts	280
	10.8.8	Electrocatalysis of Anodic Methanol Oxidation	281
	10.8.8.1	Technoeconomic Significance of the Process	281
	10.8.8.2	Self-Poisoning of Methanol Oxidising Pt-Catalyst by Oxidation Products of Methanol	281
	10.8.8.3	Anodic Methanol Oxidation at Alloy Catalysts	281
	10.8.9	Gas-Diffusion Electrodes in Membrane (PEM) Fuel Cells	282
	10.8.9.1	Rationale of Developing a Method of Internal Wetting for Membrane Fuel Cell Electrodes	282
	10.8.9.2	Improving Catalyst Utilisation by Ionomer Impregnation of Gas-Diffusion Electrodes	282
	10.8.9.3	The Preparation of Membrane Electrode Assemblies (MEAs) for Membrane Fuel Cells	283
	10.8.10	Electrodes for High-Temperature Fuel Cells	284
	10.8.10.1	Stability of Electrode Structures at High Temperatures	284

	10.8.11	Electrode Kinetics and Electrocatalysis in Molten-Carbonate Fuel Cells	285
	10.8.11.1	Anodic Hydrogen Oxidation.	285
	10.8.11.2	Cathodic Oxygen Reduction	285
	10.8.12	Electrodes in Solid-Oxide Fuel Cells (SOFC)	287
	10.8.12.1	Electrodes and Electrode Structure.	287
	10.8.12.2	The SOFC-Anode	287
	10.8.12.3	The SOFC-Cathode	288

References 289
Further Reading 289

Chapter 11
Industrial Processes

11.1	Introductory Remarks	290
11.2	Inorganic Electrolysis and Electrosynthesis	291
11.3	Chloralkali-Electrolysis	291
	11.3.1 The Electrochemical Reaction.	292
	11.3.2 Thermodynamics and Energy Demands.	292
	11.3.3 Anodic Chlorine Evolution	293
	11.3.4 The Cathodic Reaction	294
	11.3.4.1 Cathodic Sodium Deposition in the Mercury Process . .	294
	11.3.4.2 Cathodic Hydrogen Evolution in the Diaphragm and Membrane Process	295
11.4	Process Technologies	295
	11.4.1 The Amalgam Process	295
	11.4.2 The Diaphragm Process	297
	11.4.3 The Membrane Process.	298
	11.4.3.1 Process-Flow Sheets	300
	11.4.3.2 Brine Recycling	302
	11.4.4 Gas Purification and Conditioning	303
	11.4.4.1 Chlorine	303
	11.4.4.2 Hydrogen.	304
	11.4.5 Comparison of the Three Processes	304
11.5	Hypochlorite, Chlorate and Chlorine Dioxide	306
	11.5.1 Production of Sodium Hypochlorite	306
	11.5.1.1 Electrolytic Generation of Hypochlorite	306
	11.5.1.2 Current Efficiency Losses	307
	11.5.2 Production of Sodium Chlorate	307
	11.5.2.1 Balance of Plant of Chlorate Electrosynthesis	310
	11.5.2.2 Construction Materials.	311
	11.5.3 Chlorine Dioxide from Sodium Chlorate.	311
11.6	Perchloric Acid, Perchlorates, Peroxidsulfates	312
	11.6.1 Perchloric Acid	312
	11.6.2 Sodium Perchlorate	312

	11.6.3	Peroxidisulfates	313
11.7	Fluorine		315
11.8	Hydrogen by Water Electrolysis		316
	11.8.1	Technoeconomic Environment	316
	11.8.2	Thermodynamics and Technological Principles of Electrolytic Water Splitting	317
	11.8.3	Process Technologies	318
	11.8.4	Conventional Alkaline Water Electrolysis	320
	11.8.4.1	Monopolar Technology	320
	11.8.4.2	Bipolar Technology	320
	11.8.4.3	Improved Alkaline Technologies	323
	11.8.5	New Technologies	324
	11.8.5.1	Membrane Water Electrolysis	324
	11.8.5.2	Steam Electrolysis	324
	11.8.6	Economic Implications of Technical Innovations for Alkaline Water Electrolysis	325
11.9	Electrowinning and Electrorefining of Metals		326
	11.9.1	Metal Electrowinning and Refining from Aqueous Electrolytes	326
	11.9.2	Copper Electrowinning and Electrorefining	330
	11.9.3	Nickel Electrowinning	331
	11.9.4	Nickel from the Chloride Leach Process	333
	11.9.5	Nickel Refining	334
	11.9.6	Zinc Electrowinning	334
	11.9.7	Lead Electrorefining	335
11.10	Metal Electrowinning from Molten Salt Electrolytes		335
	11.10.1	General Considerations	335
	11.10.2	Aluminium Production – the Hall–Heroult Process	336
	11.10.2.1	The Melt	336
	11.10.2.2	Electrode Reactions	338
	11.10.3	The Cell	339
	11.10.4	Alkali Metals from Chloride Melts	341
	11.10.5	Magnesium Electrolysis	342
	11.10.5.1	Production of the Feed Salt	343
	11.10.5.2	Magnesium Electrolysis Cells	344
11.11	Organic Electrosynthesis Processes		345
	11.11.1	General Overview	345
	11.11.2	Cell Types Used in Commercial Electroorganic Synthesis	347
	11.11.3	Process and Reaction Techniques of Some Examples of Industrial Organic Electrosyntheses	349
	11.11.3.1	Adipodinitrile Production by the Monsanto/Baizer Process	349
	11.11.3.2	Electrosynthesis of Sebacic Diesters by Kolbe Synthesis	352

		11.11.3.3	Benzaldehydes by Direct Anodic Oxidation of Toluenes.	353

    11.11.3.3 Benzaldehydes by Direct Anodic Oxidation of Toluenes 353
    11.11.3.4 The Selective Anodic Oxidation of L-Sorbose in Commercial Vitamin C Synthesis 353
    11.11.3.5 Anodic Formation of Perfluoro-Propylene Oxide. 355
11.12 Selected Electrochemical Procedures Outside the Chemical and Metallurgical Industries 357
    11.12.1 Electrochemical Wastewater Treatment by Electrodeposition and by Electroosmosis 357
    11.12.1.1 General Considerations 357
    11.12.1.2 Particular Cells for Removal of Metal Ions from Effluents 358
    11.12.1.3 Electrodialysis 361
    11.12.2 Electrochemical Surface Treatment and Shaping of Metals 362
    11.12.2.1 Electrochemical Shaping. 362
    11.12.2.2 Electropolishing. 363
    11.12.2.3 Electrochemical Machining (ECM) 365
    11.12.2.4 Electrochemical Grinding 366
    11.12.3 Electroreforming of Microdies and Microtools by the LIGA-Process 368
References 369
Further Reading 369

Chapter 12
Fuel Cells

12.1 Fuel Cells as Gas Supplied Batteries. 370
12.2 Theoretical Efficiency of Hydrogen/Oxygen Fuel Cells. 371
12.3 Fuel Cell Types 373
    12.3.1 Low-Temperature Fuel Cells – Their Technological State 375
    12.3.1.1 Phosphoric-Acid Cells 375
    12.3.1.2 Membrane Cells 376
    12.3.1.3 Direct and Indirect Methanol-Combusting Membrane Cells 377
    12.3.1.4 Process Principles of the PAFCs and PEMFCs with Proton Conducting Electrolyte 378
    12.3.2 High-Temperature Fuel Cells 379
    12.3.2.1 Molten-Carbonate and Solid Oxide Fuel Cells 379
    12.3.2.2 Process Schemes of MCFCs and SOFCs 379
    12.3.2.3 Internal Reforming in High-Temperature Fuel Cells . . . 380
    12.3.3 Cell Technologies of MCFCs and SOFCs 381
    12.3.3.1 Molten-Carbonate Fuel Cells 381
    12.3.3.2 Solid Oxide Fuel Cells 382
    12.3.3.3 The Westinghouse Technology 382

	12.3.4	Flat-Plate Solid Oxide Cells	384
12.4		Current Voltage Curves of Different Fuel Cells	385
12.5		Fuel-Cell Systems	387
	12.5.1	Phosphoric-Acid Fuel Cell / PC 25	387
	12.5.2	Molten Carbonate Cells	390
	12.5.2.1	ERC-2 MW Plant	390
	12.5.2.2	Hot Module of MTU	390
	12.5.3	Proton Exchange Membrane Cells	391
	12.5.3.1	The Ballard Cell	392
	12.5.3.2	De Nora's Cell	394

Further Reading 394

Subject Index 395

List of Symbols and Abbreviations

a	activity	i	current density
A	area (of electrodes and cells)	i_0	exchange current density
A_e	electrode area	K	equilibrium coefficient
AE	activation energy	k	rate coefficient
a	constant value in Tafel equation	k_m	mass transfer coefficient
a	temperature conductivity	L	characteristic length
a	volume specific surface (in beds and porous electrodes)	l	length
		M	molecular weight
a_e	electrode surface per volume ratio	m	mass
B, b	breadth of electrode	m	slope of semilogarithmic current/potential correlation
b	Tafel factor		
C	capacitance	MC	maintenance costs
c	concentration	N	collection efficiency of ring at ring disc electrode
D	diffusion coefficient		
d	thickness (of interelectrodic gap of cell)	Nu	Nusselt number
		n	number of mols
d_p	diameter of particle or pore	P	amount of product
E	electrode potential – usually related to a reference	Pe	Peclet number
		Pr	Prandtl number
$\dot{E}$	potential velocity in sweep voltammetry	p	pressure
		pH	negative decadic logarithm of proton activity
E (t)	differential residence time distribution in an electrolyzer		
		p_i	partial pressure of species i
E_a	Effect of factor A (in factorial design)	Q	current quantity, also heat
		q, Q	heat
Eu	Euler number	R	gas constant
e_0	elementary charge	R	resistance
F	Faraday constant=96500 As	Re	Reynolds number
f	frequency	R^*	effective surface specific resistance
f	fugacity coefficient	r	rate
f(x,y,...)	function of x, y and ...	r, R	(inner, outer) radius
f_{osm}	osmotic coefficient	r_V	volume related chemical or electrochemical rate
FC	fixed costs		
G	Gibbs free enthalpy	S	entropy
Gr	Grashoff number	S	selectivity
H	enthalpy	Sc	Schmidt number
Ha	Hatta number	Sh	Sherwood number
g	gravity	T, t	temperature
h	height of electrodes	t	time
I	current	t_i	transference number of species i

List of Symbols and Abbreviations

TC	total costs	δ_r	reaction layer thickness
U	voltage	δ_N	thickness of Nernst's diffusion layer
U,V,W	characteristic velocity	δ_{Pr}	thickness of Prandtl-layer
u,v,w	components of linear velocity in space	ε	dielectric constant
		ε	porosity
u	catalyst utilisation	ε_0	influence constant
V	Volt	ε_g	gas voidage in two-phase flow
V	volume	Φ^e	current efficiencies
VC	variable costs	Φ_{Th}	Thiele modulus
$\dot{V}$	volumetric flow rate	γ_i	activity coefficient of species i
Wa	Wagner number	γ	yield
X	fractional conversion	γ_H	enthalpy related yield
x	mol fraction	Γ	surface concentration
x,y,z	cartesian space coordinates	η	overpotential
$z_{A,B,...}$	coordinates in factorial design	η	energy efficiency
z_i	charge number	κ	specific conductivity
(˙)	flux, quantity per unit time	λ	equivalent conductivity
( )'	point quantity	λ	heat conductivity
( )°	standard quantity	Λ	molar conductivity
α	phenomenological charge transfer coefficient	μ	dynamic viscosity
		μ	chemical potential=Gibbs free enthalpy per mol
α_a, α_c	anodic and cathodic charge transfer coefficient		
		ν	kinematic viscosity
α	heat transfer coefficient	ν_i	stoichiometrie factor of substance i
β	symmetry factor of charge transfer	Θ	degree of coverage
		Θ	wetting angle
β	Frumkin coefficient in adsorption isotherm	Θ_p	conversion related yield
		ρ	specific resistance
τ	characteristic time	ρ	density
γ_G	energy related yield	ρ	space time yield
γ	activity coefficient	σ	surface tension
δ_i	thickness of layer i	$\omega=2\pi f$	rotation speed

CHAPTER 1

The Scope and History of Electrochemical Engineering

1.1
Carl Wagner and the Beginning of Electrochemical Engineering Science

The Scope of Electrochemical Engineering was the title of a contribution of Carl Wagner in the second volume of Advances in Electrochemistry and Electrochemical Engineering in 1962. On ten pages he described some examples of his own work on particular topics in electrochemical engineering science [1]. Typical for the situation in electrochemical engineering science at that time was that he concentrated mainly on questions of mass transfer by convective diffusion on one hand and on current-density distributions on electrodes on the other hand. Also typical for the prevailing situation at that time and even more typical for his personality was his methodical approach. Carl Wagner was – compared to other physicochemists and even more compared to average chemists of this time – obsessed by mathematics. He was never content with only formulating the problem under consideration exactly and well founded by the respective set of differential equations. He was satisfied only if – at least for some well defined borderline cases – a closed solution could be found for the problem. In that way he even treated relatively complicated problems as for instance the onset of free convection driven by small density differences of the electrolyte at copper anodes and cathodes of copper refining electrolysis cells and its influence on steady state mass transfer at these electrodes.

Another example in this article was the interaction of mass transfer, crystal nucleation and crystal growth kinetics in performing the anodic formation of lead chromate under conditions which allow to avoid the formation of closed passivating deposits on the lead surface in favour of the formation of a powdery precipitate due to homogeneous nucleation in front of the electrode in the bulk of the electrolyte. In his third example he dealt with the current density distribution at flat electrodes and he found the dimensionless quantity named afterwards the Wagner number, to be the relevant quantity determining the current density distribution. Although Carl Wagner's work in the field of electrochemical engineering science was rather more occasional than abundant and systematic he met with his contributions the salient points and topics which were to keep the electrochemical engineers in academia (though much less the practitioners) busy for at least the next two decades. Therefore he and Charles W. Tobias, Fumio

Hine and Norbert Ibl are known as the founders of electrochemical engineering science.

1.2
Electrochemistry and Electrochemical Engineering Science

At the beginning of physical chemistry as a scientific discipline, electrochemistry played a central role and was often at the leading edge of this branch of chemistry. Apart from the physics and thermodynamics of at first diluted, and then of more concentrated electrolyte solutions the thermodynamics of at first simple and then complicated electrode reactions were the more important issues till the late 1940s. In the 1950s and 1960s the theory of the electrochemical double layer was completed and the advances which had been made in electronics provided the tools for electrochemists to investigate also with fast methods the kinetics of electrochemical reactions and to develop electrode kinetics as a special discipline to an extent which became comparable to what had been achieved in homogeneous chemical kinetics already two to three decades ago. After this very productive period, whose results were finally compiled in Vetter's famous book [2], and after many disappointing trials to advance the understanding of the elementary process of interfacial charge transfer which did not lead much further than the established approaches of Marcus [3], Levich [4] and Gerischer [5], very soon – beginning in the early 1970s – within ten years the physicochemical scientific community as a whole lost almost any interest in electrochemistry as a discipline of fundamental importance. Electrochemistry soon became a very diversified field of – though still also today numerous – specialists working more and more on very particular problems, which are very often multidisciplinary in character, and it moved slowly to the rim of the physicochemist's vision. In the little justified view of the majority of the academic physicochemists who today imagine electrochemistry to be mainly manifested in terms of the Nernst equation and perhaps – if it goes that far – the Butler–Volmer equation, electrochemistry is now a chapter of science history.

Indeed today electrochemistry is mainly what one would call "applied electrochemistry" a very diversified field ranging from electrode processes, electrosynthesis and electrolytic corrosion to solid state ionics, to surface processes of any kind and a lot more provided they are connected with charge transfer or ionic charge and mass transport. But this is not electrochemical engineering – on the contrary it is most often very far from electrochemical engineering and electrochemical engineering science. What then is electrochemical engineering and electrochemical science?

Electrochemical engineering science is certainly a discipline of its own rights which on the basis of the fundamental laws of electrochemistry deals with electrochemical systems and processes according to scientific principles finding its place between chemical engineering and electrochemistry. It developed according to its own rules and logic relatively independently of electrochemical science and it is essentially interdisciplinary as must be an engineering science of any kind. Established facts of electrochemistry are but the fundaments and in a way also the tools of this branch of engineering science which deals rather with systems and systematically arranged situations related to industrial electrochemical technologies in which a limited number of fundamental phenomena are interacting rather than with isolated phenomena. Its aim is analysing and describing mathematically – that means modelling – such situations in a way, which will allow to design and operate processes under full control at (financially) optimal conditions.

1.3
Electrochemical Engineering Science and Technology Since the Mid 1960s

Carl Wagner distinguishes in his article electrochemical engineering science from the "art of electrochemical engineering", being quite aware of the role of electrochemical processes in the chemical industries and knowing well that at that time there existed for instance an established technology of chloralkali electrolysis, which was developed continuously with ever improving efficiency over more than 60 years and a considerable number of other electrolysis processes in the chemical process and the metallurgical industry of comparable age and history. All of them had been developed more or less by intuition and by trial and error and almost without any need for a deeper insight into the more fundamental phenomena which contribute to controlled and steady performance of these processes.

But in the 1960s after chemical engineering science had already developed the capability to treat unit operations in a way that linked the underlying physical and physicochemical phenomena to chemical technology, it became apparent that this kind of treatment was also due in the field of electrochemical technology.

Established electrochemical technologies were at that time chloralkali electrolysis according to the mercury and diaphragm processes and related processes as hypochlorite and chlorate synthesis, electrowinning of aluminium and magnesium from molten salt electrolytes, zinc electrowinning and copper refining as branches of hydrometallurgy and a few small scale processes as for instance MnO_2-production. The contribution of electrochemical engineering science to cell and process design was at the mid-1960s practically nil – almost all that had been achieved in technical perfection was due to the art of engineering. Testemony to this are a number of textbooks published till the mid-1970s on industrial electrochemistry and dealing with established processes in a descriptive manner together with some reference to electrolyte theory and electrochemical thermodynamics but containing little or nothing about electrochemical engineering science.

In particular mass transfer considerations were not accounted for – mostly because mass transfer did not really limit the classical processes. Current density distributions were not essential and current densities were limited rather more by energy consumption than by mass transfer hindrance or uneven current distributions. Moreover the words electrode kinetics and electrocatalysis, the latter was introduced in the context of fuel cell electrochemistry in 1963 by Grubb [6], were almost unknown and at least irrelevant to the practitioners in industry of that time, who were designing, constructing and running cells and processes. No wonder, that Henry Beer, the inventor of the (at first Pt then) RuO_2-activated dimensionally stable anodes (DSAs) [7] had remarkable difficulties in convincing the owners of chloralkali electrolysis plants that electrocatalytically activated electrodes could not only save, but earn them a lot of money. Till the 1960s corrosion of electrodes, cells, troughs and cell frames and other components could only be mitigated by relatively primitive coatings made of asphalt or rubber. Titanium as a valve metal was too expensive and polymer linings were not cheap enough or could not yet be reliably applied as surface coatings on large scale. Diaphragms made of asbestos and even more polymer-bonded asbestos diaphragms were a real advance over diaphragms made of porous cement but the possibility of using porous polymers as separators were not yet fully acknowledged and exploited at the late 1950s early 1960s. The host of diverse polymers, not to speak of ionomers whose most important example is the Nafion membrane developed by Grot in the early 1970s at DuPont [8], were still unknown and so were the words polymer coating, fluoromers, polymer engineering, and little was known of materials engineering and materials science at that time. The advances which were achieved in materials development in the late 1960s and 1970s and the decreased costs of highly corrosion resistant valve metals were still to open unexpected possibilities in cell design and process technology.

Electrochemical reaction engineering was till the beginning of the 1970s also a term of no significance because the electrochemical reactions were unequivocal, that means by their very nature very selective and were in general not followed by chemical consecutive reactions – with one exception – the chlorate process. Although already in the early 1950s the chemical kinetics of hypochlorite disproportionation [9] had been investigated by several authors, it was sufficient to find empirically the reactor volume, which secured highest degree of conversion and chlorate selectivity at lowest cost. Contrary to the non-existence of electrochemical engineering science at the early 1960s, process costing and optimisation is that part of practical electrochemical engineering, which was – of course – already systematically developed and applied, though little was published on that subject as it was proprietory knowledge. Not a complete system analysis but empirical approaches based on current voltage correlations of total cells together with conservative determination of investment costs for cells and additional equipment were the approaches which were not much different from, but almost identical to the methods of today. What then is electrochemical engineering today?

1.4
What Means Electrochemical Engineering Science and Technology Today?

Obviously it is easier to say what is the aim of electrochemical engineering, than to try to define it. Its aim is to design and perform electrochemical processes-comprising all stages – that means also the non-electrochemical process steps – including preparation of raw materials up to conditioning of the products at minimum total costs but simultaneously observing the issues of resource-saving and protecting the environment.

The scope of electrochemical engineering is determined by the diversity of industrial electrochemical processes. The classical processes in the chemical process industry (CPI) as chloralkali electrolysis and related processes, the synthesis of chlorates, perchlorates, peroxoacids, chromate and manganese dioxide had been amended during the last ten years by a number of organo-electrosynthesis processes. The electrowinning and refining of metals in hydrometallurgy and molten salt electrolyses had been little extended for several decennia to new processes. The invention of pulsed copper deposition with current reversal is an exception which was mainly based on ideas of unsteady mass transfer due to improved scientific insight into the electrowinning process. Another case is lithium electrowinning from chloride melts, which became important because of the increasing demand of this metal in the battery industry. Two very important and extending fields of relevance to electrochemical engineering are electrochemical surface treatment of metals and electrochemical methods in environmental protection. In the latter field electrochemical procedures are unique by their efficiency in removing noxious transition metal and heavy metal cations from waste waters and effluents of the galvanic industries. The invention of the porous so called three-dimensional or bed electrodes together with the need for improved porous battery electrodes gave rise to developing the theory of porous electrodes. Lately the development of fuel cell technologies during the recent ten years opened a new field of activity to the electrochemical engineer and directed his interest to the field of electrochemical catalysis – electrocatalysis – a field which had been neglected by the more physicochemically or mathematically oriented electrochemical engineer and was for long time a subject treated rather by electrochemists than electrochemical engineers.

Within the last 30 years electrochemical engineering science extended far beyond the limits which had been drawn by Carl Wagner in his article of 1962. Still, occasionally special mass transfer problems are investigated. The recently reported findings of mass transfer and fluid flow in front of gas evolving electrodes are an echo of the early, intensive investigations during the 1960s and 1970s. But closed mathematical solutions are not and cannot be aimed at and the extensive tables of the correlations of the relevant adimensional quantities, Sherwood, Reynolds, Grashoff and Schmidt, cover today almost any technical situation imaginable. Current density distributions are now routinely calculated with high reliability by numerical methods even for complicated electrode structures and also incorporating two-phase flow of the electrolyte and gases between the electrodes of an electrolyzer or a separator, be it a membrane or a diaphragm, and an electrode. Even the molten cryolite and aluminium circulation in aluminium smelters due to magneto-hydrodynamics are now routinely calculated in order to optimally designing the smelter, the arrangement of current feeding and busbars and pot lines. The whole field of electrode kinetics with inclusion of electrocatalysis became essential for electrochemical engineering but the treatment of electrode kinetics is always rather more directed towards a formal description of overpotential current density correlations than to fundamental mechanisms and also more to improving electrocatalyst performance than to explain their electrocatalysis from physicochemical and chemical principles.

Moreover the concept of utilisation of porous catalysts which was state of current chemical engineering knowledge since the 1950s was introduced into the concepts of electrochemical engineering science in the late 1960s in the context of fuel cell development. It was but consequent that starting from corrosion science on one hand and from the development of technical electrocatalysts the electrochemical engineer had to become more familiar and to use creatively more and more knowledge from materials science and materials engineering. Finally in the 1980s as a consequence of dealing with complex organo-electrosynthesis processes the principles of reaction and process engineering which since long had been established in chemical engineering were also adopted by the electrochemical engineers. Is that the end of any further development in electrochemical engineering science?

If one looks on chemical engineering as an example and counterpart to electrochemical engineering one would expect that the actual work and interest in electrochemical engineering would return to problems and questions which are more chemical and electrochemical (with the stress on chemical) in nature than it used to be. In chemical engineering the interest has shifted clearly to homogeneous and even more heterogeneous catalysis. Accordingly one would expect that also electrocatalysis and the far-reaching parallelism between chemical and electrochemical catalysis would again be faced more closely. Now we have almost every engineering tool for the formal treatment of electrochemical processes at hand. It becomes clearer than ever that for developing new processes – for instance in organic electrosynthesis – new chemical ideas with respect to promising synthetic routes of high selectivity and particularly with respect to catalytic control of the reaction are decisive. It is more than obvious that in the field of electrocatalysis – if it goes beyond chlorine, oxygen and hydrogen electrochemistry – our knowledge is only rudimentary. For instance still relatively little is achieved in a relatively simple reaction like electrocatalytic methanol oxidation at close to its equilibrium potential and catalyst alternatives to the platinum metals and their alloys are not yet at hand. In contrast chemical catalysis of, for instance, the partial oxidation of organic compounds (though working at much higher oxygen pressures and therefore not directly comparable) is today much better understood. It might be possible that improving the knowledge of electrocatalysis in the context of fuel cell electrochemistry might open new ways and might enable the electrochemical engineer to understand electrocatalysis in general much better and on a wider scope than today. That might lead to the end to introduce also new electrocatalytic ideas in the field of organo-electrosynthesis. Certainly this in only a faint hope, but perhaps a promising one with respect to the future of the discipline of electrochemical engineering science and technology.

References

1. Carl Wagner, "The Scope of Electrochemical Engineering" in "Advances in Electrochemistry and Electrochemical Engineering", P. Delahaye, C.W. Tobias, editors, Interscience publishers, New York, London, Sydney, 1962, Vol. 2, p.1
2. K. Vetter, Electrochemische Kinetik, Springer Verlag, Berlin, Göttingen, Heidelberg 1961
3. R.A. Marcus, J. Chem. Phys. *24*, 966 (1956); *43*, 679 (1965)
4. V.G. Levich, Present State of the Theory of Oxidation/Reduction in solution in P. Delahay, C.W. Tobias eds., Advances in Electrochemistry and Electrochemical Engineering, Interscience Publisher/John Wiley, New York, 1966, Vol. 4, p. 249
5. H. Gerischer, Zeitschr. physik. Chemie, N.F., *26*, 223 (1960), *26*, 325 (1960), *27*, 48 (1961)
6. W.T. Grubb, Low temperature hydrocarbons, 17; Annual Power Source Conference, Atlantic City, 1963
7. H.B. Beer, Magneto Chemie, N.V., Brit. Pat. No 8,551,707 (1958)
8. W. Grot. US Appl. 178.782, 8. Sept. 1971
9. N. Ibl, H. Vogt. In: Inorganic Electrosynthesis, Comprehensive Treatise on Electrochemistry, J.O'M Bockris, B.E. Conay, E. Yeager, R.A. White, editors, Plenum Press New York and London, 1981, Vol. 2, Chapter 3, p. 167

Further Reading

J.S. Newman, Electrochemical Systems, Prentice Hall 1973
De P. Le Goff, Preface, In: F. Coeuret, A. Storck, Elements De Genie Electrochimique, Lavoisier TEC 8 DOC, Paris, 1984

CHAPTER 2

Basic Principles and Laws in Electrochemistry

2.1
Stoichiometry of Electrochemical Reactions

An electrochemical reaction, for example the cell reaction of copper electrowinning,

$$CuSO_4 + H_2O \rightarrow Cu + 1/2 O_2 + H_2SO_4 \qquad (2.1)$$

represents a redox reaction and may be considered as composed of two partial reactions, an anodic oxidation and a cathodic reduction. For the process under discussion the partial reactions are:
reduction of divalent copper ions to copper metal at the cathode

$$Cu^{2+} + 2e^- \rightarrow Cu \qquad (2.2)$$

oxidation of water at the anode

$$H_2O \rightarrow 1/2 O_2 + 2H^+ + 2e^- \qquad (2.3)$$

In general any homogeneous redox reaction

$$A + D \rightarrow A^{z-} + D^{z+} \qquad (2.4)$$

may be divided likewise into two different redox reactions

$$D \rightarrow D^{z+} + ze^- \qquad \text{(oxidation).} \qquad (2.5)$$

$$A + ze^- \rightarrow A^{z-} \qquad \text{(reduction).} \qquad (2.6)$$

In homogeneous redox reactions the electrons are transferred directly from the donor molecule or atom D to the acceptor molecule or atom A. In electrochemical reactions both partial reactions at Eqs. (2.5) and (2.6) are heterogeneous and proceed at different sites, at opposite electrodes of an electrolyzer whereas the electrons are transported via an external electrical circuit from the anode to the cathode. This is shown schematically in Figs. 2.1 a and b.

2.1 Stoichiometry of Electrochemical Reactions

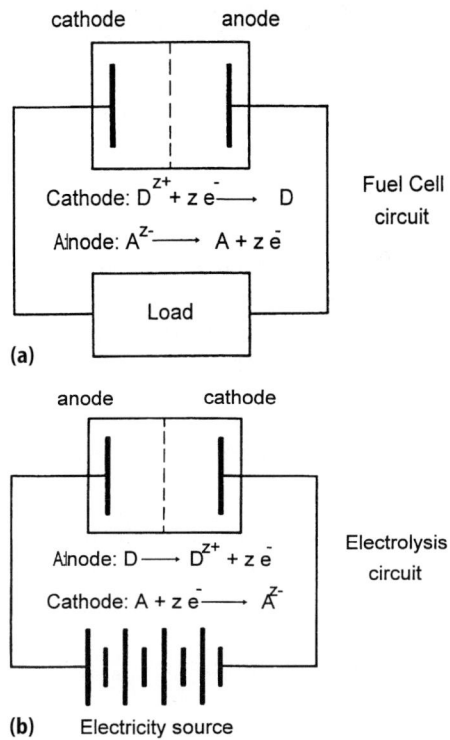

Fig. 2.1a,b. Schematic representation of electrochemical circuits: **a** fuel cell; **b** electrolyzer

According to basic rules of thermodynamics (see Chap. 3) for a spontaneous redox reaction at Eq. (2.4) a free enthalpy change of $\Delta G<0$ is required. If such a system reacts in an electrochemical cell then electrical energy is produced under consumption of the stored chemical energy which is the case in batteries and fuel cells and shown in Fig. 2.1 a. In an electrochemical cell usually a cell reaction with $\Delta G >0$ is performed by external input of electrical energy as shown in Fig. 2.1 b. The polarity of fuel cells and electrolysis cells is different: A fuel cell cathode forms the positive and the anode the negative pole whereas to the cathode and anode of an electrolysis cell the negative and positive pole of the current source is attached.

Considering an electrode reaction (Eq. 2.5 or 2.6) as a special type of a chemical reaction shows that electrons are reactants.

$$A + ze^- \rightarrow A^{z-} \qquad (2.7)$$

2.2
Faraday's Law

In the special case of reaction at Eq. (2.7) the stoichiometric coefficient of the electrons v_e is equal to the charge number z of the product. However, this is not in general the case as for instance demonstrated by Eq. (2.3). The stoichiometric consideration of reaction at Eq. (2.7) shows that for the formation of 1 mol product v_e mols of electrons are consumed. Taking the electron charge as $e_0 = 1.6021917 \cdot 10^{-19}$ C and Avogadro's number as $N = 6.022169 \cdot 10^{23}$ mol^{-1} yields for the charge of 1 mol electrons F=96,486.69 C·mol^{-1} =26.80 Ah/C·mol^{-1} which is known as one Faraday. The number of the mols of product, n, produced at constant current, I, by a charge of Q=I t is given as

$$n = \frac{Q}{v_e F} = \frac{It}{v_e F} \tag{2.8}$$

multiplication by the molar mass of the product M and substituting the Integral $\int I dt$ for Q yields the more general equation

$$m = \frac{MQ}{v_e F} = \frac{M \int I dt}{v_e F} \tag{2.9}$$

usually referred to as Faraday's law. For the more general electrode reaction scheme, Eq. (2.10)

$$v_s S + v_e e^- \to v_p P \tag{2.10}$$

Faraday's law can be formulated as

$$m_p = \frac{v_p M_p}{v_e F} \int I dt \tag{2.11}$$

or with respect to the consumed substrate, S

$$-m_s = \frac{v_s M_s}{v_e F} \int I dt \tag{2.12}$$

Division of Eq. (2.11) by M_p and differentiation with respect to t yields the point rate of production

$$\frac{dn_p}{dt} = \frac{v_p}{v_e F} I \quad \text{and} \quad \frac{dm_p}{dt} = \frac{v_p M_p}{v_e F} I \tag{2.13}$$

2.3
Production Rates and Current Densities

Equation (2.13) shows that the rate of an electrochemical reaction is simply proportional to the current flowing through the electrolyzer cell. Therefore the easiest experimental method of determining electrochemical reaction rates is to measure the electric current.

Since the rates of the heterogeneous charge-transfer reactions Eq. (2.13) are always proportional to the interfacial area, that means the electrode surface, electrochemical reaction rates usually are normalized with respect to the electrode area A_e.

$$\frac{1}{A_e}\frac{dn_p}{dt} = \frac{v_p}{v_e F}\frac{I}{A_e} = \frac{v_p}{v_e F}i \tag{2.14}$$

the current related to the electrode area is called current density, i.

A prerequisite for the validity of Eqs. (2.11)–(2.14) is that only the considered reaction takes place at the electrode. But very often several parallel electrochemical reactions occur at the electrode. For example cathodic copper deposition (Eq. 2.2) may be accompanied by a parallel formation of hydrogen or, under practical conditions more likely, by reduction of dissolved oxygen

$$Cu^{2+} + 2e^- \rightarrow Cu \tag{2.2}$$

$$2H^+ + 2e^- \rightarrow H_2 \tag{2.15a}$$

$$1/2\, O_2 + 2e^- + 2H^+ \rightarrow H_2O \tag{2.15b}$$

In such cases the individual contribution of each of these parallel reactions is given by their current efficiency Φ_i^e. Summing up the current efficiencies of all parallel reactions yields 1 by definition:

$$\sum_i \Phi_i^e = 1 \tag{2.16}$$

The current efficiency of an individual reaction may be calculated applying Faraday's law

$$\Phi_p^e = \frac{n_p \cdot v_e F}{v_p Q} \tag{2.17}$$

with

$$Q = A_e \int_0^t i\, dt \tag{2.18}$$

The quantity Φ_p^e is referred to as overall current efficiency and represents the mean value over a certain time period taking into account that the current efficiency may vary with time. The actual current efficiency at time t, the so called differential or point current efficiency $\Phi_p^{e'}$ is defined by

$$\Phi_p^{e'} = \frac{(dn_p/dt)v_e F}{v_p A_e i} \tag{2.19}$$

In Eqs. (2.17) and (2.19), instead of the stoichiometric coefficients, their absolute values are taken since these coefficients are defined as negative numbers for educts and as positive numbers for products. The product

$$\Phi_p^{e'} i(t) = i_p(t) \tag{2.20}$$

represents the partial current density, $i_p(t)$ of reaction (2.10).

2.4
Ohm's Law and Electrolyte Conductivities

Any electrochemical circuit contains ohmic resistors. There are mainly the wires composed of electronically conducting materials and the electrolyte representing an ionic conductor. In metals (wires) electrons act as charge carriers whereas in the electrolyte ions are the carriers of electric charge. Electronic and electrolytic conductivities are distinguished by their temperature coefficient. In metallic electronic conductors (not in semiconductors) the conductivity decreases with rising temperature whereas with increasing temperature the ionic conductivity of electrolytes increases due to decreasing viscosity of most solvents.

The voltage drop caused by a current flowing through a resistor is given by Ohm's law.

$$U_R = RI \tag{2.21}$$

Ohm's law is only valid if the number of charge carriers per unit volume is independent of the current. The resistance of a resistor of length l and cross-section A is given by

$$R = \frac{l}{A}\rho = \frac{l}{A} \cdot \frac{1}{\kappa} \tag{2.22}$$

where ρ represents the specific resistivity and κ the specific conductivity of the resistor material. Combination of Eqs. (2.21) and (2.22) results in the calculation of the ohmic potential drop ΔU_Ω:

$$\Delta U_\Omega = \frac{l}{A_e} \cdot \frac{1}{\kappa} \cdot I = \frac{l}{\kappa} \cdot i \tag{2.23}$$

2.4 Ohm's Law and Electrolyte Conductivities

The specific conductivity κ of an electrolyte containing several ionic species i is given as

$$\kappa = \sum_i c_i z_i \lambda_i \qquad (2.24)$$

with λ_i the equivalent conductivity of ionic species i. Some numerical values of equivalent ionic conductivities are given in Table 2.1. A more extented treatment of ionic conduction in electrolytes is given in Sect. 5.4.

Table 2.1. Equivalent conductivities of some ions at infinite dilution at 25 °C

Cation	$\lambda_i^0/\text{cm}^2\Omega^{-1}\text{val}^{-1}$	Anion	$\lambda_i^0/\text{cm}^2\Omega^{-1}\text{val}^{-1}$
Ag^+	62.9	Br^-	78.14
Al^{3+}	63	BrO_3^-	55.4
Ba^{2+}	63.6	Cl^-	76.35
Ca^{2+}	59.5	ClO_3^-	64.6
Cd^{2+}	54	ClO_4^-	67.3
Ce^{3+}	69.6	CO_3^{2-}	69.3
Co^{2+}	49	CrO_4^{2-}	83
Cr^{3+}	67	F^-	55.4
Cs^+	77.2	$Fe(CN)_6^{3-}$	99.1
Cu^{2+}	55	$Fe(CN)_6^{4-}$	111
Fe^{2+}	53.5	HCO_3^-	44.5
Fe^{3+}	68	$H_2PO_4^-$	36
H^+	349.8	HSO_4^-	52
Hg^{2+}	63.6	I^-	76.85
Hg_2^{2+}	68.6	IO_3^-	40.8
K^+	73.5	IO_4^-	54.5
La^{3+}	69.6	MnO_4^-	61.3
Li^+	38.6	NO_2^-	71.4
Mg^{2+}	53.0	NO_3^-	71.4
Mn^{2+}	53.5	OH^-	198.3
NH_4^+	73.5	SCN^-	66
$N(CH_3)_4^+$	44.9	SO_3^{2-}	72
$N(C_2H_5)_4^+$	32.6	SO_4^{2-}	80
$N(C_3H_7)_4^+$	23.4	$HCOO^-$	54.6
$N(C_4H_9)_4^+$	19.4	CH_3COO^-	40.9
Na^+	50.1	$C_2H_5COO^-$	35.8
Pb^{2+}	70	$C_6H_5COO^-$	32.3
Rb^+	77.8	$C_2O_4H^-$	40.2
Sr^{2+}	59.4		
Tl^+	74.7		
Zn^{2+}	54		

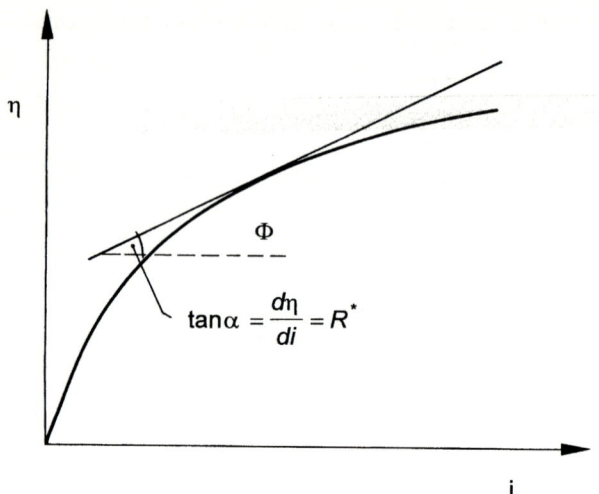

Fig. 2.2. Schematic polarisation curve (current density vs overpotential, i vs η, correlation) of an electrode process. The slope of the η/i curve is a function of current density and represents the "polarisation resistance", R*, which is also, like the resistance l/κ which is contained in Eq. (2.22), a surface specific resistance with the dimension Ωcm^2

An electrochemical circuit consists not only of ohmic resistors as treated above. Also the electrode/electrolyte interface represents a resistor. But its resistance is not ohmic since its value depends on the current or the current density resp. The dependence of electrode potential (potential drop across the interface) on current density usually is presented as a logarithmic polarisation curve, shown schematically in Fig. 2.2. The slope of this curve, dη/di

$$R_p^* = \frac{d\eta}{di} \qquad (2.25)$$

is denoted as polarisation resistance, which is a point quantity and usually decreases with current density

2.5
Parallel Circuits and Cells with Electrolytic Bypass and Kirchhoff's Rules

In electrochemical circuits several resistors may be connected in series or in parallel. Therefore some fundamental laws of electrical resistor networks should be mentioned here known as Kirchhoff's rules.
1. At any connecting point of several conductors the sum of all currents equals zero.
2. If two points of a circuit are connected by several conductors containing no power sources, then the product of current and ohmic resistance is equal for all conductors ($IR = \Delta U_\Omega$).

2.5 Parallel Circuits and Cells with Electrolytic Bypass and Kirchhoff's Rules

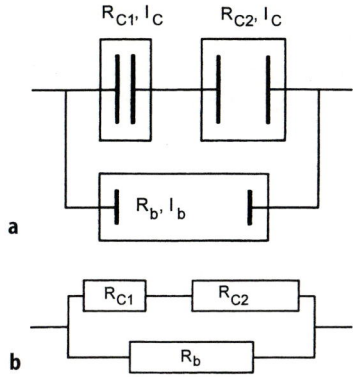

Fig. 2.3. a Circuit for two electrochemical cells in series with an electrolytic bypass parallel to the cells. **b** If for simplicity all resistors are assumed to be ohmic, then according to Kirchhoff's first rule for the currents the relation of Eq. (2.27) holds

3. Since electrochemical cells are not simple ohmic resistors condition 2. must be substituted by the postulation that the potential drop across each electrochemical element, connected to two points 1 and 2 of a more complex circuit must be equal to $U_1 - U_2$.

As an example two cells in series with a parallel electrolytic bypass resistor are considered as shown in Fig. 2.3. With respect to practical applications this demonstrates the case of two electrochemical cells in series. Whereas the currents I_c (cell) and I_b (bypass) are summing up to the total current I:

$$I = I_c + I_b \tag{2.26}$$

Kirchhoff's second rule requires the relationship

$$I_b R_b = I_c (R_{c1} + R_{c2}) \tag{2.27}$$

According to Ohm's law the cell voltages are

$$U_{c1} = I_c R_{c1} \tag{2.28}$$

$$U_{c2} = I_c R_{c2} \tag{2.29}$$

Combination of Eqs. (2.26) and (2.27) yields for the bypass current

$$I_b = I \frac{R_{c1} + R_{c2}}{R_b + R_{c1} + R_{c2}} \tag{2.30}$$

This means the bypass current which causes current and energy losses, as it does not contribute to the electrochemical reaction, will be lower the larger the bypass resistance.

Further Reading

C.H. Hamann, W. Vielstich, Elektrochemie I, II, VCH, Weinheim 1985
D. Dobos, Electrochemical Data, A Handbook for Electrochemists in Industry and University. Elsevier 1975
D.R. Stull, H. Prophet Eds. JANAF Thermochemical Tables 2nd Edition, National Standard Reference Data Systems, 1971
I. Barin, O. Knacke, Thermochemical Properties of Inorganic Substances, Springer 1973.

CHAPTER 3

Electrochemical Thermodynamics

3.1
Equilibrium Cell Potential and Gibbs Energy

As discussed in Chap. 2 any chemical reaction involving charge exchange between two different redox reactants, see Eqs. (3.1 a,b) can be performed virtually – and very often also practically – by performing the redox reactions of the two redox-couples separately at two different electrodes but jointly in a divided electrochemical cell (Fig. 3.1):

$$A^{x+} + B^{(y+1)+} \rightarrow A^{(x+1)+} + B^{y+} \quad \text{total reaction} \quad (3.1)$$

$$A^{x+} \rightarrow A^{(x+1)+} + e^- \quad \text{redox couple A} \quad (3.1\,a)$$

$$B^{(y+1)+} + e^- \rightarrow B^{y+} \quad \text{redox couple B} \quad (3.1\,b)$$

thus the oxidation of ferrous ions by ceric ions

$$Fe^{2+} + Ce^{4+} \rightarrow Fe^{3+} + Ce^{3+} \quad (3.1\,c)$$

may be performed in two half cells, one containing the ferric/ferrous system $(Fe^{2+} \rightarrow Fe^{3+} + e^-)$, the other containing the ceric/cerous system $(Ce^{4+} + e^- \rightarrow Ce^{3+})$.

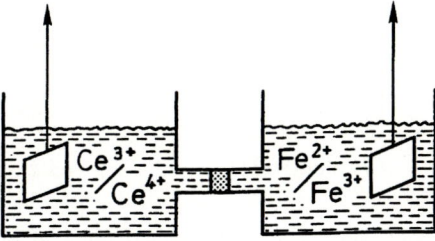

Fig. 3.1. Schematic of a cell reaction composed of two separate redox reactions which jointly would establish a homogeneous redox reaction between an oxidant and a reductant.

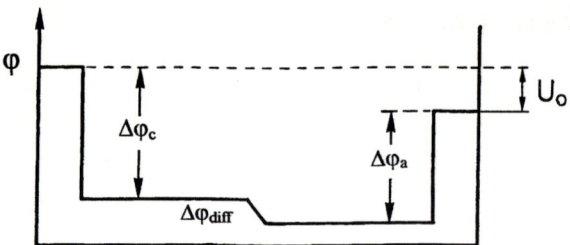

Fig. 3.2. Equilibrium cell potential U_0 in a cell divided by a diaphragm composed of two interfacial potentials $\Delta\varphi_i$ and a (mostly negligible) diffusion potential $\Delta\varphi_{diff}$. The diffusion potential $\Delta\varphi_{diff}$ vanishes if a salt bridge with an electrolyte whose cations and anions possess transference numbers of 0.5, each, connects the two separate cells. With aqueous KCl solutions this is the case

Both being connected by a salt bridge[1], they deliver an electrical potential difference U_0 between two Pt-electrodes inserted into the two separated electrolytes (Fig. 3.2).

Likewise the formation of hydrochloric acid, dissolved at a given concentration in water, from gaseous chlorine and gaseous hydrogen, realized in two half cells connected by an electrolyte bridge, generates a cell potential. If the reaction is performed electrochemically, ($H_2 + Cl_2 \rightleftarrows 2HCl$; $H_2 \rightleftarrows 2H^+ + 2e^-$ and $Cl_2 + 2e^- \rightleftarrows 2Cl^-$), by dipping a chlorine electrode (made of platinized platinum and being supplied and sparged with elemental chlorine) and a hydrogen electrode (platinized, H_2-sparged platinum electrode) into an aqueous solution of hydrochloric acid of the respective concentration, the equilibrium cell potential can be measured between the two platinum electrodes at vanishing current with a voltmeter of high internal resistance.

This cell potential is sometimes called "electromotive force". Under conditions of vanishing cell current and established reversibility of the two different electrode reactions, (which is accomplished by effective electrocatalysis of the two electrode reactions by applying for instance platinum black), the so called open cell potential becomes the equilibrium cell voltage U_0.

U_0 equals the free energy ΔG of the cell reaction per mol of product divided by the number ν_e of Faradays (1F=96,500 As) necessary for the electrochemical generation of one mol of product.

$$1/2 H_2 + 1/2 Cl_2 \rightleftarrows HCl_{diss} \qquad (3.2\ a)$$

$$1/2 H_2 \rightleftarrows H^+_{diss} + e^- \qquad (H_2\text{-electrode}) \qquad (3.2\ b)$$

$$1/2 Cl_2 + e^- \rightleftarrows Cl^-_{diss} \qquad (Cl_2\text{-electrode}) \qquad (3.2\ c)$$

---

1 The connecting electrolyte in the bridge should have equal transference numbers (0.5) for cations and anions imposing equal distribution of cationic and anionic currents on the cell.

3.1 Equilibrium Cell Potential and Gibbs Energy

U_0 is defined as the equilibrium potential difference between the respective anode and cathode.

$$U_0 = E_0 \text{ (anode)} - E_0 \text{ (cathode)} \tag{3.2 d}$$

$$U_0 = \frac{\Delta G}{v_e F} \tag{3.3}$$

where $v_e = 1$ for HCl.

If the reaction Eq. (3.2 a) proceeds spontaneously, the hydrogen electrode is the anode and if the electronic current passes through a wire or a load from the anode to the cathode, then ΔG is negative and the cell would spontaneously produce hydrochloric acid. For HCl electrolysis, however, the chlorine electrode is the anode, the hydrogen electrode becomes the cathode, and under these conditions where reaction at Eq. (3.2 a) is reversed by imposing an external potential differentially greater than U_0, hydrochloric acid would be electrolytically decomposed and ΔG becomes positive.

The Gibbs energy of any chemical reaction depends on:
(1) the temperature;
(2) the concentrations of dissolved reactants; and
(3) the partial pressures of gaseous reactants.

Because of the latter two dependencies it is reasonable to define standard molar Gibbs energies called standard chemical potentials μ_i^0 and also standard cell voltages U^0 defined by standard concentrations and standard pressures of reactants and products. Standard concentrations of dissolved substances – in particular for aqueous solutions of electrolytes are defined by solute activities of 1 mol dm^{-3} and for gaseous substances by partial fugacities of 0.1 MPa or 1 bar respectively.

Concentration related activities and fugacities are derived from actual concentrations and partial pressures by multiplying molar concentrations with activity coefficients, γ_i,

$$a_i = c_i \gamma_i \tag{3.4 a}$$

or partial pressures with fugacity coefficients f_i,

$$p_i^* = p_i f_i. \tag{3.4 b}$$

These coefficients account for deviations from ideal behaviour which is described by the two limiting equations, Eqs. (3.5 a, b), which describe chemical potentials of gases and solutes at low – or more precisely – vanishing pressures and concentrations.

$$\mu_i = \mu_i^0 + RT \ln(p_i / p^0) \tag{3.5 a}$$

where p^0 is standard fugacity (0.1 MPa)

$$\mu_j = \mu_j^0 + RT \ln(c_j / c^0) \tag{3.5 b}$$

c^0 is standard activity (1 mol dm^{-3}), μ_i and μ_j are chemical potentials or molar free enthalpies of either gaseous species i or dissolved materials j.

Reduced quantities p_i/p^0 and c_j/c^0, which are very often omitted in the literature and will be also omitted in the book from hereon, are used in order to make the argument of the logarithms adimensional.

The corrected equations for higher pressures and concentrations read

$$\mu_i = \mu_i^0 + RT \ln(p_i\, f_i) \tag{3.5 c}$$

$$\mu_j = \mu_j^0 + RT \ln(c_j \gamma_j) \tag{3.5 d}$$

Introducing Eqs. (3.5 c,d) into Eq. (3.3) and neglecting deviations from ideality yields

$$U_0 = \left\{ \sum v_{i,j} \mu_{i,j}^0 + RT \ln\left(\Pi_{i,j} c_j^{v_j} p_i^{v_i} \right) \right\} / (v_e F) \tag{3.6}$$

with $v_{i,j}$ the stoichiometric factors of gaseous, index i, and dissolved, index j, reactants and products of the cell reaction; $v_{i,j}$ is positive for products and negative for educts and v_e is the stoichiometric factor of electrons for the reaction under consideration..

For the hydrogen/chlorine cell this translates into

$$F U_0 = \Delta G_{HCl}^0 + RT \ln\left\{ \left(p_{H_2} p_{Cl_2}\right)^{1/2} c_{HCl}^{-1} / \left(mol^{-1} dm^3 bar\right) \right\} \tag{3.6 a}$$

Equation (3.6 a) defines the concentration and pressure dependence of the cell potential of the hydrochloric acid cell and simultaneously the so called standard equilibrium potential U^0 of the cell reaction

$$U_{HCl}^0 = \Delta G_{HCl}^0 / F \tag{3.6 b}$$

which is the equilibrium cell potential of a hypothetical cell in which all reactants and products possess standard activities and fugacities respectively.

ΔG^0 determines the equilibrium coefficient $K_{p,c}$ of the chemical reaction

$$\Delta G^0 = -RT \ln K_{p,c} \tag{3.6 c}$$

so that the measurement of equilibrium cell potentials combined with their extrapolation to standard condition is used to determine data for chemical equilibria, in particular for chemical equilibria which involve dissolved electrolytes.

For the chosen chlorine/hydrogen cell the corresponding relation, which allows to determine the standard Gibbs energy, reads

$$U_{HCl}^0 F = \Delta G_{HCl}^0 = -RT \ln K = -RT \ln\left[a_{HCl,\,diss} / \left(p_{H_2} \cdot p_{Cl_2}\right)^{1/2} \right]_{equil} \tag{3.6 d}$$

For the electrochemical engineer it is most important to note that the equilibrium cell potential U_0 determines the minimum amount of electrical energy which has to be expended for any given electrochemical conversion or electrolysis process ($\Delta G > 0$) accounting for any given concentration and pressure of reactants and products or the maximal amount of electrical energy which can be generated from a battery or a fuel cell ($\Delta G < 0$) which is driven by the respective reaction. Any irreversibility in a practically performed electrochemical process increases the expended electrical energy for an energy consuming electrolysis and decreases the extractable electrical energy from a fuel cell or a battery.

3.2
Electrode Potentials, Reference Electrodes, Voltage Series, Redox Schemes

For vanishing cell currents and neglecting diffusion potentials the electrical potential in the electrolyte is constant between the two electrodes (Fig. 3.2). Therefore the equilibrium cell voltage U_0 is to a good approximation equal to the difference of the two so-called interfacial potentials $\Delta\varphi_a$ and $\Delta\varphi_c$ which exist at the phase boundaries anode/electrolyte and cathode/electrolyte. Interfacial potentials mean differences of internal potentials of neighboured phases:

$$U_0 = [\Delta\varphi_a - \Delta\varphi_c]_{i=0} = [(\varphi_a - \varphi_{el}) - (\varphi_c - \varphi_{el})]_{i=0} \tag{3.7}$$

For principle reasons the internal potential differences $\Delta\varphi_a$ and $\Delta\varphi_c$ which are called interfacial potentials cannot be measured whereas the difference of interfacial potentials occurring in electrochemical cells are well defined and easily measurable quantities. If one of these interfacial potentials is fixed by a rapidly established electrochemical equilibrium which does not vary with time, a change in $\Delta\varphi_a - \Delta\varphi_c$ reflects any change in the second non-fixed interfacial potential.

This is the reason to introduce reference electrodes which possess themselves well defined and easily reproducible, though principally unknown, interfacial potentials $\Delta\varphi_{reference}$ (Fig. 3.3) in order to measure electrode potentials of single electrodes vs. such fixed electrode potential of the reference electrode. As the most popular example of a reference electrode we refer to the normal hydrogen or standard hydrogen electrode (NHE). This reference electrode, which is by no means a uniquely distinguished reference, is a hypothetical hydrogen electrode which dips into an acidic aqueous solution in which protons possess the activity (activity means activity coefficient corrected concentration) of 1 mol dm^{-3} and is sparged with hydrogen of 0.1 MPa pressure (more precisely with a fugacity of 0.1 MPa). Other practical reference electrodes are for instance the 0.01n HCl/hydrogen electrode, the KCl-saturated calomel electrode (SCE: Hg, covered by a Hg_2Cl_2/Hg mixture and being in contact with a saturated aqueous potassium chloride solution) or 1 mol l^{-1} or also a KCl-saturated silver/silver chloride electrodes 1 mol l^{-1} KCl Ag/AgCl or KCl$_{sat}$ Ag/AgCl which are connected electrolytically by means of an electrolyte bridge most often in the form of a Luggin – capillary, filled with aqueous saturated KCl-solution, Fig. 3.3, to the electrode under

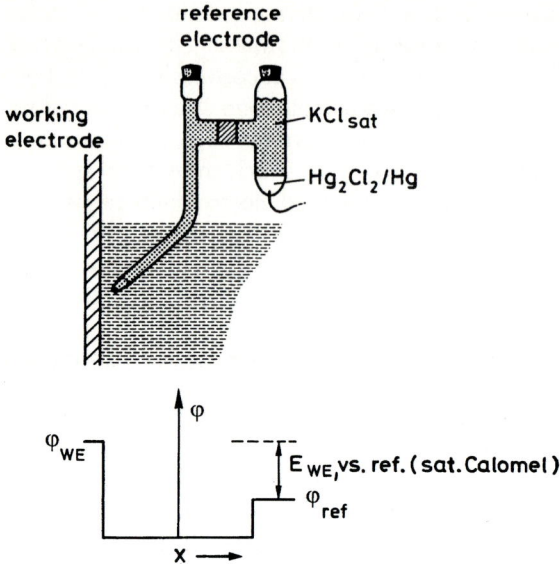

Fig. 3.3. Schematic description of the use of the interfacial potential $\Delta\varphi_{ref}$ of a reference electrode in an electrochemical cell by means of an electrolyte bridge, most often filled by sat. KCl solution. The electrolyte bridge is introduced in the form of a so called Luggin-capillary. Index $_{WE}$ means working electrode

investigation. Of practical importance is the reversible hydrogen electrode (RHE) in any respective electrolyte solution which can very often be established by dipping a Pt-black activated Pt-electrode into the electrolyte and sparging it with pure hydrogen.

The RHE-potential is defined by the proton-activity of the solution under study and is – although well defined – not a solvent/solute independent quantity since it is determined by the pH of the electrolyte under consideration. The RHE is of high practical value in those electrochemical systems in which water is decomposed (water electrolysis) or formed (fuel cells). It is therefore often used as reference for any hydrogen or oxygen evolving electrode process and the respective hydrogen or oxygen consuming fuel-cell electrodes. Likewise useful is the reversible chlorine electrode for all systems where chlorine is evolved or electrochemically reduced. Table 3.1 collects some of the more popular reference electrodes used for measurements in aqueous solutions and their potential with respect to the standard or normal hydrogen electrode (vs. NHE).

The difference of the interfacial potential of any arbitrary electrode under study and that of any selected reference electrode is called the electrode potential versus the chosen reference-electrode. (For instance: +0.2 V vs KCl sat. calomel, or in abbreviated form: +0.2 V vs SCE).

Although the normal hydrogen electrode is by no way distinguished among all possible reference electrodes it had been distinguished historically and by convention as the preferred reference for all electrode-potential measurements in aqueous

3.2 Electrode Potentials, Reference Electrodes, Voltage Series, Redox Schemes

Table 3.1. List of reference electrodes at 25 °C

Electrode	Potential/V
standard hydrogen (NHE)	0
reversible hydrogen (RHE)	$-0.059 \cdot \text{pH}$
calomel, KCl sat. (SCE)	+0.243
calomel, 1 mol l^{-1} KCl sat. (NCE)	+0.2828
silver/silver chloride, 1 mol l^{-1} KCl (NSE)	+0.2224
aqu. Quinhydrone	+0.699

solution and the (unknown) value of the internal potential difference of the standard hydrogen electrode, $\Delta\varphi^0_{H_2}$, is deliberately and per definitionem taken to be zero.

The standard potential value of any other equilibrium electrode (all participating reactants possess either unit activity, if they are pure phases or all concentration activities of dissolved reactants equal 1 mol/dm^3 or gas fugacities equal 0.1 MPa or 1 bar respectively) are referred to the potential of the standard hydrogen electrode and correspondingly ordered in the voltage series.

The voltage series for aqueous solutions of a number of metal/metal ion electrodes of practical interest is collected in Table 3.2 a, for element/element-anions electrodes in Table 3.2 b and for redox couple electrodes, where the reduced as well as the oxidised species is dissolved in solution, in Table 3.2 c.

The concentration dependence of the equilibrium potential of a redox couple

$$A + e^- \rightleftarrows A^- \qquad (3.8\text{ a})$$

is given by the Nernst equation

$$E = E^0 + \frac{RT}{F} \ln \frac{c_A}{c_{A^-}} \qquad (3.8\text{ b})$$

For the more involved and more generally formulated redox-reaction which is written with omission of the electric charge of species A and B

$$\nu_A A + \nu_e e^- \rightleftarrows \nu_B B \qquad (3.9\text{ a})$$

one obtains the Nernst equation

$$E = E^0 + \frac{RT}{\nu_e F} \ln \frac{c_A^{\nu_A}}{c_B^{\nu_B}} \qquad (3.9\text{ b})$$

Table 3.2 a. Normal potentials in Volt of selected cation-element couples[1]

Li/Li^+	−3.01	Se/Se^{3+}	−2.0	Ti/Ti^+	−0.335	Cu/Cu^{2+}	+0.34
Rb/Rb^+	−2.98	Ti/Ti^{2+}	−1.75	Co/Co^{2+}	−0.27	Os/Os^{2+}	+0.7
Cs/Cs^+	−2.92	Al/Al^{3+}	−1.66	Ni/Ni^{2+}	−0.23	Rh/Rh^{3+}	+0.7
K/K^+	−2.92	V/V^{2+}	−1.5	Sn/Sn^{2+}	−0.14	Tl/Tl^{3+}	+0.71
Ba/Ba^{2+}	−2.92	Nb/Nb^{3+}	−1.1	Pb/Pb^{2+}	−0.126	Hg_2/Hg_2^{2+}	+0.796
Sr/Sr^{2+}	−2.89	Mn/Mn^{2+}	−1.05	Fe/Fe^{3+}	−0.036	Ag/Ag^+	+0.799
Ca/Ca^{2+}	−2.84	Cr/Cr^{2+}	−0.86	$H_2/2H^+$	0.00	Pb/Pb^{4+}	+0.8
Na/Na^+	−2.713	Fe/Fe^{2+}	−0.44	Bi/Bi^{3+}	+0.2	Hg/Hg^{2+}	+0.854
La/La^{3+}	−2.4	Cd/Cd^{2+}	−0.44	Sb/Sb^{3+}	+0.24	Pt/Pt^{2+}	+1.2
Mg/Mg^{2+}	−2.38	In/In^{3+}	−0.34	As/As^{3+}	+0.3	Au/Au^{3+}	+1.42

Table 3.2 b. Normal potentials in Volt of some selected anion-element couples[1]

$Te^{2−}/Te$	−0.92	$2I^−/I_{2\,s}$	+0.536	$2Br^−/Br_{2\,g}$	+1.08	$2Cl^−/Cl_{2\,diss}$	+1.40
$Se^{2−}/Se$	−0.78	$2I^−/I_{2\,diss}$	+0.62	$2Br^−/Br_{2\,diss}$	+1.09	$OH^−/OH$	+1.4
$S^{2−}/S$	−0.51	$2CNS^−/(CNS)_2$	+0.77	$ClO_2^−/ClO_{2\,g}$	+1.15	$2F^−/F_{2\,g}$	+2.85
$4OH^−/O_2+H_2O$	−0.401	$2Br^−/Br_{2\,diss}$	+1.066	$2Cl^−/Cl_2(g)$	+1.358		

Table 3.2 c. Normal potentials in Volt of redox couples with soluble partners[1]

$[Co(CN)_6]^{4−}/[Co(CN)_6]^{3−}$	−0.83	$[Mn(CN)_6]^{4−}/[Mn(CN)_6]^{3−}$	−0.22	$Hg_2^{2+}/2Hg^{2+}$	+0.906		
Ga^{2+}/Ga^{3+}	−0.65	$[Co(NH_3)_6]^{2+}/[Co(NH_3)_6]^{3+}$	+0.1	$IrCl_6^{3−}/IrCl_6^{2−}$	+1.02		
In^{2+}/In^{3+}	−0.45	Sn^{2+}/Sn^{4+}	+0.15	$3Br^−/Br_3^−$	+1.05		
Eu^{2+}/Eu^{3+}	−0.43	Cu^+/Cu^{2+}	+0.159	Ti^+/Ti^{3+}	+1.28		
Cr^{2+}/Cr^{3+}	−0.41	$[Fe(CN)_6]^{4−}/[Fe(CN)_6]^{3−}$	+0.36	Au^+/Au^{3+}	+1.29		
$WCl_5^{2−}/WCl_5^−$	−0.4	$3I^−/I_3^−$	+0.535	Ce^{3+}/Ce^{4+} (in 0.1 M H_2SO_4)	+1.44		
Ti^{2+}/Ti^{3+}	−0.37	$MnO_4^{2−}/MnO_4^−$	+0.54	Mn^{2+}/Mn^{3+}	+1.51		
In^+/In^{2+}	−0.35	Fe^{2+}/Fe^{3+}	+0.783	Pb^{2+}/Pb^{4+}	+1.69		
V^{2+}/V^{3+}	−0.255	$OsCl_6^{3−}/OsCl_6^{2−}$	+0.85	Co^{2+}/Co^{3+}	+1.842		

Table 3.2 d. Normal potentials in Volt of oxoanions and metal oxides[1]

$Cr(OH)_{3,s}/CrO_4^{2−}$	−0.12	$IO^−/H_3IO_6^{2−},OH^−$	+0.7	$Mn^{2+}/MnO_4^−,H^+$	+1.51		
$(PbO)_s/(PbO_2)_s,OH^−$	+0.248	$VO^{2+}/HVO_3,H^+$	+1.1	$Cl^−/HClO_2,H^+$	+1.56		
$(PbO)_s/(Pb_3O_4)_s,OH^−$	+0.25	$Cr^{3+}/Cr_2O_7,H^+$	+1.33	$MnO_2/MnO_4^−,H^+$	+1.7		
$TeO_3^{2−}/TeO_4^{2−},OH^−$	+0.4	$Cl_2/ClO_4^−,H^+$	+1.34	$Fe^{3+}/FeO_4^{2−},H^+$	+1.9		
$ClO^−/ClO_2^−,OH^−$	+0.66	$Cl^−/HClO,H^+$	+1.49	$SO_4^{2−}/S_2O_8^{2−},H^+$	+2.01		

[1] Extracted from Dobos [1]

3.2 Electrode Potentials, Reference Electrodes, Voltage Series, Redox Schemes

For a metal/metal ion electrode where the redox reaction at Eq. (3.10 a) is potential determining, and the metal of the electrodes is the pure material of activity 1

$$M^{z+} + ze^- \rightleftarrows M \qquad (3.10\,a)$$

the Nernst equation reads

$$E = E^0_{M^{z+}/M} + \frac{RT}{zF} \ln a_{M^{z+}} \qquad (3.10\,b)$$

or if activity corrections may be neglected ($a_{M^{z+}} = c_{M^{z+}}$) one obtains

$$E = E^0_{M^{z+}/M} + \frac{RT}{zF} \ln c_{M^{z+}} \qquad (3.10\,c)$$

For an element/element-anion electrode as the chlorine electrode with gaseous chlorine

$$1/2\, X_2 + e^- \rightleftarrows X^- \qquad (3.11\,a)$$

one obtains with neglect of activity coefficients

$$E = E^0_{X_2/X^-} + \frac{RT}{F} \ln p^{1/2}_{X_2}/c_{X^-} \qquad (3.11\,b)$$

where in the second term of Eq. (3.11 b) ($a_{X_2} = fc_{X_2}$) is to be inserted instead of p_{X_2} if the molecular species X_2, for instance bromine, is not supplied as a gas but is dissolved with the concentration c_{X_2} in the electrolyte.

A redox couple which may involve in its redox-reaction additionally the solvent, water, like for instance with the formation of metal oxo-anions from lower valent metal cations as with the couple chromate/chromium III or permanganate/Mn II or in general terms the reaction at Eq. (3.12 a):

$$M_xO_y^{z-} + 2nH^+ + v_e e^- \rightleftarrows M_xO_{(y-n)}^{z^*-} + nH_2O \qquad (3.12\,a)$$

with $z^* = z + v_e - 2n$, imposes onto an inert indicator electrode like a Pt wire an electrode potential which is pH-dependent:

$$E = E^0_{M_xO_y^{z-}/M_xO_{y-n}^{z^*-}} + \frac{RT}{v_e F} \ln \frac{c_{M_xO_y^{z-}} \cdot c_{H^+}^{2n}}{c_{M_xO_{y-n}^{z^*-}}} \qquad (3.12\,b)$$

As an example chromate and chromic ions form a redox couple:

$$CrO_4^{2-} + 8H^+ + 3e^- \rightleftarrows Cr^{3+} + 4H_2O\,, \qquad (3.13\,a)$$

the electrode potential of which changes with $c_{Cr^{3+}}$, $c_{CrO_4^{2-}}$ and pH according to

$$E = E^0_{CrO_4^{2-}/Cr^{3+}} + \frac{RT}{3F}\ln\frac{c_{CrO_4^{2-}} \cdot c_{H^+}^8}{c_{Cr^{3+}}}$$

$$= E^0_{CrO_4^{2-}/Cr^{3+}} + \frac{RT}{3F}\ln\frac{c_{CrO_4^{2-}}}{c_{Cr^{3+}}} - \frac{RT}{3F} 2.302 \cdot 8 \cdot pH \quad (3.13\ b)$$

In acidic solutions bichromate instead of chromate is reduced

$$Cr_2O_7^{2-} + 14H^+ + 6e^- \rightleftarrows 2Cr^{3+} + 7H_2O \quad (3.13\ c)$$

and the Nernst equation reads

$$E = E^0_{Cr_2O_7^{2-}/Cr^{3+}} + \frac{RT}{6F}\ln\frac{c_{Cr_2O_7^{2-}}}{c_{Cr^{3+}}^2} - \frac{RT}{3F} 2.302 \cdot 14 \cdot pH \quad (3.13\ d)$$

Table 3.2 c lists some more prominent examples of this type of redox couples. Normal potentials refer in these cases to $a_{H^+}=1$ mol dm^{-3}, that means to a pH of 0 or if hydroxyl ions are involved in the reaction to $a_{OH^-}=1$ mol dm^{-3} or pH = 14 respectively.

The equilibrium cell potential for any process conditions may be obtained by calculating the difference of the two involved single electrode potentials. For this purpose starting from the normal potentials of Tables 3.2 a–d, single equilibrium electrode potentials are calculated by inserting concentrations (activities) and pressures (fugacities) of dissolved or gaseous reactants as they prevail under practical operating conditions into the respective Nernst equation.

In the membrane chloroalkali electrolysis, for instance, hydrogen (1 bar) is evolved at the cathode from a 30 wt% KOH solution, whereas the chlorine anode evolves chlorine (1 bar) from brine which contains approximately 200 g dm^{-3} NaCl (3.4 mol dm^{-3}). Omitting in a first approach any activity and fugacity corrections and calculating c_{NaOH} with $\rho_{30\ wt\%\ NaOH}=1.3$ g cm^{-3} to be $c_{NaOH}=6$ mol dm^{-3} one obtains for the single electrode potentials

$$E_{Cl_2} = E^0_{Cl_2} - (RT/F)\ln c_{Cl^-} = (1.40 - 0.032)\ V = +1.368\ V \quad (3\ a)$$

$$E_{H_2} = E^0_{H_2} - 2.302 \cdot RT/F \cdot pH = (0 - 0.887)\ V = -0.887\ V \quad (3\ b)$$

With this approximations one obtains:

U_0 (chloroalkali electrolysis) $\approx (1.358+0.887)$V$=2.255$ V (3 c)

This comes already quite close to the value of 2.3 V which is obtained by taking into account activity corrections – in particular for the concentrated caustic

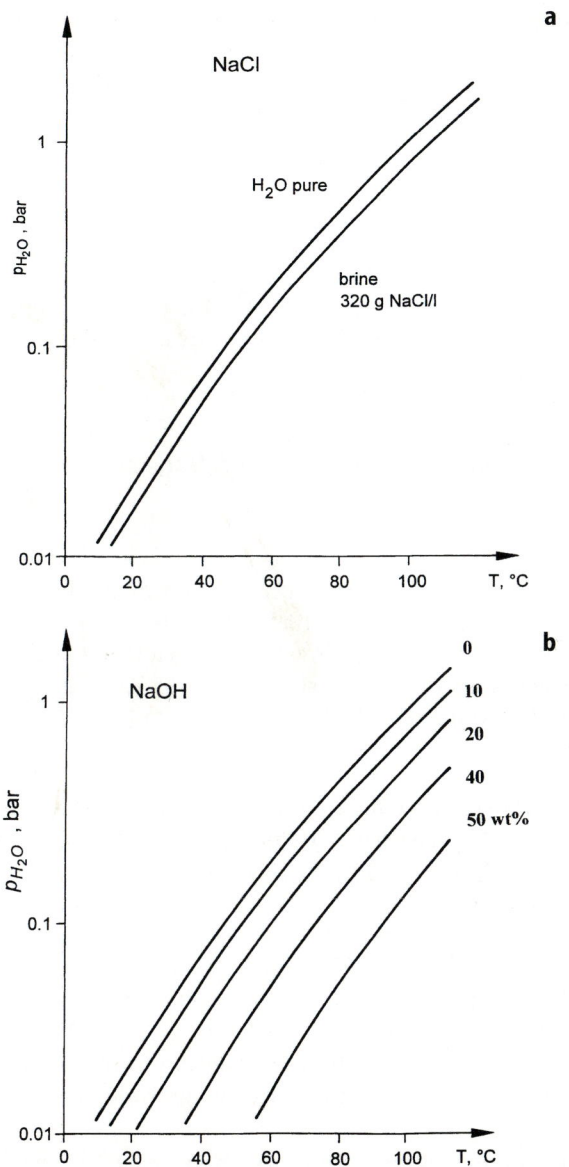

Fig. 3.4a,b. Temperature dependence of water vapour pressures of: **a** NaCl-solutions; **b** NaOH-solutions at various concentrations

soda solution of the catholyte. For technical electrolysis processes which often use highly concentrated electrolyte solutions, it is important to take into account that the activity of the solvent, which usually is water, cannot be assumed to be unity, but is very often considerably decreased.

As an example for decreased water activities in concentrated electrolytes, Fig. (3.4 a,b) plots the vapour pressures of pure water, of NaCl-solutions and of NaOH solutions of different concentrations vs temperature. The activity of the solvent at any temperature and any electrolyte concentration is calculated according to Eq. (3.14) from the ratio of the vapour pressures of the concentrated solution, p^c, and pure water p^0.

$$a^c_{H_2O} = p^c_{H_2O} / p^0_{H_2O} \qquad (3.14)$$

Obviously the decrease of water activity with increasing electrolyte concentrations is higher for NaOH solutions than for NaCl solutions due to pronounced exergetic solute–solvent interaction

3.3
Reaction Enthalpy, Reaction Entropy, Thermoneutral Cell Voltage and Heat Balances of Electrochemical Reactions

Equation (3.3) defines the Gibbs free energy ΔG of any cell reaction as the minimal amount of electrical energy per mol of converted substrate which must be expended in an electrolysis process. This equation differs from the total energy balance and the heat balance of a given cell reaction which is performed under equilibrium conditions. The total energy to be converted is the reaction enthalpy ΔH which differs from ΔG by the entropic term $T\Delta S$

$$\Delta G = \Delta H - T\Delta S \qquad (3.15)$$

For any electrolysis process ($\Delta G > 0$) which involves a positive reaction entropy, ΔG is smaller than ΔH and the entropic term $T\Delta S$ has to be supplied by introducing thermal heat from the environment into the cell, if the reaction is performed reversibly, but at constant temperature.

$$T\Delta S = \Delta H - \Delta G \qquad (3.16)$$

If ΔS is negative, $T\Delta S$ must be transferred as heat from the cell to the environment.

In general all electrolysis processes which involve gas evolution from a liquid electrolyte have a positive reaction entropy of the order of magnitude of R per mol of evolved gas.

Table (3.3 a) collects for four important electrolysis processes which involve gas evolution standard Gibbs enthalpies and reaction enthalpies together with standard entropies at the given process temperatures.

Figure 3.5 depicts the temperature dependence of ΔG^0 and ΔH^0 with the difference representing $T\Delta S^0$ of the most important water splitting/water formation reaction. For water formation ΔG^0 and ΔH^0 are negative for the formation of water. At 373 K the transition from liquid water to water vapour causes a drop in ΔH due to the heat of evaporation. Whereas ΔH^0 is almost constant from 373–1100 K, ΔG^0 decreases by approximately 20%. Only the quantity ΔG^0 can be con-

3.4 Heat Balances of Electrochemical Processes

Table 3.3. a Thermodynamic data. **b** Equilibrium voltage, U_0, and thermoneutral cell voltage U_{th} of some cell reactions with gas evolution

		(a)			(b)	
Cell reaction		ΔG^0/kJmol^{-1}	ΔH^0/kJ mol^{-1}	ΔS^0/Jmol^{-1}K^{-1}	U_0/V	U_{th}/V
$(H_2O)_l \rightarrow H_2 + 1/2 O_2$	100 °C	224	277.5	+159	1.16	1.42
$(H_2O)_g \rightarrow H_2 + 1/2 O_2$	100 °C	224	243	+50	1.16	1.25
$(Al_2O_3)_s \rightarrow 2Al_l + 3/2 O_2$	1300 K	1263	1688	+326	2.18	2.9
$2(Al_2O_3)_{diss} + 3C_s \rightarrow 3CO_2 + 4Al$	1300 K	1763	2189	+431	1.89	1.89

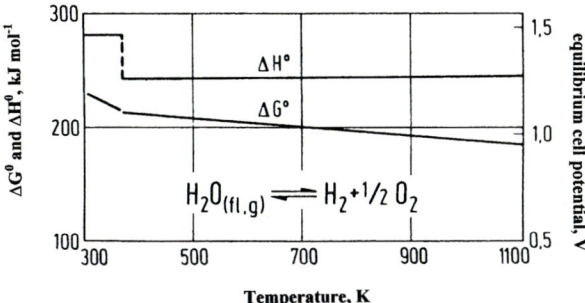

Fig. 3.5. ΔG^0, ΔH^0 and U_0 of the water splitting/water formation reaction from 300 to 1100 K

verted into electricity in a fuel cell and must at least be expended in water electrolysis.

Whenever the electrolysis process has to be assumed to be performed under adiabatic conditions, which is almost realised in large industrial electrolyzers, the total reaction enthalpy must be provided for by expending electrical energy so that instead of U_0 – the minimal or equilibrium cell voltage – the so called "thermoneutral" cell voltage U_{th} or ΔH cell voltage must be applied in order to balance the input of electrical energy with the enthalpy demand of the cell reaction.

$$U_{th} = \Delta H / (v_e F) \quad (3.17)$$

In order to perform the electrolysis process under adiabatic conditions without temperature change of the electrolyte U_{th} has to be the steady state cell voltage. Table 3.3 b compares equilibrium cell voltages U_0 and "thermoneutral" U_{th} cell voltages of the electrolysis processes which are listed in Table (3.3) under (a).

3.4
Heat Balances of Electrochemical Processes

Under technical conditions electrolysis processes are always performed irreversibly so that actual cell voltages exceed the equilibrium cell voltages often by far.

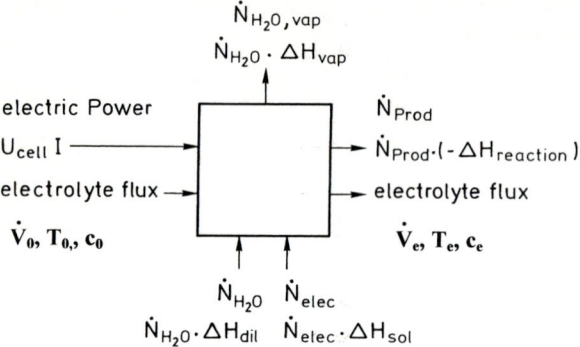

Fig. 3.6. Balance of electric power-input, electrochemical energy demand and generated heat

For a fuel cell process the fuel cell voltage is correspondingly lower than U_0. The heat Q generated in excess per mol of product is calculated by

$$Q_{molar} = U_{cell}\, \nu_e\, F - \Delta H \tag{3.18}$$

Since most technical electrolysis processes are performed at elevated temperatures, that means in the temperature range between 70 and 90 °C, the evolved gases which, being by the very nature of their generation fully saturated with water vapour, carry away substantial amounts of heat by the respective amount of latent heat of vaporisation which is eventually removed from the product gas stream in coolers by condensing of the water vapour.

In chloroalkali electrolysis according to the membrane process the catholyte is 30 wt% NaOH and the anolyte is brine with 200 g NaCl dm^{-3}. At a process temperature of 90 °C, the chlorine gas leaving the cell contains approximately 60 mol% of water vapour whereas the evolved hydrogen – due to the diminished water activity in 30 wt% NaOH solutions – contains approximately only 30 mol% of H_2O. Per mol of H_2 or Cl_2 evolved, approximately 0.9 mol of water evaporate in total, which with a molar heat of evaporation of 38 kJ mol^{-1} are able to remove approximately 34 kJ mol^{-1} of chlorine from the cell or per two mol of NaCl converted in the process.

The total energy balance of an operating cell (Fig. 3.6) comprises the following terms per unit time:

(i) the reaction enthalpy:

$$\dot{n}_p \Delta H_r$$

(ii) convective heat transport due to heating of the circulating electrolyte:

$$\dot{V}_{el} \rho c_p (T_{out} - T_{in})$$

(iii) heat removal by solvent evaporation:

$$\dot{N}_{H_2O,vap} \cdot \Delta H_{vap} = I \sum_j \frac{1}{\nu_{e,j}F} \left(\frac{p_{H_2O}}{P - p_{H_2O}} \right) \cdot \Delta H_{vap}$$

(iv) enthalpies of dissolution or dilution for those cases where pure solvent or solid or concentrated electrolyte is fed directly into the cell additionally to the recirculated electrolyte:

$$\dot{N}_{salt} \Delta H_{diss} + \dot{N}_{H_2O} \cdot \Delta H_{dil}$$

(v) consumed electrical energy

$$U_{cell} I$$

$\dot{N}_{H_2O}$ and $\dot{N}_{salt}$ are the molar fluxes of pure water and salt or other electrolytes per unit time which might be added continuously to the cell additionally to the volumetric flow rate $\dot{v}$ of the electrolyte solution, which enters and leaves the cell under steady state conditions.

Equation (3.19) adds these different terms and equates them to the dissipated electric power

(vi)
$$U_{cell} I = (\Delta H_r / \nu_e F) I + \dot{V}_{el} \rho c_p (T_{out} - T_{in}) + \dot{N}_{salt} \cdot \Delta H_{diss}$$
$$+ \dot{N}_{H_2O} \cdot \Delta H_{dil} + I \sum_j \frac{1}{\nu_{e,j}F} \left(\frac{p_{H_2O}}{P - p_{H_2O}} \right) \cdot \Delta H_{vap} \tag{3.19}$$

In Eq. (3.19) all ΔH values are molar quantities (kJ/mol^{-1}) whereas the c_p-values of the circulated electrolyte is meant to be a mass specific quantity (kJkg^{-1}k^{-1}), hence multiplication of $\dot{v}$ with density and specific heat c_p of the circulating electrolyte.

3.5
Retrieval of Thermodynamic Data and Activity Coefficients

Standard Gibbs free energies, enthalpies and entropies are collected for numerous compounds in JANAF Tables[2] and in other collections of thermodynamic data of, for instance, Baring and Knacke[3]. Both explicitly mentioned data collections and others are not consistent with each other insofar as they choose different reference states.

---

2 JANAF tables are setting the standard Gibbs energies, enthalpies and entropies of all elements in their respective thermodynamically stable modification at any given temperature to zero. [2]
3 Baring and Knacke are setting Gibbs energies and enthalpies of all elements at the reference temperature of 298 K to zero whereas entropies are calculated for the elements from the third law [3]

Since most tables are referring only to pure elements or compounds respectively, concentration or pressure corrections to calculate free enthalpies under any condition of concentration and pressure must be applied. Neglecting to a first approximation activity (fugacity) coefficients one obtains for gases:

$$\mu_i(\text{gas}) = \mu_i^0 + RT \ln p_i \qquad (3.20\ a)$$

for solutions of non-ionised compounds

$$\mu_j(\text{solute}) = \mu_j^0 + RT \ln c_j \qquad (3.20\ b)$$

and for ideal solid solutions and liquid mixtures of non-ionised components with mol fractions extending from zero to unity, where the standard state is the pure substance

$$\mu_k(\text{mixture}) = \mu_k^0 + RT \ln x_k \qquad (3.20\ c)$$

with x_k being the mole fraction of the k_{th} component. Regard that μ_j^0 and μ_k^0 are not identical as they refer to different concentration scales and reference states. Since only the logarithm of activity and fugacity coefficients enter into the calculation of more refined data of free enthalpies, the approximate equations which neglect the deviation from ideal behaviour are very often of sufficient accuracy. For solutions of ionised compounds the situation is more complex because usually solute/solvent and solute/solute (interionic) interaction are non-negligible.

Molten mixtures of salts (for instance mixed NaCl/KCl melts), however, as far as they do not exhibit miscibility gaps can often be assumed to behave as almost ideal mixtures according to Eq. (3.20 c). For more refined calculations the empirical approximation given in Eq. (3.20 d) is used for the solvent or host melt, index 1, and for the solute, index 2. α and I are adjusted parameters.

$$\lg \gamma_1 = \alpha x_2 \qquad (3.20\ d)$$

$$\lg \gamma_2 = \alpha x_1^2 + I \qquad (3\ e)$$

This approximation holds for $0.5 < x_1 < 1$.

Often there exists a strong chemical interaction between the two components of a binary salt mixture or salt melt as for instance between alkali chlorides (LiCl, NaCl or KCl) and aluminium chloride due to complex formation

$$MCl + AlCl_3 = MAlCl_4 \qquad (3.21)$$

These binary mixtures can be described to a good approximation as an ideal binary mixture of the two species MCl and $MAlCl_4$ as far as the molar ratio of aluminium chloride to alkali chloride is kept below 1.

For dilute aqueous solutions of strong electrolytes composed of two monovalent ions, for instance HCl, the concentration of which do not exceed 10^{-2} mol

3.5 Retrieval of Thermodynamic Data and Activity Coefficients

dm^{-3} the approximation of Debye and Hückel for single ion activity coefficients $\gamma_\pm$ reads

$$\lg \gamma_\pm = A\, I^{1/2}; \quad I = \text{ionic strength} = 1/2\, c_i z_i^2 \tag{3.22}$$

with c the molar concentration of the ions, and the numerical value of A given by

$$\ln A = \frac{c e_0^2}{2.303 \cdot 2 \cdot \varepsilon \cdot \varepsilon_0 RT} \tag{3.22 a}$$

e_0 is the elemental charge, and $\varepsilon, \varepsilon_0$ the dielectric constant and induction coefficient and c the molar concentration of a 1:1 electrolyte.

Equation (3.22) allows one to calculate activity coefficients of single ions and mean activity coefficients of the dissolved electrolytes for relatively diluted electrolytes $\gamma_\pm = \{(\gamma_+)\cdot(\gamma_-)\}^{1/2}$ for 1.1 electrolytes, and in general

$$\gamma_\pm = \left\{(\gamma_-)^{z+} \cdot (\gamma_-)^{z-}\right\}^{\frac{1}{z^+ + |z^-|}}.$$

But this equation fails at concentrations higher than 10^{-2} molar concentrations and is therefore only of limited value for electrochemical engineers who usually deal with concentrated electrolyte solution.

Since the electrochemical engineer has in most cases to apply his calculations to electrolyte concentrations from 1 mol dm^{-3} upwards where these simplified approximations based on calculations of the coulombic ion/ion interaction under complete neglect of solute/solvent interactions fail, he has to account for activity corrections mainly by evaluating empirical data. For those higher electrolyte concentrations it is meaningless to aim at calculating single ion activities and the experimental determination of total activities and activity coefficients of dissolved electrolytes should be performed instead. Nearly any method applied for the investigation of thermodynamic properties of higher electrolyte concentration like for instance ebulloscopy, cryoscopy, osmometry or vapour pressure determination determines solvent activities rather than solute activities and calculates solute activities by integrating the Gibbs–Duhem equation for isothermal mixing under constant pressure

$$N_i d\mu_i = 0 \tag{3.23}$$

which results in

$$\lg \gamma_{el} = \int_1^{x_{solv}} \frac{x_{solv}}{1 - x_{solv}} d\ln \gamma_{solv} \tag{3.24}$$

x_{solv} is the mole fraction of the solvent and γ_{solv} is its activity coefficient which can be measured relatively easily.

Figure 3.4 a,b is an example for the determination of water vapour pressures of concentrated electrolyte solutions which allow to determine the activity and activity coefficients respectively of water in these solutions by making use, for instance, of Eq. (3.14). Solving Eq. (3.24) by measuring γ_{solv} over an extended concentration range of the electrolyte. This meets the difficulty that for low electrolyte concentrations ($x_{solv} \to 1, (1-x_{solv}) \to 0$) the solvent activity approaches 1, log γ_{solv} approaches zero and $x_{solv}/(1-x_{solv})$ approaches ∞ so that the integral cannot be evaluated.

Using the Gibbs Duhem equation (Eq. 3.23) in the form

$$n_{solv} \, d \ln a_{solv} = n_{elyte} \, d \ln a_{elyte} \quad (3.23\,a)$$

multiplying both sides of the equation with $1000/M_{solv} n_{solv}$, applying concentration definition on the molality scale, c_m and making use of the osmotic coefficient, f_{osm}, which is easily measured e.g. by cyroscopy and defines the solvent activity at low electrolyte concentration one arrives at Eq. (3.25) [4]:

$$\ln \gamma^\pm = (f_{osm} - 1) + 2 \int_0^c \frac{f_{osm} - 1}{c_m^{1/2}} \, dc_m^{1/2} \quad (3.25)$$

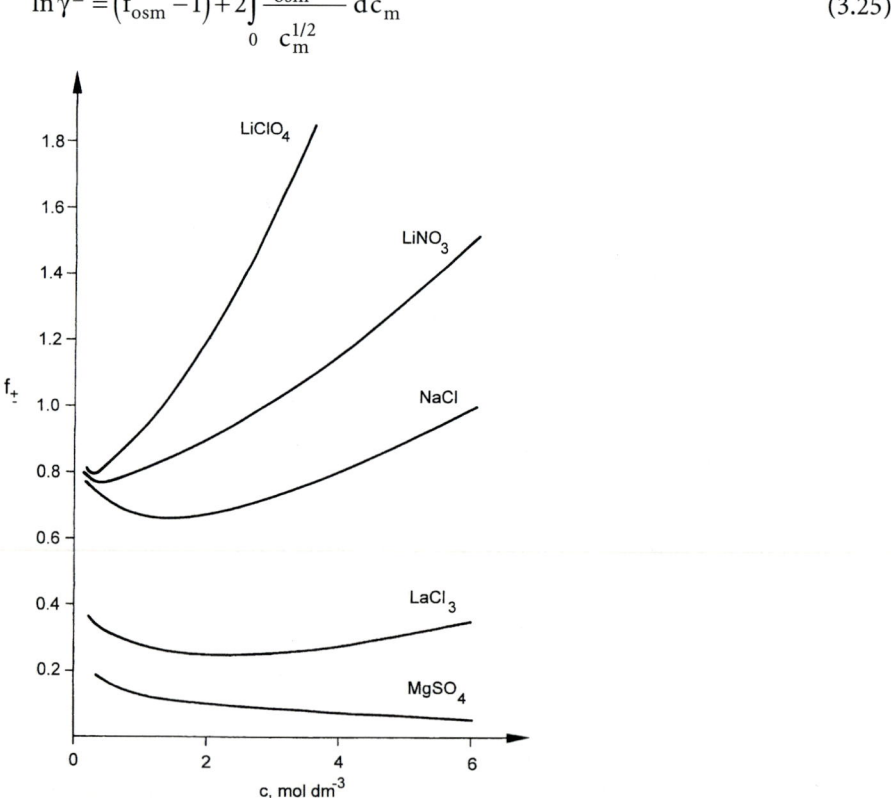

Fig. 3.7. Concentration dependence of electrolyte activity coefficients in concentrated solutions of various electrolytes

which can be integrated as the quantity $(f_{osm}-1)/c_m^{1/2}$ converges towards a finite value as $c_m^{1/2}$ approaches zero. The mean ion activity coefficient $\gamma_\pm$ is defined by Eq. (3.26):

$$(\gamma_\pm)^{z_i+|z_j|} = (\gamma_{i,+})^{z_i}(\gamma_{j,-})^{|z_j|} \tag{3.26}$$

Figure 3.7 shows for several electrolytes the dependence of experimentally determined activity coefficients vs electrolyte concentration for aqueous solutions at ambient temperature. Typically activity coefficients of dissolved electrolytes which are composed of higher valent ions – in particular cations – at first decrease steeply with increasing concentration, pass through a minimum and then increase again – very often surmounting unity at higher concentrations.

A host of activity coefficient measurements for concentrated electrolytes is found in the scientific literature. They have been collected critically by Dobos [1] in 1975.

3.6
Thermodynamics of Electrosorption

The electrosorption of reaction educts, products or reaction intermediates that means their adsorption on the electrode surface does not bear on the thermodynamics of an electrochemical process because electrosorption is always an intermediate step of the overall reaction. But electrosorption is important for electrode processes and electrocatalysis and their kinetics. Very often electrochemical reactions demand adsorption of the educt or some reaction intermediate prior to the decisive, rate determining charge transfer. Therefore as an appendix to "Electrochemical Thermodynamics" the thermodynamics of electrosorption is treated in this chapter in a general way.

The thermodynamics of electrosorption of charged particles differ fundamentally from that of uncharged particles in so far as for charged particles the Gibbs energy of adsorption ΔG_{ad} is directly influenced by the interfacial potential difference at the phase boundary or the electrode potential respectively – compare Eq. (3.29) below. For uncharged particles the electrode potential influences the free energy of adsorption more indirectly, as the difference between the actual electrode potential and the respective potential of zero charge E_{pzc} the so called rational potential, $E_r = E - E_{pzc}$, determines the magnitude of ΔG_{ad} of uncharged particles. ΔG_{ad} for uncharged particles frequently changes with the rational potential according to $(E - E_{pzc})^{-2}$. For charged particles one cannot distinguish "true adsorption" that means incorporation of the species under consideration into the inner Helmholtz layer from accumulation in the diffuse part of the double layer, where at least two sheets of solvent molecules separate the respective charged particle from the electrode surface. Uncharged particles are not subject to this type of accumulation in the diffuse layer. The formal thermodynamic description of electrosorptive processes as well as that of simple adsorption also has to take into account that the free energy of adsorption usually is influenced by the mutual interaction of the adsorbed particles, that means that the degree of coverage in contrast to the simple model of the Langmuir isotherm generally alters the adsorption enthalpy.

The Frumkin model postulates a linear dependence of the adsorption enthalpy ΔH_{ad} on the degree of coverage, Θ

$$\Delta H_{ad} = \Delta H_{ad} + \beta \Theta \tag{3.27}$$

For charged as well as for uncharged particles "true" electrosorption in the meaning of contact-adsorption or adsorption in the inner and/or outer Helmholtz layer is often described by the generalised formula (Langmuir isotherm) which holds too for adsorption on solid surface from the gas phase.

$$\Theta = \frac{Kc}{1+Kc} \quad \text{with} \quad \Delta G_{ad} = -RT \ln K \tag{3.28}$$

K is the adsorption equilibrium coefficient which depends on the electrode potential E and the coverage; c is the concentration of the adsorbate in the bulk of the electrolyte.

Correspondingly one obtains Eq. (3.29 a) for the adsorption equilibrium coefficient for charged particles:

$$K_{charged} = K_0 \exp\left\{-\frac{\Delta \mu_{ad}^0 - zF\eta}{RT}\right\} \tag{3.29 a}$$

For charged particles with charge ze and Frumkin dependence of ΔH_{ad} this equilibrium constant reads

$$K_{charged} = K_0 \exp\left\{-\frac{\Delta \mu_{ad}^0 - \beta\Theta - zF\eta}{RT}\right\} \tag{3.29 b}$$

For uncharged particles the last term vanishes whereas $\Delta \mu_{ad}^0$ very often contains implicitly the interfacial potential as $\Delta \mu_{ad}^0$ depends on the rational potential $E_r = (E - E_{pzc})$. The standard free enthalpy of adsorption $\Delta \mu_{ad}^0$ of uncharged species is often found to vary according to Eq. (3.30):

$$\mu_{ad,uncharged}^0 = \frac{A}{\left(E - E_{pzc}\right)^2} \tag{3.30}$$

So for uncharged particles frequently the Gibbs adsorption enthalpy decreases strongly with increasing and decreasing electrode potential starting from the point of zero charge.

Therefore it is a rule that uncharged particles are adsorbed most strongly at the potential of zero charge and its vicinity, whereas anion adsorption increases with positive and cation adsorption with negative rational potential.

3.6 Thermodynamics of Electrosorption

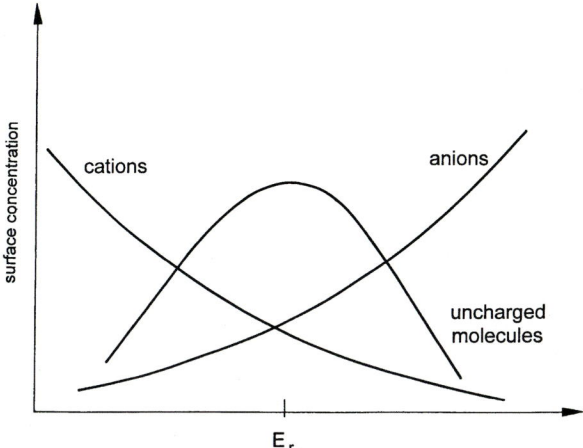

Fig. 3.8. Schematic representation of the dependence of surface concentrations on rational potential of anions, cations and uncharged molecules

Figure 3.8 depicts schematically the dependence of the surface concentration on the rational potential for different dissolved species. For non-surface-active ions the "surface concentrations" which means surface related excess in moles cm^{-2} and which must not be misunderstood as concentration of contact-adsorbed species vanish at potentials which are significantly more negative (anions) or positive (cations) than the point of zero charge and increase to very high virtually infinite values at potentials which deviate significantly in opposite direction because these ions are pushed out or attracted into the diffuse double layer by coulombic interaction with the electrode charge depending on the sign of the surface charge and the ion respectively. Surface active ions or neutral species, however, have easiest access to the Helmholtz layer at potentials close to the potential of zero charge where the Coulomb-interaction between the dipolar solvent molecules and the electrode charge vanishes as the charge vanishes. The solvent-electrode interaction increases strongly as the electrode charge increases with increasing deviation from the point of zero charge, that means with increasing positive or negative rational potential. Therefore at higher anodic or cathodic rational potentials surfactive bulky molecules are pushed out of the Helmholtz layer and replaced by solvent molecules.

References

1. D. Dobos, Electrochemical Data, Elsevier, Amsterdam, Oxford, New York, 1975
2. JANAF Thermochemical Tables, 2nd Edition, D.R. Stull. H. Prophet eds., Ntl. Bur. Stand. (U.S.) NSRDS - NBS 37, 1971
3. I. Barin, O. Knacke, Thermochemical Properties of Inorganic Substances, Springer-Verlag/Verlag Stahleisen, Berlin/Düsseldorf, 1973
4. G. Kortüm, Treatise on Electrochemistry, Elsevier Publishing Comp., Amsterdam, London, New York, 1965, p. 169

CHAPTER 4

Electrode Kinetics and Electrocatalysis

4.1
The Electrochemical Double Layer

Charge transfer between an electrode and the electrolyte proceeds by electron tunnelling between the electrode metal and a suitable electron acceptor or donor on the solution side of the electrode/electrolyte interface across very small distances of the order of fractions of nanometers. In this immediate vicinity of the electrode surface, matter is neither isotropic nor homogeneous. Considering the electrode/electrolyte interface, there is on neither side of it a well defined phase boundary. On the side of the electrode metal there is in atomar dimensions a certain change of electrical potential and charge density, that means an accumulation of negative charge at potentials negative to the potential of zero charge, $E<E_{pzc}$ or positive charge at potentials $E>E_{pzc}$. What follows on the solvent side in the direction perpendicular to the electrode surface is a relatively ordered but mobile sheet of solvent molecules and/or adsorbed species (ions or surfactive molecules) which is designated as the inner Helmholtz layer (IHL). A second less ordered sheet of solvent molecules which is called the outer Helmholtz layer (OHL) forms a smooth transition towards the fully established structure of the bulk solvent which is usually a mix of close range ordered clusters of solvent molecules being in long range statistical disorder (Fig. 4.1 a). The electronic charge on the metal side of the electrode is balanced by ionic counter-charges in the vicinity of the electrode in the IHL and OHL and particularly in the so called diffuse part of the electrochemical double layer. In this diffuse part of the double layer a diffuse cloud of counter ions accumulates due to the coulombic interaction between the charged electrode and the charges of the different dissolved ions.

Figure 4.1 b gives a schematic picture of the charge distribution in the different parts of the double layer and Fig. 4.1 c describes schematically the resulting spatial distribution of the electric potential which in essence establishes the – principally non-measurable – interfacial potential difference

$$\Delta\varphi=\varphi_M-\varphi_S \tag{4a}$$

Although $\Delta\varphi$ cannot be measured, any change in $\Delta\varphi$ can by measured accurately and easily by measuring the electrode potential against any suitable reference electrode which establishes a well defined interfacial potential $\Delta\varphi_{ref}$ by itself.

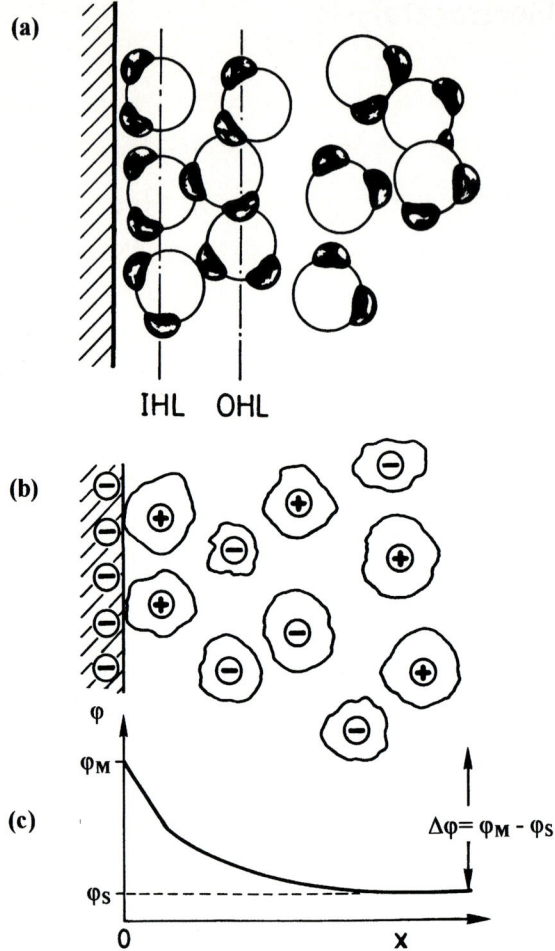

Fig. 4.1. a Schematic presentation of the water structure of inner Helmholtz layer (IHL), and the outer Helmholtz layer (OHL) at the electrode/electrolyte interface. **b** Schematic of charge distribution at the electrode/electrolyte interface determined by the spatial distribution of electrolyte ions, which are depicted here schematically as being embedded in a more or less well defined solvation sheet. **c** Potential distribution in the compact (IHL and OHL) and diffuse part of the electrochemical double layer

It is in particular the part of $\Delta\varphi$ which extends spatially from the outermost part of the metal to the OHL, which exerts the driving potential difference for interfacial charge transfer.

Under conditions of electrochemical equilibrium, the electrode potential (interfacial potential difference) of any reversible redox electrode is ruled by the Nernst equation and is a well defined quantity which can be determined in reference to an equally well defined interfacial potential difference of a suitable reference electrode – compare Chap. 3. Therefore a given electrode potential E_i is always assigned to the reference electrode which has been used.

$$E_i \text{ (vs reference)} = \Delta\varphi_i - \Delta\varphi_{reference} \tag{4b}$$

4.2
Kinetics of Interfacial Charge Transfer

For an electrochemical reaction (charges of species A and B are omitted)

$$A + v_e e^- \rightleftarrows B \tag{4.1}$$

the rate of electrochemical conversion is determined by the applied current

$$r = -dN_A/dt = \Phi^e I / v_e F \tag{4.2}$$

v_e is the stoichiometric number of electrons and Φ^e is the current efficiency for the considered reaction.

It is reasonable to relate electrochemical reaction rates and currents to unit surface in order to obtain rate equations and currents which are independent of the surface area of the electrode which actually has been used in an experiment. Thus one derives at surface specific reaction rates and current densities i as a measure for electrochemical reactions rates.

$$i = I/A = \frac{1}{A}(dN/dt)v_e F \; \Phi^{e-1} \tag{4.2 a}$$

Integrating Eq. (4.2 a) over time and multiplying dN/dt with the molecular weight leads to Faraday's law which allows to calculate the mass of an electrochemically generated product from the amount of consumed electricity

$$m_{j,(product)} = \{A \cdot M_j \Phi^e / F v_{e,j}\} \int i \, dt \tag{4.2 b}$$

It is an experimental fact that whenever mass transfer limitations are excluded, the rate of charge transfer for a given electrochemical reaction varies exponentially with the so called overpotential η which is the difference between the equilibrium potential E_0 and the actual electrode potential E, ($\eta = E - E_0$). Since for the electrode reaction Eq. (4.1) there exists a forward and a back reaction which both are changed by the applied overpotential in exponential fashion, but in opposite sense, one obtains as the effective total current density the difference between anodic and cathodic partial current densities according to the BUTLER VOLMER equation:

$$i = i_0 \left(\exp\left\{\frac{\alpha_a F \eta}{RT}\right\} - \exp\left\{\frac{-\alpha_c F \eta}{RT}\right\} \right) \tag{4.3}$$

i_0 is called the "exchange current density", that means the current density exchanged back and forth under equilibrium potential. α_a and α_c are called the anodic and cathodic charge transfer coefficients which determine the relative response of the respective partial current densities towards a change in overpotential.

For one-electron charge-transfer reactions α equals the so called symmetry factor ß. Usually symmetry factors β_a and β_c are close to a value of 0.5. There does not exist, however, any fundamental physical law which demands a value of 0.5 for symmetry factors of any arbitrary electrochemical reaction nor is there any law which demands that the sum of α_a and α_c is unity. The sum of α_a and α_c, however, has to equal the stoichiometric number of electrons v_e of the potential-determining electrode reaction. Therefore only for simple one-electron charge transfer reactions the sum of α_a and α_c is 1. This is a rule rather than a law. Further it must be stressed, that the condition $\alpha_a + \alpha_c = v_e$ is only expected to be fulfilled if the α values are determined at potentials close to the equilibrium potential E_0 where reversibility prevails. For high cathodic and anodic overpotentials, however, the respective forward and backward reactions may proceed according to very different mechanisms and stoichiometries – resulting in different i_0 values for the forward and backward reaction which differ from each other.

The electrochemical engineer rather would like to measure rate laws, that means current density/potential correlations under well defined conditions of concentrations, electrode potential, mass transfer conditions and homogeneously distributed current density than to adopt without any measurements a generally assumed or believed value for the charge transfer coefficient. In particular he would like to find out whether charge transfer coefficients for a given reaction keep constant or whether they change whenever measurements of current–voltage curves are extended over a wider range of current densities and electrode potentials, or other, non-electrical process parameters, like temperature or concentrations of reactants.

For a given overpotential the effective current density depends as well on the magnitude of the charge transfer coefficient as on the exchange current density i_0. If the overpotential is high enough – that means if either – $\alpha_c \eta F/RT$ or $\alpha_a \eta F/RT \gg 1$, then one of the partial current densities in the Butler Volmer equation overrules the other:

$$i = i_0 \exp\{\alpha_a F\eta/RT\} \quad \text{for } \alpha_a F\eta/RT \gg 1 \tag{4.4}$$

or

$$i = -i_0 \exp\{-\alpha_c F\eta/RT\} \quad \text{for } -\alpha_c F\eta/RT \gg 1 \tag{4.5}$$

From these simplified equations the so called Tafel equation is derived

$$\eta = a + b \lg i \tag{4.5 a}$$

with

$$a = -2.302 \cdot (RT/\alpha F) \lg i_0 \quad b = 2.302 \cdot (RT/\alpha F) \tag{4.5 b}$$

4.2 Kinetics of Interfacial Charge Transfer

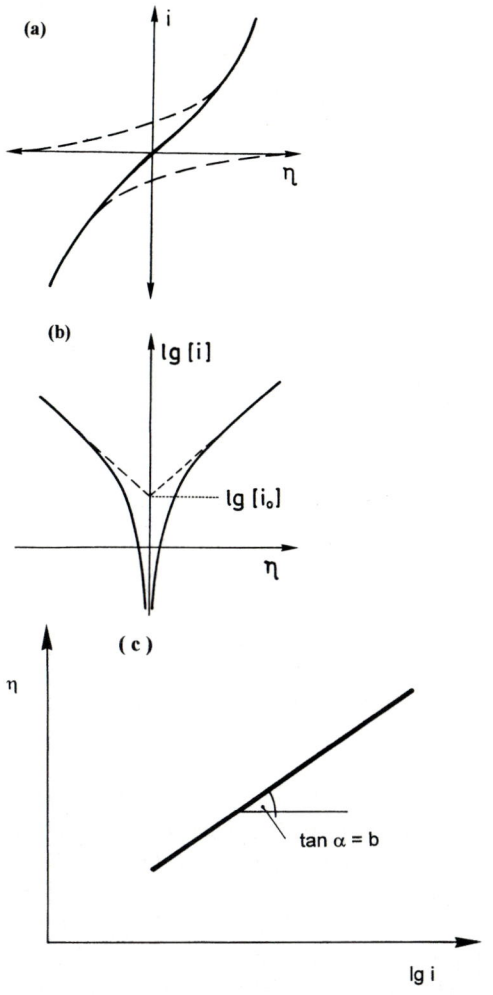

Fig. 4.2. a Plot of current-overpotential correlation. *Dotted lines* signify partial anodic and cathodic branches of current voltage correlation. The slope at $\eta=0$, $(d\eta/di)_{\eta=0}$ is the charge transfer resistance. **b** Semilogarithmic plot for high overpotential allows to determine the exchange current density i_0. **c** Semilogarithmic current voltage curves with Tafel slope $d\eta/d\lg bi = b$

The value b is called the "Tafel slope" which for $\alpha=0.5$ and 300 K amounts to 120 mV per decade.

Figure 4.2 explains schematically in a linear, Fig. 4.2 a, and semilogarithmic plots, Fig. 4.2 b,c of the anodic and cathodic current densities, i, against overpotential, how the so called charge transfer resistance $(d\eta/di)_{i=0}$, Fig. 4.2a, and the charge transfer coefficients, α, Fig. 4.2 c, are determined by extrapolating the slopes of the semilogarithmic current voltage curves (lgi against η) towards high

Table 4.1. Exchange current densities of the hydrogen reaction at different electrode materials in aqueous 1 N H_2SO_4 solution at ambient temperature

Metal	i_0 [A cm^{-2}]
palladium	1.0×10^{-3}
platinum	8.0×10^{-4}
rhodium	2.5×10^{-4}
iridium	2.0×10^{-4}
nickel	7.0×10^{-6}
gold	4.0×10^{-6}
tungsten	1.3×10^{-6}
niobium	1.5×10^{-7}
titanium	7.0×10^{-7}
cadmium	1.5×10^{-11}
manganese	1.3×10^{-11}
thallium	1.0×10^{-11}
lead	1.0×10^{-12}
mercury	5.0×10^{-13}

cathodic and anodic current densities where either the cathodic or the anodic current alone determines i. By extrapolating these slopes to the equilibrium potential ($\eta = 0$) lg i_0 is obtained, Fig. 4.2 b.

Table 4.1 collects some data of exchange current densities for the technically important reaction of cathodic hydrogen evolution as measured at different electrode materials and in 1 N sulfuric acid. As can be seen from these data the exchange current densities i_0 vary a lot. Highest observed exchange current densities for cathodic hydrogen evolution are close to 10^{-3} A cm^{-2} (at Pd and Pt) but the lowest values are as low as 10^{-13} A cm^{-2} (at mercury). For the b-values (which are a measure for the charge transfer coefficients) one determines at room temperature at many metals values close to 120 mV per dec. – and values close to 0.5 for α – but b-values close to 30 mV (α_c close to 2) are observed, too – for instance at Pt electrodes. Exchange current densities for anodic oxygen and chlorine evolution are similarly very sensitive towards the chosen anode material and so are the Tafel slopes of these reactions.

Figure 4.3 shows for cathodic H_2 evolution at nickel electrodes which are covered by coatings of a Ni/Mo/Cd alloy that one observes a relatively low Tafel slope of b=30 to 35 mV at low current densities whereas at higher current densities the slope increases and becomes eventually 120 mV. Such changes in Tafel slope are experienced frequently and are often explained to be due to a change in the reaction mechanism on surmounting a critical overpotential or a critical current density. Very often, however, also the influence of variations in surface coverage of an adsorbed reaction intermediate may be the explanation. Fig. 4.3

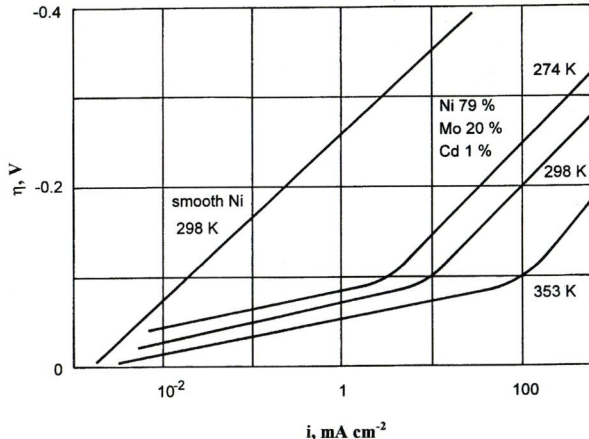

Fig. 4.3. Comparison of current voltage correlations of cathodic H_2 evolution at smooth nickel cathodes and nickel cathodes covered by Ni/Mo/Cd alloy coating. The latter are measured at 1, 23 and 80 °C

shows also the temperature effect on electrode kinetics. Increasing the temperature accelerates the charge transfer and decreases the overpotential.

4.3
Electrode Kinetics of Multielectron Charge Transfer Reaction

Multielectron reactions like cathodic hydrogen evolution or anodic hydrogen oxidation in acidic (a) and alkaline (b) solutions in Eqs (4.6)

$$2H_3O^+ + 2e^- \rightleftarrows H_2 + 2H_2O \tag{4.6 a}$$

$$2H_2O + 2e^- \rightleftarrows H_2 + 2OH^- \tag{4.6 b}$$

or the anodic oxygen evolution or cathodic oxygen reduction respectively, written for acidic and alkaline aqueous electrolytes in Eqs. (4.7)

$$6H_2O \rightleftarrows O_2 + 4H_3O^+ + 4e^- \tag{4.7 a}$$

$$4OH^- \rightleftarrows O_2 + 2H_2O + 4e^- \tag{4.7 b}$$

involve always several different and consecutive charge transfer steps coupled to each other by chemical reactions. The chemical reactions are often slow and kinetically hindered and only one of the different charge transfer or chemical steps will become rate limiting for the electrochemical reaction as a whole whereas the other, faster, steps usually may be assumed to be in fast established chemical and/or electrochemical equilibrium.

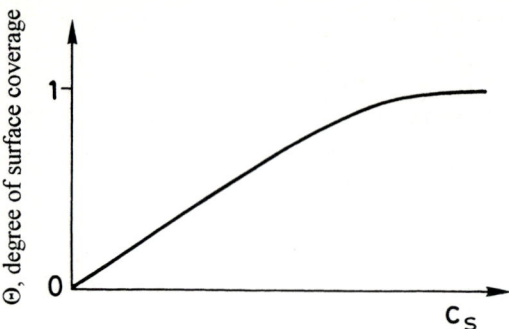

Fig. 4.4. Schematic presentation of Langmuir isotherm

For instance it shall be assumed that the cathodic hydrogen evolution proceeds via a fast established electrochemical adsorption equilibrium of H atoms followed by a slow second charge-transfer step. This rate-limiting step involves adsorbed hydrogen which reacts with protons and electrons under H_2 release. (Volmer–Heyrovsky mechanism)

$$H_2O + e^- \rightleftarrows H_{ad} + OH^- \qquad (4.8\text{ a})$$

fast established equilibrium

$$H_2O_{ad} + H_{ad} + e^- \rightarrow H_2 + OH^- \qquad (4.8\text{ b})$$

slow rate determining step

Neglecting any back-reaction, that means assuming high cathodic overpotentials, the rate of the slow and rate determining reaction at Eq. (4.8 b) would read

$$dn_{H_2}/dt = k_{H,\,ad} \cdot \Gamma_{H,\,ad} = \frac{1}{2} i/F \qquad (4.9)$$

In Eq. (4.9), k would mean to include the surface concentration of water molecules. Hence $k_{H,ad}$ would be a pseudo-first-order rate constant for the surface reaction in Eq. (4.8 b).

Since Eq. (4.8 b) is a charge transfer reaction, k would be assumed to depend on the electrode potential according to the Butler Volmer equation

$$k = k_0 \exp\{-\beta_c \eta F/RT\} \qquad (4.9\text{ a})$$

where β_c is inserted for the cathodic charge transfer coefficient of reaction Eq. (4.8 b) which is meant to be the rate determining one-electron charge transfer. $\Gamma_{H,ad}$ is determined by the fast established electrochemical adsorption equilibrium, Eq. (4.8 a). The simple model of adsorption ($A_{diss} + S \rightleftarrows AS$) is based on the Langmuir isotherm: $\Gamma_{AS} = \Gamma^S K_{ad} c_A / (1 + K_{ad} c_A)$ which is schematically de-

4.3 Electrode Kinetics of Multielectron Charge Transfer Reaction

picted in Fig. 4.4. Γ^S is the saturation surface concentration obtained at full coverage ($\Theta=1$).

With $\Gamma^s_{H,ad}$ the saturation surface concentration of adsorbed hydrogen and $K_{H,ad}$ the adsorption equilibrium coefficient one obtains for prevailing Langmuir adsorption according to reaction at Eq. (4.8 a)

$$\Gamma_{H,\,ad} = \Gamma^s_{H,\,ad} \frac{K_{H,\,ad} \frac{c_{H_2O}}{c_{OH^-}}}{1 + K_{H,\,ad} \frac{c_{H_2O}}{c_{OH^-}}} \qquad (4.10\ a)$$

which can be simplified for small degrees of coverage by:

$$\Gamma_{H,\,ad} = \Gamma^s_{H,\,ad} K_{ad} \frac{c_{H_2O}}{c_{OH^-}} = \Gamma^s_{H,\,ad} \cdot K^*_{H,\,ad} c^{-1}_{OH^-} \qquad (4.10\ b)$$

(with $K^*_{H,\,ad} = K_{H,\,ad} c_{H_2O}$).

K_{ad} varies exponentially with the electrode potential because this adsorption equilibrium involves two charged species, the hydroxyl ion and the electron which both exist in two separate phases, the electrolyte and the electrode metal, which are at different potentials φ_{elode} and φ_{elyte}. Relating the interfacial potential, $\Delta\varphi$, of $\varphi_{elode} - \varphi_{elyte}$ to the interfacial potential at equilibrium $\Delta\varphi_0$ and defining the overpotential $\eta = \Delta\varphi - \Delta\varphi_0$ leads to

$$K_{ad} = K_{ad,\,0} \exp\{-\eta F/RT\} \qquad (4.10\ c)$$

Inserting Eq. (4.10 b,c) into Eq. (4.9) yields

$$k = k_0 \cdot K_{ad,\,H,\,0} \Gamma^s_{H,\,ad} \exp\{-(1+\beta_c)\,\eta F/RT\} \qquad (4.10\ d)$$

Converting surface specific reaction rates dn/dt into current densities i results in

$$dn_{H_2}/dt = i/2F = i_0 \exp\{-\alpha_e F\eta/RT\}/2F \qquad (4c)$$

and one obtains eventually

$$i = 2Fk_0 K^*_{ad,\,H,\,0} \Gamma^s_{H,\,ad} c^{-1}_{OH^-} \exp\{-(1+\beta_c)\eta F/RT\} \qquad (4.11)$$

Equation (4.11) allows identification of the quantities i_0 and α_c.

$$i_0 = 2Fk_0 K^*_{ad,\,H,\,0} \Gamma^s_{H,\,ad} c^{-1}_{OH^-} \qquad (4.11\ a)$$

and

$$\alpha_c = 1 + \beta_c \qquad (4.11\ b)$$

As mentioned above for many one-electron charge transfer reactions the symmetry factor β_c comes close to 0.5, so that one obtains

$$\alpha_c \approx 1.5 \tag{4.11 c}$$

for the assumed reaction mechanism under the limiting condition that the surface coverage by adsorbed hydrogen is small, the Tafel slope would be b = 2.302 (RT/1.5 F) equal to 40 mV per dec. at ambient temperature.

This very simplified hypothetical treatment of the hydrogen evolution reaction is intended, to show that Tafel slopes which are significantly smaller than 120 mV and in particular Tafel slopes which are below 60 mV are indicative for rapidly established electrochemical adsorption/desorption equilibria preceding the rate limiting charge transfer step of multielectron reactions. If the hydrogen reaction is treated in full generality the Volmer-, Heyrovsky- and Tafel reactions have to be considered simultaneously and also in forward and backward direction according to Eq. 4.11 d–f. (MH has the meaning of adsorbed hydrogen):

$$M + H_2O + e^- \underset{k_{-1}}{\overset{k_1}{\rightleftarrows}} MH + OH^- \quad \text{(Volmer)} \tag{4.11 d}$$

$$MH + H_2O + e^- \underset{k_{-2}}{\overset{k_2}{\rightleftarrows}} M + H_2 + OH^- \quad \text{(Heyrovsky)} \tag{4.11 e}$$

$$2MHe^- \underset{k_{-3}}{\overset{k_3}{\rightleftarrows}} M + H_2 \quad \text{(Tafel)} \tag{4.11 f}$$

The rate expressions read:

$$r_1 = k_{1,0}(1-\Theta)e^{-\beta_1 \eta F/RT} - k_{-1,0}\Theta e^{(1-\beta_1)\eta F/RT} \tag{4.11 g}$$

$$r_2 = k_{2,0}\Theta e^{-\beta_2 \eta F/RT} - k_{-2,0}(1-\Theta)e^{(1-\beta_2)\eta F/RT} \tag{4.11 h}$$

$$r_3 = k_3 \Theta^2 - k_{-3}(1-\Theta)^2 \tag{4.11 i}$$

By elimination of the surface coverage θ by the relation $r_1 = r_2 + r_3$ and making use of the relation $\dfrac{k_{1,0}k_{2,0}}{k_{-1,0}k_{-2,0}} = \dfrac{k_{1,0}^2 k_{3,0}}{k_{-1,0}^2 k_{-3,0}} = 1$, which is obtained from $r_1 = r_2 = r_3 = 0$ at $\eta = 0$ one can eliminate two rate coefficients out of six – say k_3 and k_{-3}. This leads with the assumption $\beta_1 = \beta_2 = \beta_3 = 0.5 = \beta$ to the more involved kinetic equation at Eq. (4.11 k) in which only the exchange current densities of two of the three separate processes are defined [1].

$$i = 2F \frac{k_{1,0}k_{2,0}\exp\{-\beta\eta F/RT\}\left(1-\exp\{2F\eta/RT\}\right)}{k_{1,0}+k_{2,0}+\left(k_{-1,0}+k_{-2,0}\right)\exp\{-\beta\eta F/RT\}} \tag{4.11 k}$$

4.4
Thermal Activation and Activation Energies of Electrochemical Reactions

Each electrochemical reaction – even simple redox reactions like the oxidation of hexacyanoferrate II to hexacyanoferrate III, $(Fe(CN)_6^{3+}/Fe(CN)_6^{4+})$, or the reverse reduction are kinetically hindered to a certain degree and can therefore be accelerated by thermal activation. That means that because i_0 is temperature dependent and usually increases as the temperature increases, it is possible to decrease the overpotential for a given current density by increasing the temperature. This is so, because there exists an activation energy for electrochemical reactions as well as for chemical reactions.

Referring to the simplified Butler Volmer equation (for high overpotentials) one obtains by differentiation of ln i with respect to d(1/T) at constant overpotential

$$\left(d\ln i / d(1/T)\right)_\eta = d\ln i_0 / d(1/T) \pm d(\alpha_{a,c} \eta F/RT)_\eta / d(1/T) \quad (4.12)$$

$$= -AE_{i_0}^* / R \pm d(\alpha_{a,c} \eta F / RT) / d(1/T) \quad (4.12\text{ a})$$

$$= -AE_{i_0}^* / R \pm \alpha_{a,c} \eta F / R \quad (4.12\text{ b})$$

$AE_{i_0}^*$ is the activation energy determined for i_0. The second term on the right hand side of Eq. (4.12 a) holds for the anodic and cathodic branch of the current potential correlation. The second term would amount to a potential determined increment of the effective activation energy of approximately 5 kJ mol^{-1} at 0.1 V overpotential at ambient temperature. Plus and minus in the second term on the right hand side of Eq. (4.12 a and b) holds for the anodic and cathodic branch of the current potential correlation. Clearly the effect of temperature on the current density at a given overpotential is not only determined by the activation energy of the electrochemical process $AE_{i_0}^*$ but additionally by the charge transfer coefficient and the overpotential. Figure 4.3 gives an example of the decrease of overpotential with temperature as a consequence of the activation energy of cathodic hydrogen evolution.

4.5
Electrochemical Reaction Orders

The exchange current density i_0 depends in general on the concentration of the species which undergoes charge transfer and it very often depends also on the concentrations of other constituents of the electrolyte. For instance one observes for the anodic oxidation of chloride anions at RuO_2-activated TiO_2 anodes a dependence of i_0 on the pH, that means on the proton concentration.

Therefore i_0 in general must be expressed as a function of the concentrations of different reactants and their respective reaction orders n_i.

$$i_0 = i_0^0 \prod_i c_i^{n_i} \tag{4.13}$$

i_0^0 has the meaning of a standard exchange current density for all reactants being present at unity concentration. The reaction order n_i is defined experimentally by

$$n_i(i_0) = \left(d\ln i_0 / d\ln c_i\right)_{c_j} \tag{4.13 a}$$

or for high overpotentials by

$$n_{i,j} = \left(\frac{d\ln i}{d\ln c_i}\right)_{c_j} \tag{4.13 b}$$

If one chooses simple physical models including established adsorption equilibria to describe electrochemical kinetics, integer or rational numbers n_i for the reaction order of the i-th species are not necessarily to be postulated nor is it to be expected, that reaction orders keep constant across an extended range of reactant concentrations c_i or electrode potentials E. In particular, whenever the i-th species contributing to the charge transfer or to a preestablished equilibrium is electrosorbed, then depending on the degree of surface coverage Θ_i, n_i might change from one towards zero. This is evident from Fig. 4.4, depicting schematically a Langmuir adsorption isotherm. If one assumes a linear, first order dependence of the reaction rate (rate of charge transfer) on the surface concentration Γ_j:

$$i = i_0' \Gamma_j \tag{4.13 c}$$

with i_0' the exchange current density standardised with respect to surface concentrations.

For a simple Langmuir isotherm Γ_j reads:

$$\Gamma_j = \Gamma_j^S \left(K_{ad,j} c_j / (1 + K_{ad,j} c_j)\right) \tag{4.14 a}$$

with Γ_j^S equalling the maximal or saturated surface concentration.

For low degrees of coverage one obtains

$$i = i_0' \Gamma_j^S K_{ad,j} c_j \tag{4.14 b}$$

That means, the formal reaction order is unity. For nearly complete coverage one obtains

$$i = i_0 \Gamma_j^S \tag{4.14 c}$$

That means, the reaction order becomes zero for sufficiently high surface concentration of the j-th species.

The adsorption equilibrium constant K_{ad} depends in general on the electrode potential E, Eq. (4.10 c), and for charged species in particular K_{ad} varies exponentially with E. Therefore one can envisage that changing the electrode potential by 0.15–0.2 V will change at room temperature K_{ad} by from two to three orders of magnitude for an adsorbed species of unit charge. Accordingly the electrochemical reaction order may change substantially also with potential. Therefore one is not allowed, to extrapolate simply exchange current densities from low reactant concentrations to high concentrations or from one potential to another, significantly different one. Instead one is forced to check the validity of established rate equations and reaction orders by additional experiments at the respective overpotential and concentration range relevant for the envisaged technical process.

4.6
Current Density/Potential Correlations for Different Limiting Condition

4.6.1
Micro- and Macrokinetics of Electrochemical Reactions

For chemical reactors one distinguishes microkinetics – as measured under well defined conditions of mass transfer, concentration, temperature and pressure – from macrokinetics which describes the global conversion rate, selectivity and degree of conversion in a chemical reactor – whose physical conditions (concentrations, temperature, pressure) might vary a lot from the reactor inlet to the reactor outlet or from the central part of a packed bed reactor to its peripheral parts. The same distinction of micro and macrokinetics holds for electrochemical reactors. For them there exist not only non-uniform conditions of the physical parameters as concentrations, temperature or pressure but additionally non-uniform electrode potential, mass transfer conditions and current density distributions along the electrode surface. But a thorough investigation of the microkinetics as measured in the form of current voltage curves at electrodes with well defined mass transfer conditions enables the electrochemical engineer to model his reactor appropriately describing its macrokinetic behaviour.

Microkinetic current density voltage correlations for electrochemical reactions can be described by two different limiting cases, namely mass transfer controlled and charge transfer controlled current potential correlations. The charge transfer reactions can be so fast that the potential of the electrode obeys Nernst equation

$$E = E^0 + \frac{RT}{v_e F} \ln\{c_{OX}/c_{Red}\} \tag{4d}$$

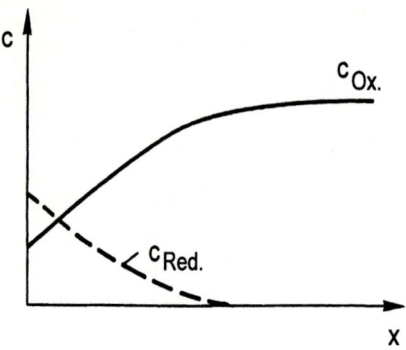

Fig. 4.5. Mass transfer induced concentration profile in front of an electrode due to electrochemical consumption of the substrate Ox and generation of product Red due to cathodic reduction of Ox

At the electrode surface under the condition of non-negligible current density the concentrations c_{ox} and c_{Red} differ from the respective values in the electrolyte bulk due to mass transport (see Fig 4.5) and the electrode potential is ruled by the ratio c_{ox}/c_{Red} at the electrode surface. Therefore one speaks in such cases of mass transfer controlled current voltage curves. If mass transfer induced concentration changes are definitely negligible but the charge transfer proper is hindered and so slow, that it becomes rate determining, charge transfer controlled current voltage curves are measured. If neither mass transfer limitations nor charge transfer hindrance can be neglected one speaks of mixed control.

4.6.2
Mass Transfer Controlled Current Potential Curves

The rate of mass transfer towards an electrode is dictated by the driving concentration difference $\Delta c = c_\infty - c_0$ (which exists between the bulk of the electrolyte, ($x=\infty$, c_∞), and at the surface of an electrode, ($x=0$, c_0), and the mass transfer coefficient, k_m. One obtains

$$i = k_m(c_\infty - c_0)Fv_e \qquad (4.15)$$

Correspondingly the mass transfer controlled cathodic current for the cathodic deposition of a metal according to Eq. (4.15 a):

$$M^{z+} + v_e e^- \rightarrow M; (v_e = z) \qquad (4.15\ a)$$

reads:

$$i^c = -v_e F k_m (c_\infty - c_0)_M \qquad (4.15\ b)$$

4.6 Current Density/Potential Correlations for Different Limiting Condition

and the mass transfer limited current density, i_l (index l for limited), is defined by Eq. (4.15 c).

$$i_l = v_e F k_m \cdot c_\infty \qquad (4.15\ c)$$

From both equations one obtains

$$c_0(i) = c_\infty(1 - i/i_l) \qquad (4.15\ d)$$

Inserting this value into Nernst equation and defining the overpotential by $E(i) - E_0 = \eta$ one obtains

$$E(i) = E_0 + \frac{RT}{v_e F} \ln c_\infty (1 - i/i_l) \qquad (4.16\ a)$$

$$\eta = E(i) - E(i=0) = \frac{RT}{v_e F} \ln(1 - i/i_l) \qquad (4.16\ b)$$

$$i = i_l \left(1 - \exp\left\{-\frac{v_e \eta F}{RT}\right\}\right) \qquad (4.16\ c)$$

Figure (4.6 a) depicts schematically the respective current potential correlation. For a redox system with two soluble redox partners

$$A_s + v_e e^- \rightleftarrows B_s \qquad (4.17)$$

cathodic and anodic limiting currents i_l^c and i_l^a are defined by

$$i_l^c = -k_m c_{A,\infty} v_e F \qquad (4.17\ a)$$

and

$$i_l^a = k_m c_{B,\infty} v_e F \qquad (4.17\ b)$$

Equal values of the mass transfer coefficient k_m being assumed for A and B, one obtains for this case a current potential correlation:

$$i = i_l^{a,c}\left(1 - \exp\{\pm v_e \eta F/RT\}\right) / \left(1 + \exp\{\pm v_e \eta F/RT\}\right) \qquad (4.17\ c)$$

In Eq. (4.17 c) the index + holds for anodic and – for cathodic current densities and overpotentials. Figure 4.6 b describes schematically the current voltage curve of a redox couple with two soluble redox partners for instance $Fe(CN)_6^{3-}/Fe(CN)_6^{4-}$; obeying Eq. (4.17 c).

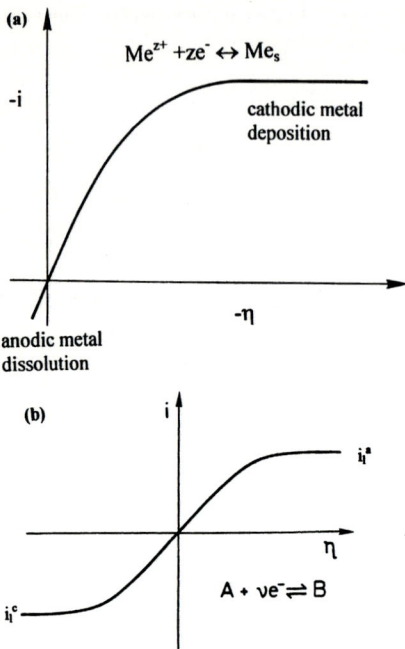

Fig. 4.6a,b. Current-potential correlation: **a** for mass transfer determined metal deposition; **b** for a redox system with two soluble redox partners and well defined mass transfer conditions

4.6.2.1
Reaction Controlled Current Voltage Curves

A slow homogeneous chemical reaction

$$A \xrightarrow{slow} B \tag{4.18 a}$$

$$B + \nu_e e^- \rightarrow C \tag{4.18 b}$$

which precedes the charge transfer may limit the current density to a value i_r which may be remarkably smaller than the mass transfer limited current density.

$$i_r < i_l \tag{4e}$$

Whenever the electrochemical charge transfer is so fast, that the potential-concentration response of the electrode still obeys Nernst law, then one obtains for this case:

$$i = i_r \left(1 + \exp\{\pm 2\nu_e \eta F/RT\} - 2\exp\{\pm \nu_e \eta F/RT\} \right)^{1/2} \tag{4.18 c}$$

4.6 Current Density/Potential Correlations for Different Limiting Condition

(Current voltage correlations which are ruled by homogeneous reactions and the calculation of i_r will be treated in more detail in Sect. 4.7)

4.6.3
Charge Transfer Controlled Current Voltage Correlation

Exclusive charge transfer control of electrochemical reaction rates are observed for slow charge transfer in absence of mass transfer limitations that means if c_0 still approximately equals c_∞. Then charge transfer kinetics are described by the Butler Volmer equation with the assumption that reactant concentrations at the electrode surface (x = 0) equal bulk concentrations (x = ∞). Writing the Butler–Volmer equation and taking account of the concentration dependence of i_0 of Eq. (4.13) one obtains Eq. (4.19):

$$i = i_0^0 \left[\Pi c_i^{n_i} \exp\left\{\frac{\alpha_a \eta F}{RT}\right\} - \Pi c_j^{n_j} \exp\left\{-\frac{\alpha_c \eta F}{RT}\right\} \right] \quad (4.19)$$

Figure 4.2 b depicted schematically in a semilogarithmic plot the potential dependence of the cathodic and anodic partial current for the purely charge transfer determined current. In practice very often only one of both branches (cathodic or anodic) and only one term of Eq. (4.19) can be observed and measured because the chemistry of the system does not allow a reversal of the electrode reaction. In particular the electrochemical conversion of organic compounds often has such an irreversible character. Figure 4.7 shows as a typical ex-

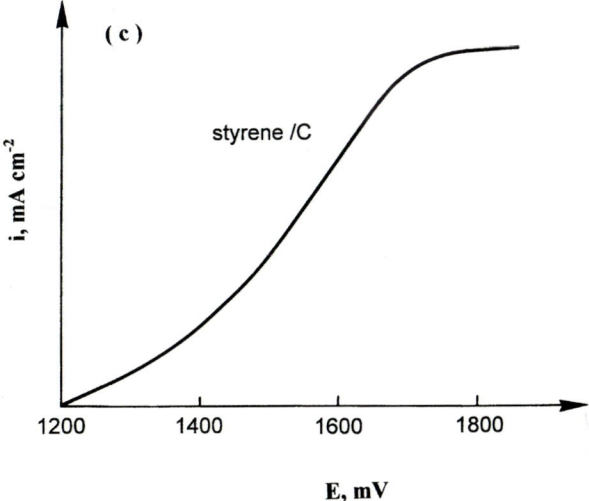

Fig. 4.7. Current-potential correlation of the anodic styrene oxidation at carbon electrodes as an example of a totally irreversible reaction; reference: SCE.

ample the current voltage curve of the anodic oxidation of styrene in methanol where the cathodic branch can never be observed because the initially formed styrene radical cation B^* is very short living and reacts chemically almost immediately to form a non-reducible intermediate C:

$$A + e^- \rightleftarrows B^* \rightarrow C \tag{4.20}$$

so that appreciable steady state concentrations of the radical cation, B^*, can never be established at the electrode and hence its reduction cannot be observed at all.

4.6.4
Combined Activation and Mass Transport Control

Combined activation and mass transfer control is accounted for in the Butler Volmer equation, reaction scheme at Eq. (4.20), for an electrochemical reaction order of 1 for the species A and B (a very common case) by calculating the concentrations at the electrode surface c_0, for instance, from Eq. (4.15 d). So one obtains the concentration-corrected Butler Volmer equation

$$i = i_0 \left(\frac{c_{B,0}}{c_{B,\infty}} \exp\left\{\frac{\alpha_a \eta F}{RT}\right\} - \frac{c_{A,0}}{c_{A,\infty}} \exp\left\{-\frac{\alpha_c \eta F}{RT}\right\} \right) \tag{4.21 a}$$

$$i = i_0 \left[\left(1 - \frac{i}{i_l^a}\right) \exp\left\{\frac{\alpha_a \eta F}{RT}\right\} - \left(1 - \frac{i}{i_l^c}\right) \exp\left\{-\frac{\alpha_c \eta F}{RT}\right\} \right] \tag{4.21 b}$$

$$i = \frac{i_0 \left[\exp\left(\frac{\alpha_a \eta F}{RT}\right) - \exp\left\{-\frac{\alpha_c \eta F}{RT}\right\} \right]}{1 + \frac{i}{i_l^a} \exp\left\{\frac{\alpha_a \eta F}{RT}\right\} - \frac{i}{i_l^c} \exp\left\{-\frac{\alpha_c \eta F}{RT}\right\}} \tag{4.21 c}$$

for high overpotentials one obtains

$$i = \frac{\pm i_0 \exp\left\{\pm\frac{\alpha \eta F}{RT}\right\}}{1 + \frac{i}{i_l^{a,c}} \exp\left\{\frac{\alpha |\eta| F}{RT}\right\}} \tag{4.21 d}$$

and calculates the overpotential according to Eq. (4.21 d)

$$\eta = \left(\frac{RT}{F}\right) \ln\left\{\frac{i}{i_l^{a,c} - i}\right\} - \left(\frac{RT}{F}\right) \ln i_0 \tag{4.21 e}$$

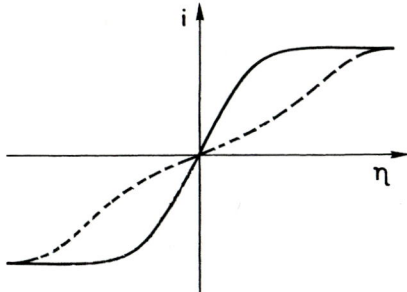

Fig. 4.8. Schematic presentation of mass transfer control, *full line*, and mixed control, *dotted line*, of the reaction Ox+e⁻ ⇌ Red with Ox and Red two soluble species. Mixed control is observed if mass transfer is fast compared to charge transfer, that is $i_0/c_\infty F \ll k_m$ but i becomes comparable to i_l

Equation 4.21 e can be rewritten in a more generalised form by introducing the so called half wave potential $E_{1/2}$

$$E = E_{1/2} \pm m \lg \frac{i_l - i}{i} \tag{4.21 f}$$

where m is the slope of the mass transfer corrected semilogarithmic current voltage curve. The case of metal deposition/dissolution see Fig. 4.6a under combined mass and charge transfer control is obtained from Eq. (4.21 b) by inserting $i^a_l = \infty$

$$i_{depos.} = i_0 \left(1 - \exp\{v_e F\eta/RT\} / \left(i_0 / i_l^c\right) - \exp\{\alpha_c F\eta/RT\}\right) \tag{4.22}$$

Figure 4.8 compares schematically the different cases of mass transfer control, charge transfer control and of mixed control for reaction Ox + e⁻ ⇌ R with dissolved partners Ox and Red of a one electron redox couple of equal concentrations.($c_{O,\infty} = c_{R,\infty}$). Charge transfer control is prevalent in the dotted line for low overpotential.

4.7
Reaction Controlled Current Voltage Curves

4.7.1
Introductory Remarks

Fast homogeneous chemical reactions may precede or may follow the charge transfer step. In both cases the kinetics of the homogeneous reaction influences the current-voltage correlation. In preceding reactions an electrochemically active minority species which is present at very low equilibrium concentrations may be formed from an electrochemically non-active abundant substance in a relatively fast chemical reaction. In another case in a fast consecutive reaction, an electrochemically formed product might be consumed.

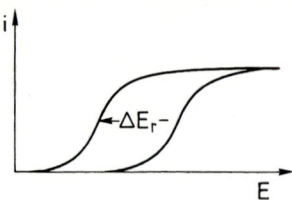

Fig. 4.9. A fast chemical reaction following charge transfer shifts the current voltage curves according to Nernst's law to lower anodic potentials for anodic oxidations and lower cathodic potentials for cathodic reductions

If a fast chemical step precedes electrochemical conversion, one observes a reaction limited current density i_r which is sizeably smaller than the hypothetical mass transfer limited current density i_l, for the non-reactive majority species. If on the other hand the charge transfer step is followed by a fast chemical step, the magnitude of the limiting current density is still the mass transfer limited value. But the potential is affected. Anodic current voltage curves are shifted to more cathodic and cathodic current voltage curves to more anodic potentials (See Fig. 4.9). For the effect of preceding as well as consecutive reactions the Hatta number is the characteristic adimensional quantity which changes current voltage correlations.

4.7.2
Fast Preceding Reaction of an Electroactive Minority Species

Assume a chemical equilibrium between an electroinactive species A and an electroactive minority species B.

$$A \underset{k_{-1}}{\overset{k_1}{\rightleftarrows}} B \tag{4.23 a}$$

with the equilibrium constant $K = k_1 / k_{-1} = c_B / c_A$. If the equilibrium concentration of the species B which undergoes electrochemical conversion:

$$B + v_e e^- \rightarrow C \tag{4.23 b}$$

is low but the reaction rate k_1 is high, the observed reaction rate limited current density i_r may be greater by orders of magnitude than would be expected from the low equilibrium concentration of B.

$$i_r > k_m c_B v_e F \tag{4.23 c}$$

Figure 4.10 explains the situation. Any species of B with a given lifetime τ_r is able to reach the surface of the electrode by diffusion, provided it is generated within the diffusion length δ_r which defines the thickness of the reaction layer.

$$\delta_r = (D\tau_r)^{1/2} \tag{4.24 a}$$

4.7 Reaction Controlled Current Voltage Curves

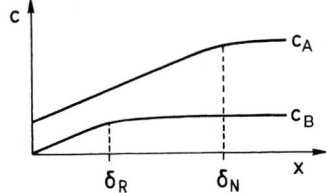

Fig. 4.10. Reaction layer thickness δ_R defined by the diffusion coefficient and the life time, τ_r, of the species B

In Fig. 4.10 the reaction layer thickness δ_r is schematically depicted and compared with δ_N the thickness of the Nernst diffusion layer.

For the chemical first order reaction of Eq. (4.23 a) one obtains

$$\tau_r \approx (k_1 + k_{-1})^{-1} \tag{4.24 b}$$

and from Eq. (4.24 a)

$$\delta_r = \{D/(k_1 + k_{-1})\}^{1/2} \tag{4.24 c}$$

According to Fig. 4.10 one obtains

$$(dc_B/dx)_{max} = c_{B,\infty}/\delta_r = c_{A,\infty} \frac{k_1}{k_{-1}} \left\{\frac{k_1 + k_{-1}}{D}\right\}^{1/2} \tag{4.24 d}$$

For the case $k_1 \ll k_{-1}$ (very low equilibrium concentration of species B) one can calculate the reaction limited current density i_r and the ratio of i_r to i_l, the hypothetical mass transfer limited current density of species A.

$$i_r = c_{A,\infty}(k_1 D K)^{1/2} v_e F \tag{4.25 a}$$

with hypothetical mass transfer limited current density i_l

$$i_l = c_{A,\infty} k_m v_e F \tag{4.25 b}$$

$$i_r/i_l = \left(\frac{k_1 D K}{k_m^2}\right)^{1/2} = Ha K^{1/2} \tag{4.25 c}$$

The adimensional quantity $(k_1 D/k_m^2)^{1/2}$ is called the Hatta-number, Ha, which is a well known characteristic adimensional quantity in unit operations dealing with mass transfer and superimposed chemical reactions. Equation (4.25 c) holds for the simplified case $k_1 \ll k_{-1}$ and would become much more involved if both rate coefficients are of the same order of magnitude.

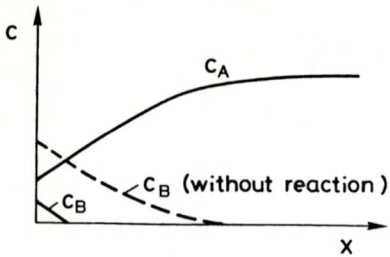

Fig. 4.11. Decrease of product concentration c_B close to the electrode as a consequence of a fast consecutive reaction following the initial charge transfer

4.7.3
Fast Consecutive Reactions

Assume the reaction sequence

$$A + \nu_e e^- \rightleftarrows B \xrightarrow{k} C \qquad (4.26\ a)$$

where the intermediate B is consumed in a fast, irreversible first order chemical step.

A reaction layer thickness δ_r can be calculated with life time τ_r ($\tau_{r,B} = k^{-1}$) of the species B

$$\delta_r = (D/k)^{1/2} \qquad (4.26\ b)$$

The fast consecutive reaction is also characterised by its particular Hatta number

$$Ha_{cons} = \left(\frac{kD}{k_m^2}\right)^{1/2} \qquad (4.26\ c)$$

A shift of the current potential curves to lower electrode potentials is observed as a consequence of the consumption of the species B by chemical conversion to C. In assumed absence of the consecutive reaction the concentration of B at the electrode surface would be $c_{B,0}$ according to the dotted line in Fig. 4.11 and the fast consecutive reaction decreases the concentration of B at the electrode surface to $c^*_{B,0}$. The shift of the potential curve is given by Nernst's law:

$$\Delta E = \frac{RT}{\nu_e F}\ln\frac{c^*_{B,0}}{c_{B,0}} \qquad (4.26\ d)$$

Again the Hatta number, $\left(\dfrac{kD}{k_m^2}\right)^{1/2}$ – now being defined for the consecutive reaction – is relevant for the quantitative treatment of this problem as it determines the potential shift according to Eq. (4.26 e)

$$\Delta E_r = \left(\dfrac{RT}{v_e F}\right) \ln\left(kD/k_m^2\right)^{1/2} = \dfrac{RT}{v_e F} \ln Ha \qquad (4.26\ e)$$

This equation is restricted to those cases, where charge transfer is fast enough, so that the Nernst equation still governs the electrode potential. If the consecutive reaction becomes still faster, then the current voltage curve will become definitely irreversible with the slope approaching $RT/\alpha F$ instead of $RT/v_e F$ for the Nernstian, reversible case.

4.8
Electrocatalysis

4.8.1
Principles of Electrocatalysis

Simple charge-transfer reactions, like the one-electron oxidation or reduction of a dissolved solvated metal ion to a solvated ion of higher or lower oxidation state, most commonly exhibit charge-transfer coefficients of ca. 0.5 and often possess only moderately low activation energies so that the exchange current densities i_0 are high and may range from 10^{-3} up to 10 A/cm². Whenever the electrochemical reaction is more involved, that is, when it is composed of at least two charge-transfer steps with at least one intermediate chemical reaction, then the reaction is likely to be kinetically hindered to a significant degree. Such kinetic hindrance results in low exchange current densities or enhanced overpotentials at given current densities. Simple charge transfer being relatively fast is therefore only moderately or not sensitive to catalytic acceleration. Hindered intermediate chemical reactions in multielectron electrode reactions however, can often be accelerated very strongly by means of heterogeneous catalysis. In these cases electrocatalysis means heterogeneous catalysis of chemical reaction steps by electrode surfaces.

4.8.2
Heterogeneous Electrocatalysis in Cathodic Evolution and Anodic Oxidation of Hydrogen

The relatively simple reaction of cathodic hydrogen evolution and anodic oxidation of dihydrogen in aqueous electrolytes is hampered by the slow reaction of discharged (but absorbed) hydrogen atoms to form the hydrogen molecule and the slow splitting of the molecules respectively. At different electrode materials the kinetic hindrance of cathodic hydrogen evolution may be so different that experimental exchange current densities differ by ten orders of magnitude:

$i_0 = 10^{-3}$ A cm^{-2} at Pt and 10^{-13} A cm^{-2} at Hg   (see Table 4.1)

Catalysis evidently is decisive for lowering the overpotential of the electrochemical hydrogen evolution and oxidation reaction and thus for the energy consumption or dissipation respectively of any process in which hydrogen is electrochemically evolved or consumed, for example in chloralkali electrolysis or fuel cell processes.

4.8.2.1
The Volcano Curve

The enthalpy of adsorption of hydrogen on a given electrode metal is related to the ease with which the hydrogen reactions at Eq. (4.27 a–c) for alkaline and acidic solutions can be performed at the respective metal/electrolyte interface.

The detailed electrode kinetic mechanisms of these reactions depend on the chosen electrode metal, electrolyte and potential or current density respectively, but quite generally three different reaction steps can be distinguished and may become rate limiting.

(a) Electrochemical adsorption/desorption of electrosorbed hydrogen, the so called Volmer reaction:

$$H^+ + e^- \rightleftarrows H_{ad} \qquad (4.27\ a)$$

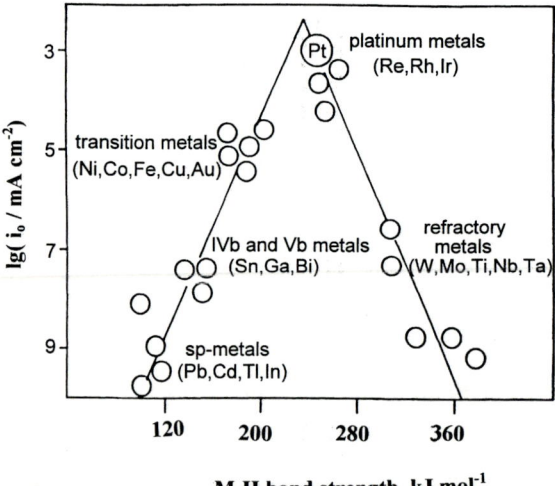

Fig. 4.12. Volcano curve obtained by plotting the log i_0 if the H$_2$ evolution reaction measured at various cathode metals against the molar enthalpy of adsorption of hydrogen or metal hydrogen bond strength respectively

(b) Chemical desorption/adsorption of adsorbed hydrogen atoms – Tafel reaction:

$$2H_{ad} \rightleftarrows H_{2,\,ad} \rightleftarrows H_2 \qquad (4.27\,b)$$

(c) Electrochemical desorption/adsorption of adsorbed hydrogen – Heyrovsky reaction:

$$H_{ad} + H^+ + e^- \rightleftarrows H_2 \qquad (4.27\,c)$$

On nickel electrodes at high current densities for instance all three reactions steps are taking actively part in controlling the overall rate of hydrogen release; see also the treatment in Eqs. (4.9)–(4.11). The overpotential for a given cathodic current density or the logarithm of the cathodic current density at a given overpotential is a measure of catalytic activity of the cathode material. Plotting these values vs the heat of hydrogen adsorption at the respective electrode material, the obtained so called volcano curve depicted in Fig. 4.12 is obtained. This curve shows that an intermediate adsorption enthalpy for hydrogen is optimal because adsorption ($H_2O + e^- \rightleftarrows H_{ad} + OH^-$) is favoured, but associative desorption ($2H_{ad} \rightleftarrows H_2$) or electrochemical desorption is not yet hindered for moderate adsorption enthalpies of approximately 50 kJ mol^{-1} or M-H bond strengths of roughly 240 kJ mol^{-1} respectively. This condition is optimally fulfilled at platinum as the best H_2 catalyst known. It should be stressed that the volcano-curve relation is by no means unique for electrochemical reactions of hydrogen but that one observes the same dependence also for heterogeneous catalysis of typical chemical reactions of hydrogen for instance for hydrogenation or dehydrosulfurization reactions. Figure 4.13 demonstrates the high electrocatalytic activity of Pt metals by comparing the current voltage curve of cathodic hydrogen evolution from alkaline electrolytes at smooth nickel and ruthenium-coated nickel cathodes.

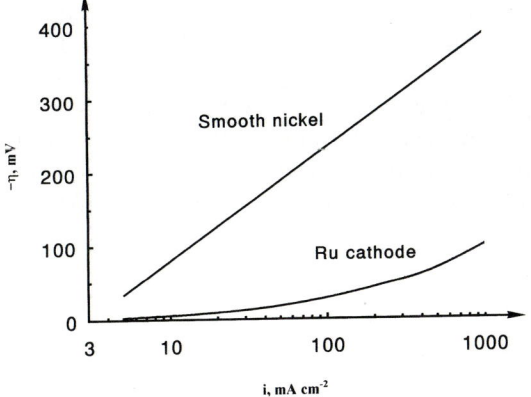

Fig. 4.13. Comparison of the current voltage curves of cathodic hydrogen evolution at smooth nickel electrodes and Ru-coated nickel electrodes in 30 wt% KOH at 80 °C

4.8.3
Electrocatalysis in Anodic Oxygen Evolution and Cathodic Oxygen Reduction

Anodic oxygen evolution and cathodic oxygen reduction (Eq. 4.28 a,b) are much more complex than H_2 evolution and oxidation since they comprise the transfer of a total of four electrons, which involves at least two chemical reactions between at least two charge-transfer steps.

$$2H_2O \rightleftarrows O_2 + 4H^+ + 4e^- \quad \text{(in acid electrolyte)} \tag{4.28 a}$$

or

$$4OH^- \rightleftarrows O_2 + H_2O + 4e^- \quad \text{(in alkaline electrolyte)} \tag{4.28 b}$$

Therefore, oxygen evolution and reduction would be expected to be kinetically hindered more strongly than the simpler hydrogen evolution. Indeed, anodic oxygen evolution and cathodic oxygen reduction, under otherwise comparable conditions, such as temperature and current densities, are usually performed at comparably higher overpotential than hydrogen evolution or oxidation. Electrochemical oxygen reactions exhibit even with the best catalysts known (Ag, Pt, RuO_2, Co_3O_4, and mixed metal oxides containing cobalt oxide) distinctly higher overpotential than cathodic hydrogen evolution or anodic hydrogen oxidation carried out at electrodes supplied with respectively best electrocatalysts (as platinum, Raney nickel, or coatings of Ni/Mo and Ni/Mo/Cd) at ambient temperature.

Another very important more fundamental kinetic difference between hydrogen and oxygen electrochemistry is that anodic oxygen evolution and cathodic oxygen reduction even at moderate overpotentials follow different reaction pathways whereas generally for H_2 evolution and H_2 oxidation the same reaction pathway, though in opposite direction, prevails. Oxygen reduction almost under all practical conditions yields in intermediate peroxide formation, whereas H_2O_2 usually is not an intermediate in anodic oxygen evolution. Since the chemistry of both reactions is so different the catalytic principles for anodic O_2 evolution and cathodic oxygen reduction are usually fundamentally different.

According to a generally accepted model, anodic O_2 evolution is catalysed by the electrochemical oxidation of multivalent metal atoms at the surface of oxide-covered electrodes toward more highly oxidised metal oxide surface sites, which in a subsequent chemical reaction release oxygen molecules (Krasilch'shikov mechanism)

$$M_xO_y + H_2O \rightleftarrows M_xO_yOH + H^+ + e^- \tag{4.29 a}$$

$$M_xO_yOH \rightleftarrows M_xO_{y+1} + H^+ + e^- \tag{4.29 b}$$

$$2M_xO_{y+1} \rightarrow M_xO_y + O_2 \tag{4.29 c}$$

4.8 Electrocatalysis

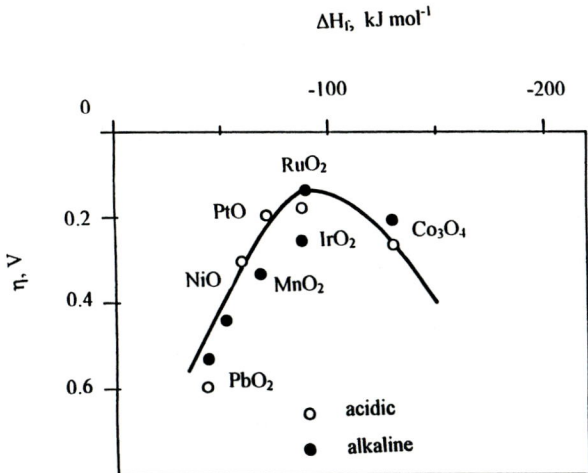

Fig. 4.14. Volcano plot for anodic oxygen evolution obtained by plotting lg i at constant η against ΔH_{Ox} for formation reaction of high valent oxide from lower valent oxide

For anodic oxygen evolution, a volcano curve, similar to that for electrochemical hydrogen evolution, can be obtained by plotting, for instance, log i (for fixed η) or η (for fixed i) against the heat of formation of the higher valent oxide from the lower valent metal oxide (Fig. 4.14). This type of correlation is equivalent to correlating the catalytic activities of various metal oxides with the standard potential, or the Gibbs free enthalpy, for the formation of the higher valency oxide from the lower valency oxide. The maximal catalytic activity is obtained for metal oxides with redox potentials close to the equilibrium potential of the O_2/H_2O or O_2/OH^- electrode.

Similar is the now generally accepted mechanism of electrocatalysis of anodic chlorine evolution by ruthenium dioxide at RuO_2-coated titanium electrodes. This process also involves a change in the valency of surface groups of RuO_2; from the resulting pentavalent, (Ru V), ruthenium oxide chlorides, chlorine is released by a chemical reaction that returns the ruthenium to its original oxidation state as written schematically at Eq. (4.30 a,b).

$$RuO_2 + Cl^- \rightleftarrows RuO_2Cl + e^- \tag{4.30 a}$$

$$2RuO_2Cl \rightarrow 2RuO_2 + Cl_2 \tag{4.30 b}$$

Figure 4.15 demonstrates the effectivity of this type of electrocatalysis in comparing the anodic chlorine evolution at carbon anodes and RuO_2 activated titanium anodes. With RuO_2 coatings the anodic overpotential keeps relatively low even at the highest current density. Comparing the catalytic activity of different anodic electrocatalysts for oxygen evolution from caustic alkaline solutions one takes it for certain that RuO_2 is one of the most efficient catalysts. It is

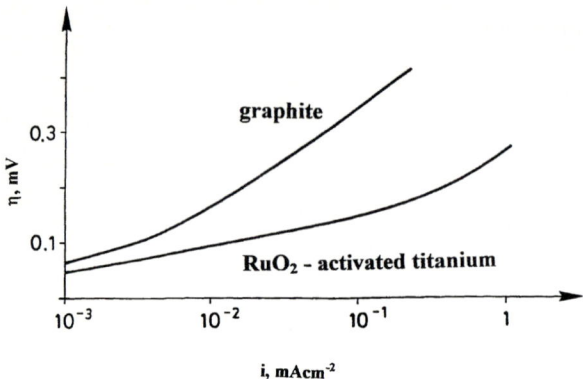

Fig. 4.15. Comparing current voltage curves of anodic evolution of chlorine on carbon anodes and RuO_2-coated titanium anodes

however not stable under electrolysis conditions and its dissolution is a matter of hours in the alkaline electrolyte. Therefore, today, coatings of Co_3O_4, stabilised by formation of mixed crystals with Fe_3O_4, which are less active but much more stable are, practically spoken, better oxygen evolution electrocatalysts.

Cathodic oxygen reduction starts usually with a two-electron reduction of oxygen to hydrogen peroxide ($O_2+H_2O+2e^-\rightarrow OH^-+O_2H^-$) or, in alkaline solution, its anion respectively. Electrocatalysis of cathodic O_2 reduction has mainly to deal with the chemical conversion of the intermediately formed H_2O_2 in particular by accelerating its disproportionation ($2H_2O_2\rightarrow 2H_2O+O_2$) by heterogeneous catalysis. This is brought about by suitable (often quite conventional) heterogeneous catalysts for this reaction which have to be chemically and electrochemically stable under the working potential of O_2-reducing cathodes. In fuel cells, for instance, where the cathodic overpotential should possibly not exceed –250 to –350 mV, such catalysts are Pt, Ag, MnO_2 and cobalt containing mixed oxides like Perovskites (e.g. $La_{0,5}Sr_{0,5}CoO_3$) or spinells (e.g. $NiCo_3O_4$). Today active carbon supported finely dispersed Pt is the catalyst of choice for cathodic oxygen reduction in low temperature fuel cells with acid electrolytes, that means in phosphoric acid and membrane fuel cells. This catalyst is operated at potentials so cathodic, that the Pt-surface is definitely not covered by PtO.

4.8.4
Redox Catalysis

The heterogeneous electrocatalysis of anodic evolution of oxygen or chlorine are only two special cases of heterogeneous redox catalysis which is based on the initial oxidation of appropriate redox systems which undergo rapid and easy electrochemical conversion or recuperation followed by a secondary chemical redox reaction with a substrate. This reaction technique is known as redox catalysis and is used since long in two completely different procedures: namely heteroge-

4.8 Electrocatalysis

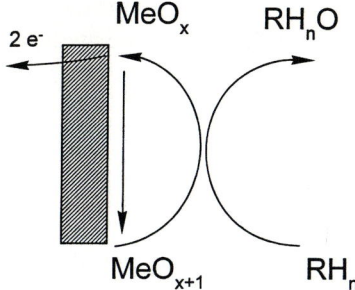

Fig. 4.16. Schematic of heterogeneous redox catalysis mediated by a solid state metal oxide redox couple at the electrode surface

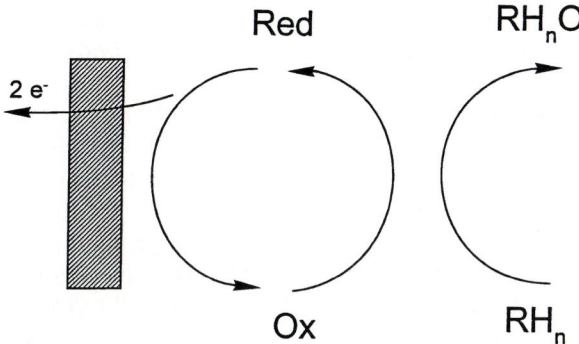

Fig. 4.17. Schematic of homogeneous redox catalysis brought about by electrochemical production of a dissolved oxidant (or reductant) which reacts with an organic substrate – for instance in a stirred tank reactor. Converted oxidants or reductants are subsequently recuperated electrochemically

neous redox catalysis on one hand and mediated electrochemical conversion on the other hand.

In heterogeneous redox catalysis surface groups at the electrode surface are mediating the final heterogeneous chemical oxidation or reduction of the target substrate. The converted surface groups are continuously recovered by electrochemical oxidation or reduction. Homogeneous redox catalysis which also is labelled "mediated electrochemical conversion" makes use of the initial oxidation or reduction of a soluble redox couple as, for instance Fe II/Fe III, Mn II/Mn III or Ce III/Ce IV which react in homogeneous reactions with a soluble substrate.

Heterogeneous and homogeneous redox catalysis are frequently used methods in the field of organo-electrosynthesis where well known chemical redox methods are adopted for the electrochemical practice in closing the redox loop by introducing the continuous electrochemical recuperation of redox reactants (Fig. 4.16 heterogeneous and Fig. 4.17 homogeneous redox catalysis). Examples for heterogeneous redox catalysis are the anodic oxidation of benzene to benzoquinone and hexafluoropropylene to hexafluoropropylene oxide by PbO_2, the anodic oxidation of alcohols to aldehydes, ketones and carboxylic acids by continuously generated NiOOH on Ni(OH)$_2$ covered Ni-anodes in alkaline medium and the previously demonstrated Cr VI-mediated anodic oxidation of organic substrates on Ti/TiO$_2$ electrodes the surface of which is doped by chromate.

The heterogeneous hydrogenation of unsaturated organic compounds on Raney-nickel coated electrodes is another, cathodic example for heterogeneous redox catalysis. At such Raney-nickel cathodes hydrogen is evolved and simultaneously they are applied for catalytic hydrogenation reactions.

Mediated homogeneous redox conversion is performed in a number of commercial organic syntheses processes for instance in the Mn^{3+}-mediated oxidation of toluenes to benzaldehydes, or the oxidation of alcohols to carboxylic acids by anodically recuperated chromate/bichromate. From the process engineering point of view these two reaction techniques, heterogeneous and homogeneous redox catalysis, are quite different. Heterogeneous redox catalysis is applied like electrocatalysis in any other normal electrolysis process and allows us to perform the electrochemical process in the usual way but with lower cell voltage, higher yield and sometimes higher selectivity than without applying electrocatalysis. One has, however, to match the current density to the velocity of the heterogeneous redox reaction, so that highest current efficiencies are obtained.

In the case of continuously operated mediated homogeneous redox conversion the electrochemical reactor – the cell – and the chemical redox reactor very often are physically separated in different vessels in order to match the electrochemical reaction rate for redox recuperation as defined by the applied current to the volume related rate r_V of the homogeneous redox reaction properly according to Eq. 4.31:

$$I\Phi^e/v_e F = V_R r_V \qquad (4.31)$$

V_R=volume of the reactor, r_V=volume specific homogeneous conversion rate, v_e is the stoichiometric number of electrons for (mediated) conversion of the substrate. Detailed examples of electroorganosynthesis processes are given below in Sect. 4.10.

4.9
Catalyst Morphology and Utilisation

4.9.1
Structural Features and Catalyst Morphology of Electrocatalysts for Gas Evolving and Gas Consuming Electrodes

Electrocatalysts for the production and electrochemical conversion of gases must exhibit the following structural features (given in hierarchical order).
(a) *Primary* structure as defined by the chemical composition and catalytic capability of the electrocatalyst.

(b) *Secondary* structure comprising the true inner surface of the highly porous catalyst, which might be for instance a Raney metal (pore diameters of Raney nickel: 10–30 nm and mass specific inner surface of e.g. 50–70 $m^2 g^{-1}$) or platinum-doped active carbon.
(c) *Tertiary* structure made up of coarse cracks and pores in a metallic carbonaceous or oxidic catalyst layer or realised by the intergranular space of more or less toughly bounded manoporous catalyst granules as in PTFE-bonded fuel cell gas diffusion electrodes.

The tertiary structure is important for gas evolving and gas consuming electrodes as it provides short diffusion paths in the electrolyte which floods the nanoporous electrocatalyst particles comprised in catalytic layers and electrode coatings.

4.9.2
Utilisation of Porous Electrocatalyst Particles

In chemical heterogeneous catalysis of gas phase reactions it is common to use highly porous catalysts which come in particles of millimetre to centimetre size in order to increase the effective catalyst surface by making use of the internal surface of the catalyst particle or pellet. In electrocatalysis, in particular applying electrocatalysts in fuel cells and on the surface of hydrogen evolving cathodes, it is also usual to use highly porous – though accounting for the low diffusion coefficients in liquid electrolytes, 10^{-5} $cm^2 s^{-1}$ vs 1 $cm^2 s^{-1}$ in gases, much smaller – catalyst particles, which come rather in micrometer than in centimetre size. For hydrogen evolution also thin nanoporous Raney nickel coatings on smooth nickel supporting electrodes are state of the art with a thickness of several tens of micrometers.

A question of concern common to heterogeneous catalysts in porous chemical catalysist as well as in porous electrocatalytic coatings is the problem of effective catalyst utilisation. Whenever electrodes coated by nanoporous electrocatalyst layers are used, the electrochemical engineer is confronted with mass transfer hindrance of reactants or products into or out of the porous catalyst layer or catalyst particle respectively. If the thickness of the porous catalytic coating or the diameter of the porous particle does not properly match the mean diffusion length of consumed or produced species, then the inner surface of the catalyst is no longer supplied sufficiently with substrate, or becomes electrochemically inactive due to product accumulation and concentration polarisation in the pores. The chemical engineer applies the Thiele-modulus, Eq. (4.32 a) as a characteristic adimensional quantity defining the degree of utilisation of a porous catalyst particle, which is obtained from the adimensionalized differential equation describing the competition of reaction and diffusion in a pore or a porous catalyst particle – see Chap. 7.

$$\text{Thiele modulus} = \Phi_{Th} = L(k/D)^{1/2} \tag{4.32 a}$$

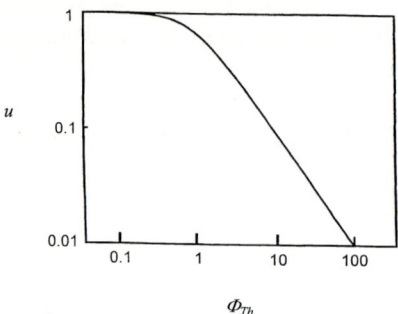

Fig. 4.18. Double logarithmic plot showing the dependence of catalyst utilisation u on Thiele modulus Φ_{Th}. The catalyst is fully utilised up to a Thiele modulus of approximately 1. At high Φ_{Th} values η varies with Φ^{-1}_{Th}

L=pore length or coating thickness or radius of porous particle; k=volume related rate coefficient of first order reaction; D=effective diffusion coefficient of the reactant in the porous matrix.

Figure 4.18 demonstrates in a double logarithmic plot the dependence of the catalyst utilisation u on Φ_{Th} showing that for Φ_{Th} >3 the utilisation decreases according to a limiting Φ^{-1}-law.

The electrochemical Thiele modulus in analogy to the normal Thiele modulus is defined by:

$$\Phi_{Th,\,el} = L\left[i(\eta)a/Dv_eFc\right]^{1/2} \qquad (4.32\ b)$$

with i(η)=current density at overpotential η ; a=inner electrode surface per unit volume of porous electrocatalyst; c=concentration of reactant in the electrolyte bulk; v_e=stoichiometric number of electrons.

The definition of the electrochemical Thiele modulus (Eq. 4.32 b) characterising the degree of electrocatalyst utilisation is a prerequisite for properly tailoring the micromorphology of porous electrocatalytic electrode coatings or fuel cell electrodes, as it allows to match the coating or catalyst particle dimension to the catalytic activity of the material. Accordingly it is necessary to keep the diameter of the nanoporous particles constituting a fuel-cell electrode or the thickness of homogeneously nanoporous fuel-cell electrodes below a critical value given by the condition $\Phi_{Th}\approx 1$ in order to assure complete utilisation of the expensive electrocatalyst. For a detailed treatment of porous catalyst utilisation see Chap. 7.

4.10
Electrocatalysis in Electroorganic Synthesis

4.10.1
Introduction into the Field of Electroorganic Synthesis

In electrochemical conversion of organic substrates one distinguishes so-called mediated from direct electrochemical conversion. Mediated conversion is more related to conventional oxidation or reduction as a synthesis-tool brought about by reacting organic substrates with inorganic oxidants or reductants. In mediated anodic oxidation or cathodic reduction the reagent is electrochemically recuperated. In contrast direct anodic oxidation or cathodic reduction rests on immediate electron transfer between dissolved organic molecules and electrodes.

4.10.1.1
Mediated Electrochemical Conversions of Organic Substrates

The anodic and cathodic conversion of organic compounds might be mediated by electrochemically regeneratable redox systems. In these cases of mediated electrochemical conversion one partner of a redox couple undergoes oxidation or reduction respectively at the electrode and reacts with the substrate in the bulk of the solution shuttling the electronic charge from the electrode to the substrate. Examples for anodic mediators are the redox couples $Cr_2O_7^{2-}/Cr^{3+}$, Mn^{3+}/Mn^{2+}, Ce^{4+}/Ce^{3+}, examples for cathodic mediators are Ti^{3+}/Ti^{4+} or V^{3+}/V^{5+}, for instance.

If the anode or cathode surface is coated with an insoluble redox system which is able to oxidise or reduce organic substrates by a heterogeneous reaction one may call this type of mediated conversion "heterogeneously catalysed" anodic oxidation or cathodic reduction respectively. Examples are oxidation at lead electrodes by PbO_2 or by MnO_2-coatings. With this type of electrochemical conversion the organic substrate interacts chemically with the redox mediator contained in the coating. For instance alcohols which are to be oxidised by Cr VI species contained in the catalytic coating are initially bound to this species as a chromate-ester before they undergo charge exchange and oxidation by the mediator, a mechanism which is known by the name of Wiberg mechanism and which is operative in homogeneous chromate oxidation of alcohols. It is this chemical interaction which justifies the expression heterogeneously catalysed oxidative or reductive conversion.

4.10.1.2
Direct Anodic and Cathodic Electrochemical Conversions of Organic Substrates

Some anodic and cathodic conversions are essentially initiated by a one-electron charge transfer between the electrode and the organic molecule generating radical ions or radicals as first, reactive intermediates. (Eq. 4.33 a–d)

$$A + e^- \rightleftarrows A^{-\bullet} \qquad \text{radical anion} \qquad (4.33\text{ a})$$

$$B^+ + e^- \rightleftarrows B^{\bullet} \qquad \text{radical} \qquad (4.33\text{ b})$$

$$C \rightleftarrows C^{+\bullet} + e^- \qquad \text{radical cation} \qquad (4.33\text{ c})$$

$$D^- \rightleftarrows D^{\bullet} + e^- \qquad \text{radical} \qquad (4.33\text{ d})$$

These intermediates, reactive as they are, may react with different competing reactants yielding different reaction products. Very reactive and short lived intermediates which do not live long enough to leave the electrode surface by diffusion and intermediates, which are adsorbed at the electrode surface do react in truely heterogeneous reactions. As the electrode surface according to its very nature – e.g. metallic vs oxidic, lipophilic vs lyophilic etc. – adsorbs different reactant molecules to a different degree it may be anticipated, that the main products obtained by direct electrochemical conversion of the same molecule may be quite different at different electrode materials. In this respect carbon electrodes – composed mainly of a polycrystalline matrix of "Acheson graphite" – play an important role. Their surface is non-polar and favours the adsorption of non-polar organic molecules like olefins or arenes. Adsorption enhances their surface concentrations and enhances the reaction rate of adsorbed radicals and radical ions with the more strongly adsorbed non-polar molecules compared to their rate of reaction with polar molecules which are less adsorbed. Thus hydrophobic carbon electrodes are catalysing for instance addition reactions of radicals and radical ions to unsaturated hydrocarbon, versus their solvolytic reactions involving the polar solvent molecules, which on the other hand would be favoured and catalysed at hydrophilic electrodes.

4.10.2
Electrocatalytic Oxidations by Oxides of Multiply-Valent Metals

Redox couples, as defined by metal oxides of metals exhibiting different valencies, (e.g. $PbO_2/PbSO_4$, $MnO_2/MnOOH$, $NiOOH/NiO$) are used as oxidants in preparative organic syntheses and may be used as heterogeneous mediators for the anodic oxidation of organics provided
(i) the respective redox potential is positive enough,
(ii) the anodic reoxidation of the respective lower valent oxide is fast,

(iii) the frequent chemical reduction/anodic reoxidation does not lead to corrosive deterioration of the electrode,
(iv) the heterogeneous oxidation of the organic substrate is fast enough to guarantee technically acceptable current densities.

Table 4.2 collects some metal oxide couples and their redox potentials together with the usual base metal. Table 4.3 compares the rate data for the oxidation of some aliphatic amines and alcohols on nickel, silver, copper and cobalt oxide anodes. Homologous alcohols always react ten times slower than amines. The reaction rates on silver oxide (0.85 V vs 0.6 and 0.7 V) and on cobalt oxide are sizeably slower than on copper and nickel oxide, so that silver and cobalt anodes could be ruled out for technically applied anodic oxidation. A rate of 10^{-4} cm s^{-1} in 1 molar solution, yields in current densities of several 10^{-1} Acm^{-2} so that lower rates would be technically unacceptable.

Table 4.2. Metal/metal-oxide couples capable to act as heterogeneous mediators for the oxidation of organic substrates dissolved in aqueous solution

Redox couple[a]		Base metal	E^0/V^b vs RHE
Cu_2O/CuO	(OH$^-$)	Cu	0.74
MnO_2/Mn_2O_3	(OH$^-$)	Ti	0.94
Co_3O_4/CoO	(OH$^-$)	Co, Fe	1.04
Ag_2O/Ag	(OH$^-$)	Ag	1.18
$NiOOH/NiO$	(OH$^-$)	Ni	1.29
$PbO_2/PbSO_4$	H$^+$	Pb	1.69[c]
Cr^{VI}/Cr^{III}	H$^+$	Ti/TiO$_2$	1.7

[a] (OH$^-$): in alkaline solution, (H$^+$) in acidic solution
[b] Recalculated with reference to the reversible hydrogen electrode potential from D. Dobos, Electrochemical Data, Elsevier, Amsterdam, 1975
[c] $H_2SO_4 = 1$ val dm^{-3}

Table 4.3. Heterogeneous rate data (cm^{-1}s^{-1}) for oxidation of amines and alcohols at different anodes (25 °C, 1 mol dm^{-3} aqueous KOH) from M. Fleischmann, K. Korinek, D. Pletcher, J. Chem. Soc., Perkin Trans. 2, 1396 (1972)

Electrode Substrate:	Nickel	Silver	Copper	Cobalt
Electrode-potential (V vs NHE):	0.6	0.85	0.7	0.6
Methylamine	1.1×10^{-4}	1.9×10^{-4}	2.8×10^{-5}	1.8×10^{-5}
Methanol	7×10^{-6}	4.5×10^{-6}	6.2×10^{-5}	$<10^{-6}$
n-Propylamine	4.2×10^{-5}	8.2×10^{-5}	1.1×10^{-4}	5.5×10^{-6}
n-Propanol	5.2×10^{-6}	2×10^{-6}	2.4×10^{-5}	$<10^{-8}$
Isopropanol	4×10^{-6}	$<10^{-8}$	1.1×10^{-5}	$<10^{-6}$

4.10.2.1
The Heterogeneously Catalysed Benzene Oxidation at Pb/PbO₂ Electrodes in Sulfuric Acid

This reaction was worked out on semitechnical scale for quinone/hydroquinone production, Eq. (4.34 a, b).

$$PbSO_4 + 2H_2O \rightarrow PbO_2 + H_2SO_4 + 2e^- \qquad (4.34\ a)$$

$$3PbO_2 + C_6H_6 + 2H_2SO_4 \rightarrow 3PbSO_4 + C_6H_4O_2 + 4H_2O \qquad (4.34\ b)$$

As shown in Fig. 4.19 a,b the current density is critically determining the current yield of the reaction sequence Eq. (4.34 a)→Eq. (4.34 b). Raising the current density above 10 mA cm^{-2} leads to increasing current efficiency losses as more and more oxygen is evolved as the rate of the chemical surface reaction, Eq. (4.34 b), can no longer follow the faster rate of charge transfer, Eq. (4.34 a). The heterogeneously catalysed NiOOH-mediated oxidation of diacetone-L sorbose to diacetone-2-keto-L-sorbic acid, the latter being a precursor to vitamin C, at nickel anodes and based on the chemical oxidation of the substrate by NiOOH is of technical relevance and is a similar case. The limiting current density in 1 mol l^{-1} KOH solution for sorbose oxidation is under operating conditions only 10 mA cm^{-2} leading to relatively poor space time yields, but cannot be enhanced without significant losses in current efficiency.

4.10.3
Electrocatalytic Hydrogenation and Electrocatalyzed Mediated Reduction

Commercial catalytic hydrogenation of unsaturated compounds use Raney nickel or – less commonly – Pt-catalyst supported on active carbon. Electrocatalytic hydrogenation can be performed at platinized platinum or other platinum-metal electrodes. Adsorbed atomic hydrogen is the active reactant in catalytic as well as in electrocatalytic hydrogenation. In general cathodic electrocatalytic hydrogenation and catalytic hydrogenation are supposed not to differ with respect to mechanism, yield and selectivity. Since the adsorbed hydrogen as well as the adsorbed substrate participate in the reaction and the adsorption equilibrium coefficients of both reactants vary independently and often in adverse manner with the electrode potential, the effective hydrogenation rate may not change – as expected – exponentially with increasing cathodic polarisation but might show maximal hydrogenation rates at intermediate potentials. Using Raney nickel coated cathodes for electrocatalytic hydrogenation would promise to take advantage of the enhanced hydrogen concentration (and surface concentration) in the nanopores of the catalyst. As will be discussed in Chap. 7 in the context of cathodic hydrogen evolution and utilisation of nanoporous catalyst coatings it is possible to achieve hydrogen concentrations in nanopores which are by a factor of up to 1000 higher than the equilibrium concentration at 1 bar hydrogen.

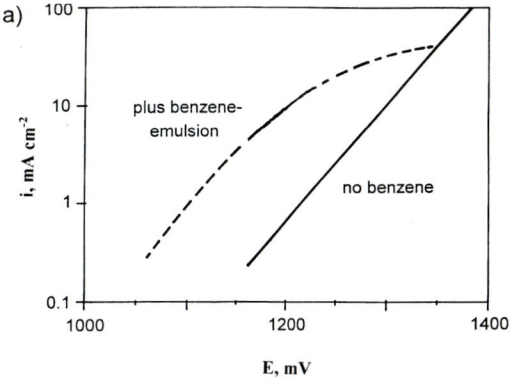

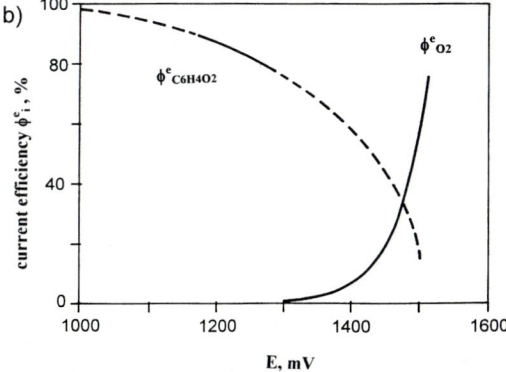

Fig. 4.19a,b. Anodic oxidation of aqueous dispersions of benzene at PbO_2 anodes: **a** current voltage curve of H_2SO_4-solution with (*broken line*) and without (*full line*) benzene; **b** current yield of benzoquinone (*broken line*) and oxygen. The *fully drawn parts of the broken lines* signify operating conditions of benzene oxidation. From I.S. Clarke, R.S. Ehigamusoe, A.T. Kuhn, J. electroanal. chem. *70* 333, (1976)

4.10.4
The Electrode Surface as Medium Catalysing Chemical Reactions of Electrogenerated Reactive Organic Intermediates

The electrosorption of reactive intermediates and of organic molecules at the surface of carbon electrodes is generally weak, and only due to physical adsorption. Nonetheless, in particular if the reactive intermediates are so reactive that they do not survive for much longer than 10^{-9} s and therefore cannot escape from the electrode surface, the chemical composition of the adsorbate layer being different from that of the bulk electrolyte composition influences catalytically the course of consecutive reactions and their yields and selectivities decisively.

The one-electron oxidation or reduction of e.g. unsaturated hydrocarbons or other electroactive organic compounds and ions creates, according to Eq. (4.33 a–d), radical cations, radical anions or radicals with an energy content, which surpasses that of the original substrate by at least 100 kJ mol^{-1}, provided their oxidation or reduction potential amounts to more than +1 V or –1 V vs aqu. SCE. Correspondingly the primarily generated oxidation or reduction products of organic substrates are so reactive, that they do not need any significant further catalytic activation e.g. by chemisorption at the electrode surface looked at as an heterogeneous chemical catalyst. Nonetheless a number of electroorganic synthesis reactions are known, whose outcome i.e. whose yield and selectivity is decisively determined by the nature of the electrode so that heterogeneous acceleration of at least one of several competitive reactions of the electrogenerated reactive intermediates might be anticipated. A famous case is the Kolbe reaction, which is essentially the anodic dimerization of alkyl radicals which are generated at platinum anodes by anodic oxidation of the anions of carboxylic acids – Eqs. (4.35)–(4.37)

$$R-CH_2-COO^- \rightarrow R-CH_2-COO^{\cdot} + e^- \tag{4.35}$$

$$R-CH_2-COO^{\cdot} \rightarrow R-CH_2^{\cdot} + CO_2 \tag{4.36}$$

$$2R-CH_2^{\cdot} \rightarrow R-CH_2-CH_2-R \tag{4.37}$$

If carboxylate anions are oxidised at carbon anodes instead of Pt anodes the main product is not the Kolbe-dimer but an ester. This reaction (Hofer-Moest) is explained by the inherent instability of C-radicals at highly anodic potentials which are necessary for the anodic oxidation of carboxylate anions. At these potentials the C-radicals are oxidised to carbonium ions which react with carboxylate anions forming esters.

$$RCH_2^{\cdot} \rightarrow RCH_2^+ + e^- \tag{4.38}$$

$$RCH_2^+ + RCH_2COO^- \rightarrow RCH_2-OOCCH_2R \tag{4.39}$$

Thermodynamically the latter reaction is to be expected, as C-radicals exhibit an oxidation potential which is at least 1 V more negative than the oxidation potential of carboxylate anions. It is therefore intriguing to understand what is the particular role of the platinum/electrolyte interface in the Kolbe synthesis favouring that reaction path – Eqs.(4.35)–(4.37) – which is thermodynamically disfavoured and unlikely to occur. It can be shown, that the expected oxidation of the C-radicals is prohibited at platinum anodes because the carboxylate radicals are so strongly adsorbed at platinum anodes that their dissociative decarboxylation is prevented on one hand and they block the electrode surface for C-radical oxidation on the other hand. The consequence is, that across the ad-

4.10 Electrocatalysis in Electroorganic Synthesis

sorbed layer of carboxylate radicals non-adsorbed carboxylate anions are oxidised to solvated radials which decarboxylate immediately forming radicals in solution. Thus contrary to the naive understanding Pt does not catalyze but inhibits the thermodynamically favoured oxidation of alkyl radicals, allowing for homogenous dimerization of the Kolbe radicals in the immediate neighbourhood of the electrode. In contrast, C-anodes allow for physisorption of these radicals with subsequent immediate anodic formation of carbenium cations, which constitutes a completely different reaction path to form esters. The Kolbe synthesis may therefore be named a non-catalytic rather than an electrocatalyzed reaction and the Hofer–Moest reaction, Eqs. (4.38) and (4.39), though in most cases undesired, is heterogeneously catalysed.

A comparable case would be the cathodic hydrogenation of the carbonyl compound to the respective alcohol according to Eq. (4.40) on one hand,

$$R_1R_2C=O+2H^+ +2e^- \to R_1R_2CHOH \tag{4.40}$$

and the formation of a pinacol according to radical dimerization on the other, according to Eqs. (4.41) and (4.42):

$$R_1R_2C=O+H^+ +e^- \to R_1R_2\dot{C}OH \tag{4.41}$$

$$2R_1R_2\dot{C}OH \to R_1R_2C(OH)-C(OH)R_1R_2 \tag{4.42}$$

Particular enhancement of the radical dimerization reaction at Eq. (4.42) is observed on hydrophobic metals like Hg, Pb and Sn at potentials close to their respective points of zero charge where electrosorption is strongest but diminishes at more negative and positive potential. If the reduction potential of the oxo compound which is pH dependent (60 mV per pH unit) is shifted by more than 100 mV away from the point of zero charge, then the ketyl radicals ($R_1R_2\dot{C}OH$) are no longer adsorbed strongly enough and they are displaced from the surface by solvent molecules so that due to homogeneous fast protonation and further reduction only the alcohol instead of the pinacol is produced according to homogeneous kinetics i.e. protonation vs heterogeneous dimerization leading to alcohol formation. This type of reaction exemplifies that preferential adsorption of reactive intermediates catalyses heterogeneously reactions of the intermediate with reaction orders of greater than one because adsorption of the reactant provides under otherwise comparable kinetic conditions a locally enhanced (surface) concentration of the intermediate.

4.10.4.1
Electrocatalytic Action of Electrosorbed Non-Reactant Species – Electrocatalysis of the Second Kind

Electrosorbed species, which by themselves are not electrochemically converted but may enhance catalytically the rate of a particular reaction of an electrochemically generated reactive intermediate, would be assumed to catalyse this particular reaction and to retard another one. One could name this type of electrocatalysis which is due to the catalytic action of adsorbed species "electrocatalysis of the second kind". Most remarkably the selectivity and commercial success of the Monsanto process – Eq. (4.43), the hydrodimerisation of arylonitrile to adipodinitrile:

$$2CH_2=CH-CN+2e^- -2H^+ \rightarrow NC-CH_2-CH_2-CH_2-CH_2-CN \quad (4.43)$$

is founded on electrocatalysis of the second kind. The hydrodimerisation of acrylonitrile can only effectively compete with and outperform the undesired reduction of acrylonitrile to propionitrile – Eq. (4.44):

$$CH_2=CH-CN+2e^- +2H^+ \rightarrow CH_3-CH_2-CN \quad (4.44)$$

if strongly surface active tetraalkyl ammonium cations are present in the electrolyte. By electrode impedance measurements it had been shown that into an adsorbate layer of tetraalkylammonium ions acrylonitrile is coadsorbed and that it and any intermediate formed by cathodic reduction of acrylonitrile embedded in a surface adlayer of these cations experiences an environment which is far less protic than the bulk of the electrolyte which is mainly water.

$$(CH_2=CH-CN)_{ad} + H^+ + 2e^- \rightarrow (\overline{C}H_2-CH_2-CN)_{ad} \quad (4.45)$$

If due to sufficiently high acrylonitrile concentrations in the bulk of the electrolyte its surface concentration in the adsorbate layer is sufficiently enhanced, the Michael addition, Eq. (4.46), of the primarily formed radical anion to coadsorbed acrylonitrile can compete effectively with (retarded) protonation – Eq. (4.44). Eventually further reduction and protonation of the Michael dimer according to Eq. (4.47) yields the desired hydrodimer.

$$(CH_2-CH_2-CN)_{ad}^- + (CH=CH-CN)_{ad} \rightarrow NC-\overline{C}H-CH_2-CH_2-CH_2-CN \quad (4.46)$$

$$NC-\overline{C}H-CH_2-CH_2-CH_2-CN + e^- + H^+ \rightarrow NC-CH_2-CH_2-CH_2-CH_2-CN \quad (4.47)$$

4.10.5
Kinetics and Selectivity of Homogeneous Chemical Consecutive Reactions Following Charge Transfer

Figure 4.11 shows schematically in the upper dotted trace of a concentration-distance diagram the steady state concentration profile of a stable electrochemically generated species which is established under well defined mass transfer conditions at a given current density. The mass transfer conditions define the extension of the Nernst-diffusion layer δ_N and the current density together with the mass transfer coefficient.

$$k_m(c_\infty - c_0) = i/v_e F \tag{4.48 a}$$

$$k_m = D/\delta_N; \tag{4.48 b}$$

as a particular solution of the general steady state equation

$$0 = \text{div}(wc) - D \, \text{div grad } c \tag{4.49}$$

Taking into account a fast chemical reaction which consumes the electrochemically generated intermediate with volume related and concentration dependent rate r modifies Eq. (4.49) to Eq. (4.50):

$$0 = \text{div}(wc) - D \, \text{div grad } c - r \tag{4.50}$$

In Fig. 4.11 the lower trace is meant to describe schematically one particular solution of Eq. (4.50) with sizably reduced c_0 and a reaction layer thickness δ_R which is remarkably smaller than δ_N. Very often δ_R is by orders of magnitudes smaller than δ_N. In this case the convective terms div(wc) in Eq. (4.49) can be neglected so that Eq. (4.50) reduces to

$$-r = D \, \text{div grad } c \tag{4.51}$$

A typical reaction-determined concentration profile can be manipulated by changing the process variable "current density". The aim is that in the term r which is a sum, comprising the rates of all different competing reactions of the reactive intermediate only the rate of one particular reaction – averaged over the extension of the reaction layer – overwhelmingly surmounts the rates of all other competing reactions to an extent which allows to define a sufficiently high selectivity. Using the space averaged rates $\bar{r}$ the selectivity for product i is defined by Eq. (4.52).

$$S_i = \frac{\bar{r}_i}{\sum \bar{r}_j} \tag{4.52}$$

For reactions of reaction orders $n_i > 1$ the space close to the electrodes where relatively highest concentrations of the intermediate are prevailing, is representing the most important part of the reaction layer. Highest concentrations induced by highest current densities by which these intermediates are generated, would therefore enhance the selectivity for the reaction path of higher reaction order. Generally speaking, if the reaction order of competing consecutive reactions differ from each other, the current density is an easy means to influence relative rates and selectivities, as its increase always favours the reaction with highest reaction order. The selectivity problem is additionally treated in Chapter 6.5.7 for consecutive electrochemical conversion steps.

References

1. A. Lasia and A. Rami, J. Electroanal. Chem. *294* 123 (1990)

Further Reading

A.J. Appleby, Electrocatalysis, Chapter 4 in Vol. 7, Kinetics and Mechanism of Electrode Processes in Comprehensive Treatise of Electrochemistry, B.E. Conway, J.O'M Bockris, E. Yeager, S.U.M. Khan, R. E. White Eds., Plenum Press, New York and London, 1983

M. Enyo, Hydrogen Electrode Reaction on Electrocatalytically Active Metals, in: Vol. 7 Kinetics and Mechanism of Electrode Processes of Comprehensive Treatise of Electrochemistry, B.E. Conway, J.O'M Bockris, E. Yeager, S.U.M. Khan, Ralph E. White Eds., Plenum Press, New York and London, 1983

S. Trasatti, Electrocatalysis of Hydrogen Evolution: Progress in Cathode Activation in H. Gerischer, C.W. Tobias eds., Advances in Electrochemical Science and Engineering. VCH Verlagsgesellschaft, Weinheim, New York, 1992

K. Vetter, Elektrochemische Kinetik, Springer Verlag, Berlin, Göttingen, Heidelberg, 1961

CHAPTER 5

Mass Transfer by Fluid Flow, Convective Diffusion and Ionic Electricity Transport in Electrolytes and Cells

5.1
Introduction

The performance of electrochemical processes is not only determined by charge transfer and electrode kinetics, but a number of additional phenomena cause and rule the electrode kinetics (in the microkinetic as well as macrokinetic sense), the heat balance of the cell and the mass balances of all process streams (electrolyte, gases, solid products). Among these factors the fluid dynamic conditions, under which the electrolyte enters, passes and leaves the electrolysis cell or moves in it under free convection, is the most powerful process parameter since hydrodynamics rule mass and heat transport.

Heat transport is important as it controls together with heat generation the temperature distribution in the cell. Another condition typical for electrochemical processes is charge transport through the electrolyte, which toghether with electrode kinetics determines in particular the current density distribution across the electrode surface. It may determine the overall current efficiency, conversion selectivity and space time yield of the process by local inhomogeneities of the current density for instance by locally too high current density which might exceed the mass transfer limited current density.

The proper handling of these characteristic process determinants: fluid dynamics, mass transport and heat transport together with proper management of ionic charge transport are the main subjects of electrochemical process engineering as far as the reactor i.e. the electrolyzer is concerned. But the electrochemical engineer has to be aware that the cell alone is only one part of the whole process. Electrolyte make up, product isolation and recycling of the electrolyte have to be taken into consideration, too.

5.2
Fluid Dynamics and Convective Diffusion

Diffusive mass transport in an electrolyte solution without superimposed convection is relatively ineffective because of the low diffusion coefficients experienced in solvents of viscosities comparable to that of water. In usual dielectric solvents the diffusion coefficient of a low molecular weight species is similar to that in aqueous solution, which is of the order of approx. 10^{-5} cm^{2}s^{-1}.

Moreover, diffusion alone is typically a non-steady process that means on extension of time it proceeds with a steadily decreasing velocity because concentration gradients tend to level off. Figure 5.1 a demonstrates this, depicting schematically the temporal development of concentration profiles of metal ions in front of a cathode on which the metal ions are deposited with time-independent current density, i, in presence of supporting electrolyte of high concentration, which allows for metal ion diffusion alone, excluding mass transport by migration.

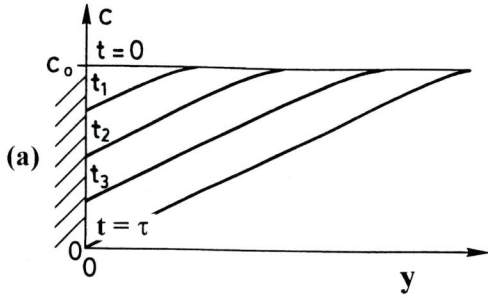

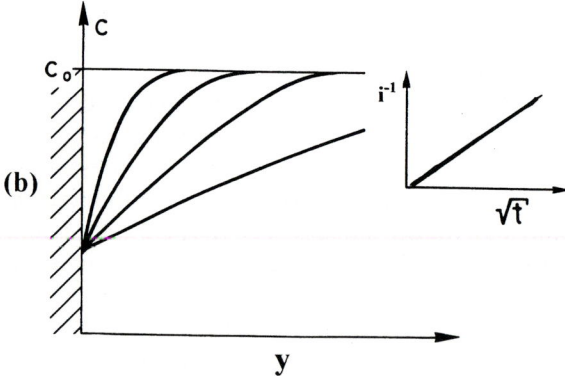

Fig. 5.1.a Schematic showing the development of concentration profiles of metal ions in a stagnant electrolyte upon galvanostatic cathodic metal deposition ($M^{z+} + ze^- \rightarrow M$). Note that $(dc/dx)_{x=0}$ remains unchanged because i is constant **b** Development of concentration profile of metal ions in stagnant electrolyte upon imposing a constant potential $\varphi < \varphi_{equil}$ on the electrode. Declining values $(dc/dy)_{x=0}$ reflect in steadily declining current densities according to $t^{-1/2}$ – law (see insert).

5.2 Fluid Dynamics and Convective Diffusion

For a given current density i the gradient (dc/dy) is fixed at the electrode surface (y=0).
For metal ion reduction,

$$M^{z+} + \nu_e e^- \rightarrow M \tag{5a}$$

one obtains

$$D(dc/dy)_{y=0} = i/\nu_e F \tag{5.1}$$

and Fick's second law describes the change of the concentration profile in front of the electrode with time:

$$\partial c/\partial t = D(\partial c^2/\partial y^2) \tag{5.2}$$

After the critical transition time τ the concentration of the metal at the electrode is depleted to zero and the mass flow $i/\nu_e F$ can no longer be maintained. Then the current will be consumed additionally by another electrode process, for instance by H_2 evolution with cathodic and O_2 evolution with anodic processes.

Therefore the electrode potential at t = τ will become so negative that hydrogen is evolved $H^+ + e^- \rightarrow 1/2 H_2$ parallel to metal deposition and hydrogen evolution would consume then larger and larger fractions of the applied current density with further extension of time.

Figure 5.2 b demonstrates, what occurs if a potential is applied to the cathode establishing according to Nernst's law a metal ion concentration c_0 being lower than c_∞. Then a steadily decreasing current density for metal deposition will be observed. The developing concentration profile becomes flatter with time, $(dc/dy)_{y=0}$ and i is decreasing with time according to $t^{-0,5}$.

Imposed steady convection, and steady rates of electrochemical conversion expressed as time-independent current densities, generate steady concentration profiles and accordingly the one-dimensional Fick's second law, Eq. (5.2.), is substituted by the two dimensional steady state Eq. (5.3):

$$\partial c/\partial t = 0 = D(\partial^2 c/\partial y^2) - \partial(wc)/\partial x \tag{5.3}$$

y and x are coordinates perpendicular and parallel to the electrode and w is the flow velocity parallel to the electrode surface. Equation (5.3) is a simplified two-dimensional expression, which holds for one-dimensional flow parallel to the electrode (in x direction) and diffusive mass transfer due to metal deposition towards the electrode in y direction, i.e. perpendicular to the electrode surface (compare Fig. 5.2 a and b). Steady viscous flow along a planar surface generates a time independent velocity boundary layer and this, together with electrochemical conversion at the electrode, rules the spatial concentration distribution depicted in the

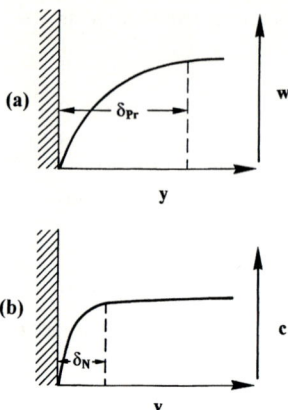

Fig. 5.2. Time independent concentration profiles are generated by forced flow along the electrode with steady velocity distribution **a** and steady concentration distribution **b**. A diffusion layer of thickness δ_N defines steady state mass transfer conditions with $k_m = D/\delta_N$

schematic concentration profile, see Fig. 5.2 b. Equation (5.3) can only be solved after solution of the Navier/Stokes equation discussed below, leading to formulation of the velocity field in front of the electrode.

5.3
Fluid Dynamics of Viscous, Incompressible Media

The treatment of fluid flow in electrolyzers can in general be confined to virtually incompressible viscous media. Moreover, viscous flow along electrodes or between parallel-plate electrodes can very often, as done above, be treated as unidirectional one-dimensional with velocity profiles extending perpendicular to the electrode.

The simplified two-dimensional Navier–Stokes equation neglecting gravity forces for steady flow in xy-plane(x parallel and y perpendicular to a planar electrode), which describes the balance of all forces (acceleration and friction forces), acting on a volume increment of the fluid, reads:

$$u\frac{\partial u}{\partial x} + v\frac{\partial u}{\partial y} = -\frac{1}{\rho}\frac{\partial p}{\partial x} + \nu\left(\frac{\partial^2 u}{\partial x^2} + \frac{\partial^2 u}{\partial y^2}\right) \quad (5.4\ \text{a})$$

$$u\frac{\partial v}{\partial x} + v\frac{\partial v}{\partial y} = -\frac{1}{\rho}\frac{\partial p}{\partial y} + \nu\left(\frac{\partial^2 v}{\partial x^2} + \frac{\partial^2 v}{\partial y^2}\right) \quad (5.4\ \text{b})$$

u and v are velocities in x and y directions respectively and ν is the kinematic viscosity $\nu = \eta/\rho$. In most cases of practical electrochemical relevance, that means in flow, along, or between parallel plates v is to a good approximation zero.

5.3 Fluid Dynamics of Viscous, Incompressible Media

One can introduce adimensional quantities

$$x^* = \frac{x}{L}; \quad y^* = \frac{y}{L}; \quad u^* = \frac{u}{U}; \quad v^* = \frac{v}{U}; \quad p^* = \frac{p}{\rho U^2} \tag{5.4 c}$$

where L and U are characteristic quantities of the system under consideration. L is for instance the total length of the plate or the distance between two plates, U may mean the mean flow velocity or it is the maximal flow velocity. ρ is the density of the fluid and p is the pressure.

The use of adimensional quantities simplifies the differential equations and allows to define important adimensional numbers, the magnitude of which characterizes the flow behaviour of the whole system. One obtains the Navier/Stokes equations, for instance Eq. (5.4 a), in adimensional form:

$$u^* \frac{\partial u^*}{\partial x^*} + v^* \frac{\partial u^*}{\partial y^*} = -\frac{\partial p^*}{\partial x^*} + \frac{1}{Re}\left(\frac{\partial^2 u^*}{\partial x^{*2}} + \frac{\partial^2 u^*}{\partial y^{*2}} \right) \tag{5.4 a*}$$

For one-dimensional steady flow between two plates in x direction this reduces still further to

$$0 = -\frac{1}{\rho}\frac{\partial p}{\partial x} + v \frac{\partial^2 u}{\partial y^2} \tag{5.5}$$

and in adimensional form to

$$0 = -\frac{\partial p^*}{\partial x^*} + \frac{1}{Re}\frac{\partial^2 u^*}{\partial y^{*2}} \tag{5.5*}$$

From Eqs. (5.4 a*) and (5.5*) the Reynolds number

$$Re = \frac{UL}{v} \tag{5.6}$$

and the Euler number

$$Eu = p/\rho U^2 \tag{5.7}$$

can be extracted.

The Reynolds number can be interpreted as the ratio of shear to acceleration forces in the system, whereas the Euler number gives the ratio of pressure versus acceleration forces.

The Reynolds number defines the character of fluid flow in the considered system with forced convection. The magnitude of the Reynolds number is indicative of the type of flow – laminar or turbulent –, which prevails under given flow conditions in a given device. For instance for channel flow in ducts or between plates a limiting Reynolds number of Re=2000–3000 determines transition from laminar at lower to turbulent flow at higher Reynolds number. The Reynolds number is by far the most important adimensional number for most cases of forced viscous fluid flow dealt with in the context of electrochemical engineering.

5.3.1
Laminar vs Turbulent Flow

Laminar flow – the expression is self explaining – means fluid flow without any velocity component perpendicular to the main direction of flow, that means without vortices or eddies. The flow can be divided up into lamella, between which there is no mixing and which interact with each other only by transmitting and exchanging momentum by viscous forces. Figure 5.3 a,b exemplifies this situation for unidirectional laminar flow between two parallel, rectangular plates and for divergent flow from a central bore outward between two parallel cylindrical plates.

Laminar flow generates smooth velocity profiles at the walls of the respective systems, which for channel flow and flow between parallel plates is typically parabolic (Fig. 5.4 a). Double integration of Eq. (5.5) leads to parabolic velocity distributions – for instance for Hagen–Poiseulle flow through circular ducts and pipes.

$$w = w_{max}\left(1 - \left(\frac{r}{R}\right)^2\right) \tag{5.8}$$

Increasing flow velocities strains the velocity profile established by viscous interaction of the flow lamellas more and more until it becomes unstable so that it

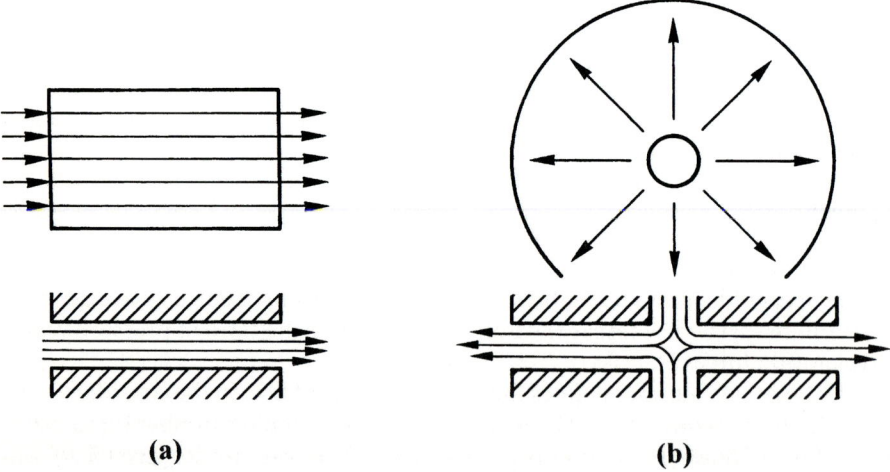

(a)   (b)

Fig. 5.3.a, b. Schematic of laminar flow between plates: **a** unidirectional flow in a duct; **b** radial flow from a central bore between two circular disks

5.3 Fluid Dynamics of Viscous, Incompressible Media

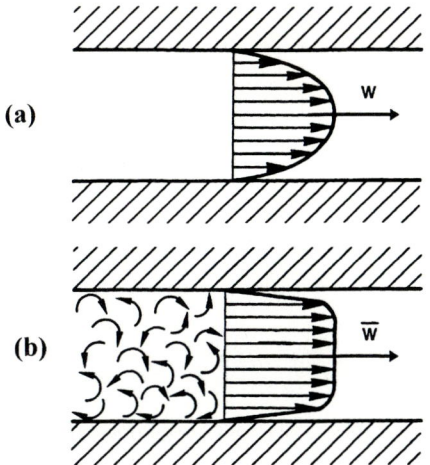

Fig. 5.4. a Parabolic velocity profile for laminar flow between plates, in tubes and rectangular ducts. **b** Steep, almost piston-like profile of mean velocities in turbulent flow between plates, in tubes and ducts

breaks down at its steepest part that means at the wall by formation of eddies. These eddies move stochastically in the direction away from the wall being carried along with the main stream of the fluid. Coming from the wall they reach the bulk of the flow and are thereby strongly accelerated in flow direction, while eddies from the bulk, which are approaching the walls (where they replace the volume of freshly generated eddies), are decelerated. This means that as soon as eddies are generated under turbulent flow conditions momentum exchange between bulk and boundary of the fluid flow becomes much more efficient than under laminar flow so that the distribution of mean flow velocities perpendicular to the main flow direction becomes remarkably more equalised – but since under any flow condition the flow velocity at the wall is zero, the slope of the velocity profile close to the wall becomes much steeper (Fig. 5.4 b) than for laminar flow. This means that with turbulent flux drag at the wall is greatly enhanced. Three examples follow, which demonstrate velocity distributions in front of a plate or between plate electrodes.

5.3.2
Velocity Distributions for Laminar Flow

5.3.2.1
Singular Electrode: Unidirectional Laminar Flow Along a Plate

For a plate electrode far away from a counter electrode in an electrolysis cell or for the initial development of the velocity profile at the inlet of a gap between a pair of parallel plate electrodes double integration of Eq. (5.5) applying a power series model for the velocity distribution results in the definition of an effective

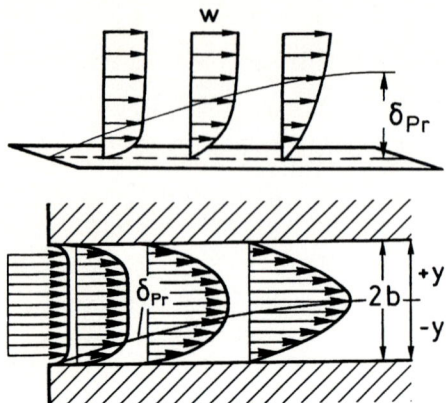

Fig. 5.5. a Extending velocity profiles for laminar flow along a plate. **b** Development and closure of velocity profile at the entrance of a duct or tube.

thickness of the velocity profile δ_{Pr} (thickness of Prandtl's boundary layer) which increases with increasing distance Z from the leading edge (Fig. 5.5 a) [1][1)]

$$w(x) = w_{max}\left(1.5(y/\delta_w) - 0.5(y/\delta_w)^3\right) \qquad (5.9\ a)$$

with

$$\delta_w = 4.64(xv/w_{max})^{0.5} = \frac{3}{2}\delta_{Pr} \qquad (5.9\ b)$$

If these equations apply for the development of the velocity profile between two parallel plates with distance 2b (Fig. 5.5 b), then the condition δ_w=b means complete closure of the velocity profile in the gap. Further downstream this closed velocity profile between two parallel electrodes does not change any longer.

5.3.2.2
Pair of Planar Electrodes

Fully established laminar flow between two plates with distance 2b yields in the parabolic velocity profile depicted schematically in Fig. (5.5 b) and written in Eq. (5.9 c):

$$w = w_{max}\left(1 - (y/b)^2\right) \qquad (5.9\ c)$$

---

1 For mathematical details compare E.R. G. Eckert, Heat and mass transfer, 2nd, ed. Mac Graw-Hill, New York, 1959 The notation δ_u defines a boundary layer thickness, which differs from the definition of the thickness of the Prandtl layer $\delta_{Pr}=w_\infty(\delta y/\delta w)_{y=0}$; δ_w= 3/2δ_{Pr}; δ_w is defined by the simultaneous boundary conditions: w=0 for y=0; w=w_∞ for y>δ_w and ($\delta w/\delta y$)=0 for y≥δ_w

5.3 Fluid Dynamics of Viscous, Incompressible Media

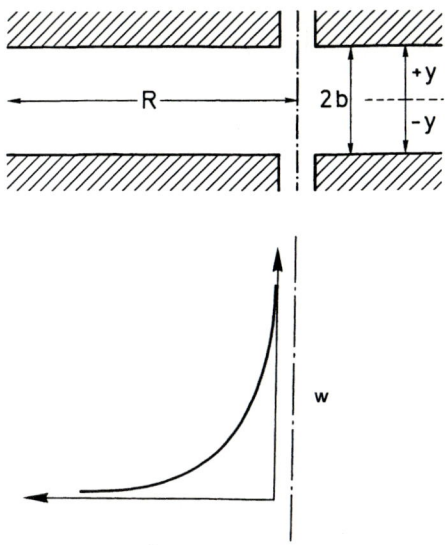

Fig. 5.5.c Radial flow from a central bore in a gap between circular discs results in decrease of mean velocities according to r^{-1}

one calculates the volumetric flow velocity

$$\dot{V} = 2 B w_{max}(2/3)b \qquad (5.9\ d)$$

with B = width of the electrodes

and the total pressure drop

$$\Delta p = 3\dot{V}\mu L / (Bb^3) \qquad (5.9\ e)$$

with L = length of the electrodes, and μ = dynamic viscosity

5.3.2.3
Circular Capillary Gap Cell

Divergent laminar flow between two plates from a central bore with radius r_0 outward through the gap between two circular parallel plate electrodes with outer radius R (see Figs. 5.3 b and 5.5 c) – in the so called circular-capillary gap cell – generates steadily decreasing linear flow velocity in going from the central bore to the periphery.

The Navier–Stokes equations at Eqs. (5.4 a) and (5.4 b) are expressed in cylindrical coordinates (r and y) and transform to:

$$w\frac{\partial w}{\partial r} = \frac{p}{\rho r} - \frac{\partial p}{\partial r}\frac{1}{\rho} + \nu\left(\frac{\partial^2 w}{\partial r^2} + \frac{1}{r}\frac{\partial w}{\partial r} - \frac{w}{r^2} + \frac{\partial^2 w}{\partial y^2}\right) \qquad (5.10)$$

Neglecting the acceleration/deceleration term du/dr for so called creeping flow and setting $d^2w/dr^2 + 1/r(dw/dr) - w/r^2$ according to the continuity equation[2] equal to zero yields:

$$0 = \frac{P}{\rho r} - \frac{\partial p}{\partial r}\frac{1}{\rho} + v\frac{\partial^2 w}{\partial r^2} \qquad (5.10\ a)$$

Due to the continuity condition one obtains for any given distance y from the middle plane and radius r:

$$w_y(r) = C_y / r \qquad (5.11)$$

One arrives at the parabolic velocity profile and radial velocity distribution of Eq. (5.11)

$$w(r,y) = \frac{3\dot{V}}{8\pi b r}\left(1 - y^2/b^2\right) \qquad (5.12)$$

with the pressure drop:

$$\Delta p_r^R = \frac{3\dot{V}v\rho}{4\pi b^3}\left(1 - r/R\right) \qquad (5.13)$$

5.4
Mass Transport by Convective Diffusion

5.4.1
Fundamentals

Mass flux density of species dissolved in stagnant or flowing electrolyte generated by spatial concentration gradients is described by Fick's law:

$$\dot{n}_{diff} = -D\ \text{grad}\ c \qquad (5.14)$$

Fick's first law is the most general description of diffusional mass transfer in the absence of additional convective mass transfer and reads in one-dimensional form as applied to diffusive mass transport to large, flat electrodes:

$$\dot{n} = -D\left(\frac{\partial c}{\partial y}\right)_{y=0} \qquad (5.14\ a)$$

---

2 In Cartesian coordinates the continuity equation reads: $\frac{d^2u}{dx^2} + \frac{d^2v}{dy^2} + \frac{d^2w}{dz^2} = 0$

5.4 Mass Transport by Convective Diffusion

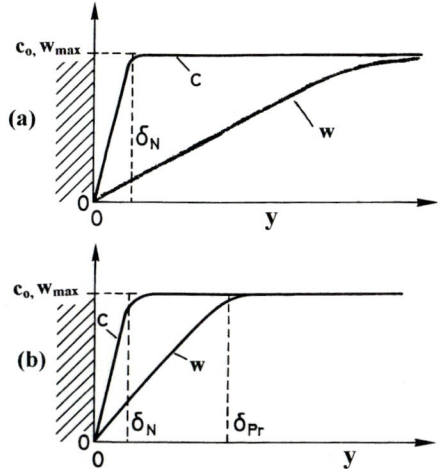

Fig. 5.6. a Defining the Nernst diffusion layer thickness δ_N by linearizing the concentration profile. **b** Schematic of relative extension of velocity and concentration boundary layers δ_{Pr} and δ_N for freely developing laminar and fully developed turbulent flow in fluids possessing Schmidt numbers of from several hundreds to thousands

The current density of any electrochemical process consuming solute species by electrochemical conversion couples the rate of electrochemical conversion with diffusive mass transport at and towards the electrode surface:

$$i = \dot{n}_0 v_e F = D(\partial c/\partial y)_{y=0} v_e F \tag{5.15}$$

The index $_0$ refers to y=0 that means at the electrode surface.

Mass transport in technical electrolyzers is always caused by the interaction of convection and diffusion. The calculation of concentration profiles and hence current densities demands the simultaneous solution of Navier Stoke's equations and Fick's first and second laws, which can be achieved for systems with laminar flow but not for turbulent flow. But under any such conditions the mass transfer rate and associated current densities are given by the product of the mass transfer coefficient k_m, which is a quantity which can be easily measured, and the driving concentration difference Δc:

$$i = \dot{n}_0 v_e F = k_m (c_\infty - c_0) v_e F = k_m \Delta c v_e F \tag{5.15 a}$$

Linearizing the concentration profiles in front of a working electrode (see Fig. 5.6 a, b) defines the thickness δ_N of the Nernst diffusion layer, which allows to interpret the mass transfer coefficient k_m as the ratio of D and δ_N[3]:

$$k_m = D/\delta_N \tag{5.16 a}$$

---

3 δ_N is similarly defined as δ_{Pr}; $\delta_N = c_\infty (\delta y/\delta c)_{y=0}$

Setting $c_0=0$ defines the mass transfer limited current density i_{lim}:

$$i_{lim} = k_m c_\infty v_e F \tag{5.16 b}$$

5.4.2
Dimensionless Numbers Defining Mass Transport Towards Electrodes by Convective Diffusion

To find an explicit equation for the mass transfer coefficient, k_m, is only possible for laminar flow. Under turbulent flow one can only measure mass transport coefficients by measuring mass transport limited current densities. But this is a tedious affair as mass transfer is influenced often by a great number of variables. Dimensional analysis of the problem allows to reduce considerably the number of variables which have to be taken into account for mass transfer determinations by introducing dimensionless groups which comprise several different characteristic quantities of the respective system like for instance mass transfer coefficient, velocity, density and characteristic lengths which might for instance be the interelectrodic distance or the electrode length in case of virtually singular electrode.

The reduction in number of variables thus obtained amounts exactly to the number of different fundamental dimensional quantities (length, time, mass, charge, voltage) which are used in total to define the complete set of variables. By introducing dimensionless groups one obtains then equations, "adimensional correlations", which relate these quantities to each other for a given flow geometry. For laminar flow these correlations are obtained by algebraic calculation. For turbulent flow such correlations have to be determined experimentally.

For mass transfer under forced convection there exist at least three different dimensionless groups: The Sherwood number, Sh, which contains the mass transfer coefficient, the Reynolds number, Re, which contains the flow velocity and defines the flow condition (laminar/turbulent) and the Schmidt number, Sc, which characterizes the diffusive and viscous properties of the respective fluid and which describes the relative extension of fluid-dynamic and concentration boundary layer:

$$Sh = \frac{k_m L}{D} \tag{5.17 a}$$

$$Re = \frac{wL}{v} \tag{5.17 b}$$

$$Sc = v/D \tag{5.17 c}$$

If forced convection does not apply, density differences between the electrolyte close to the electrodes and the bulk of the solution may be generated for instance by metal deposition and dissolution or gas evolution. Then free or so called natural convection is generated and becomes very often much more intense than forced convection driven from outside by pumping the electrolyte through the cell. Fluid flow under free convection is characterised by the Grashoff number, Gr, which contains the gravity constant, g, the density difference of the electrolyte at the electrode (ρ_0) and the electrolyte bulk (ρ_∞).

$$Gr = gL^3 \left(\rho_{e,\infty} - \rho_{e,0}\right)/\rho_\infty v^2 \tag{5.17 d}$$

In general one tries to present the dependence of Sh on Re (or Gr) and Sc – and possibly on other adimensional quantities like the ratio l/L of characteristic lengths in form of a power series:

$$Sh = C Re^m \left(\text{or } Gr^n\right) Sc^p \left(l_1/L\right)^q \left(l_2/L\right)^r \tag{5.18}$$

The experimental determination of Sh is quite easy in electrochemical systems. Measuring the mass transfer limited current density and using Eq. (5.16 b) one obtains immediately the surface averaged Sh from the numerical value of i_{lim}:

$$Sh = \frac{i_{lim} L}{c_\infty v_e FD} \tag{5.18 a}$$

5.4.3
Hydrodynamic Boundary Layer and Nernst Diffusion Layer: Planar Electrodes

Figure 5.6 a,b depicts schematically the relative extension of velocity and concentration profiles in front of a planar electrode for (a) laminar and (b) turbulent flow. (Assumption: potential conditions set for mass transfer limited current density and freely developing, that means not yet closed, velocity boundary layer for laminar flow). For both cases at sufficiently high distance from the electrode surface the maximal fluid velocity w_{max} is acquired.

Turbulent flow is distinguished from laminar flow by two main differences:
(a) Due to the eddies, which characterise the turbulent flow condition, the flow in the bulk of the fluid is stochastically fluctuating in all directions around w_{max} – which means that fluctuating velocity contributions not only parallel but also perpendicular to the main velocity vector are generated by eddies and may be observed. The intensity of these fluctuations is decaying more and more towards the electrode surface very close to the electrode surface in the region δ_0 of the so called viscous sublayer.
(b) Under comparable conditions (plate dimensions, kinematic viscosity of the fluid) the extension of the viscous sublayer of a turbulent boundary layer is remarkably less extended than the velocity boundary layer for a freely developing laminar flow. This difference bears also on mass transfer: Quite similarly the extensions of the Nernst diffusion layers connected to mass transfer

for turbulent flow is remarkably smaller and hence mass transfer is much more efficient for turbulent flow than for laminar flow.

A quantitative treatment [1] of the relative extension δ_{Pr} of the velocity boundary layer for freely developing laminar flow and the associated Nernst diffusion layer thickness δ_N shows that it is governed by the Schmidt number $(Sc = v/D)$, which is indicative of the relative effectiveness of "momentum diffusion" (described by the kinematic viscosity $v = \eta/\rho$) and mass diffusion (characterised by D). For aqueous electrolytes, with $v \approx 10^{-2}$ cm² s⁻¹ and $D \approx 10^{-5}$ cm² s⁻¹, the Schmidt number is approximately 1000.

For *freely developing laminar flow* along a plate one obtains

$$(\delta_N/\delta_{Pr})_{x,lam} \approx Sc^{-1/2} \tag{5.19 a}$$

so that for aqueous electrolytes this ratio of the extension of both boundary layers amounts to approximately 30 everywhere on the plate.

For *turbulent flow* theoretical considerations arrive at a ratio of the thickness of the Nernst layer to that of the laminar sublayer of

$$(\delta_N/\delta_0)_{turb} \approx Sc^{-1/2} \text{ to } Sc^{-1/3} \tag{5.19 b}$$

which would be approximately from 30 to 10 for aqueous electrolytes.

Figure 5.7 demonstrates schematically for free flow along a plate, compare Fig. 5.5 a, how the laminar (with thickness δ_{Pr}) and (after arriving at a critical length) the turbulent boundary layer and laminar sublayer (with thickness δ_0) are developing. According to Eq. (5.19 a,b) the Nernst diffusion layer (with thickness δ_N) would develop below δ_{Pr} and δ_0. In quantitative terms one obtains Sh (the mean Sherwood number) for laminar flow along a plate electrode of length L

$$\overline{Sh}_L = 0.67\, Re_L^{1/2}\, Sc^{1/3}; Re < 3 \cdot 10^5 \tag{5.20 a}$$

and for turbulent flow the local Sherwood number is given by

$$Sh_L = 0.03\, Re_L^{0.8}\, Sc^{1/3}; \quad Re > 3 \cdot 10^5 \tag{5.20 b}$$

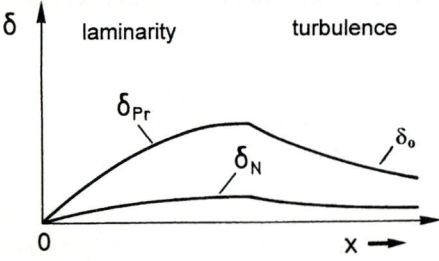

Fig. 5.7. Change from laminarity to turbulence for freely developing parallel flow along a plate induces compression of the thickness of the velocity profile or Prandtl layer thickness

5.4.4
Mass Transport Towards a Singular Planar Electrode[4] Under Laminar Forced Flow

Calculation of mass transfer towards an isolated electrode under conditions of forced convection along its surface under steady state conditions and laminar flow starts with the solution of the Navier/Stokes equation, which describes the velocity distribution in y-direction (perpendicular to the electrode surface) and the growth of the fluid dynamic boundary layer in x direction that means downstream, Eq. (5.9 a,b). One then applies the mass transfer equation neglecting diffusive mass transfer in x direction:

$$0 = D(\partial^2 c/\partial y^2) - w(\partial c/\partial x) \qquad (5.21)$$

Integration with respect to y yields

$$D(\partial c/\partial y)_0 = \frac{\partial}{\partial x} \int_0^\infty w(c_\infty - c) dy \qquad (5.22)$$

with the boundary conditions:

$$c_0 = 0 \text{ (means } i = i_{lim}\text{)}, \quad c_{y=\infty} = c_\infty \qquad (5.23)$$

and with $w_{x,z}$ equal to the already obtained velocity distribution of the laminar boundary layer, Eq. (5.8 ,b), one obtains:

$$D(\partial c/\partial y)_0 = w_\infty \frac{\partial}{\partial x} \int_0^\infty \left(\frac{3}{2} \frac{y}{\delta_w} - \frac{1}{2} \left(\frac{y}{\delta_w} \right)^3 \right) (c_\infty - c) dy \qquad (5.24)$$

In the next step the assumption of the development of a spatially similar boundary layer profile which independent of the extension of the Nernst diffusion layer satisfies in principle the same (reduced) mathematical description which is taken to be a polynomial of third order with coefficients a, b and c leads to[5]:

$$c(x,y) = c_\infty + \Delta c \left(b(y/\delta_c) + c(y/\delta_c)^2 + d(y/\delta_c)^3 \right) \text{ and } \delta_c = \delta_c(x) \qquad (5.25)$$

---

4 The counter electrode is so far away that is does not influence the velocity profile. This holds for the very entrance of a gap or duct between any two planar electrodes, too. As the fluid enters this duct the fluid at the leading edge of any of the two electrode does not yet "feel" the presence of the counter electrode. The following example is treated more extensively in order to convey to the reader an impression and better understanding that, and how, mass transfer under laminar conditions can be calculated.

Introducing the boundary conditions

$$(\partial c/\partial y)_{\delta_c} = 0; \quad (\partial^2 c/\partial^2 y)_0 = 0; \quad c_{\delta_c} = c_\infty; \quad c_0 = 0 \tag{5.25 a}$$

results in:

$$c(x,y) = c_\infty \left[\frac{3}{2}\left(\frac{y}{\delta_c}\right) - \frac{1}{2}\left(\frac{y}{\delta_c}\right)^2 \right] \text{ and } \delta_c = \delta_c(x) \tag{5.26}$$

According to mathematically similar profiles and boundary conditions the mathematical formulations for the velocity distribution in the Prandtl layer and the concentration profile of the Nernst layer are similar only being distinguished by different extensions δ_{Pr}, δ_w and δ_N, δ_c of the velocity and the concentration boundary layers. By introducing Eq. (5.26) into Eq. (5.24) and limiting the integration to δ_c one obtains

$$\frac{3}{2} D/\delta_c = w_{max} c_\infty \frac{\partial}{\partial x} \int_0^{\delta_c} \left\{ \frac{3}{2}\left(\frac{y}{\delta_w}\right) - \frac{1}{2}\left(\frac{y}{\delta_w}\right)^2 \right\} \left\{ 1 - \frac{3}{2}\left(\frac{y}{\delta_c}\right) + \frac{1}{2}\left(\frac{y}{\delta_c}\right)^2 \right\} dy \tag{5.27}$$

Solving the integral and introducing δ_w for the isolated planar electrode, compare Eq. (5.96), one arrives with $\delta_c/\delta_w = \left(\frac{13}{14}\right)^{1/3} Sc^{-1/3}$, $\delta_N = \frac{2}{3}\delta_c$ and

$$\delta_w = \left(\frac{280 \nu x}{13 w_{max}}\right)^{1/2} \text{ at}$$

$$-(\partial c/\partial y)_{y=0} = c_\infty/\delta_N = c_\infty \frac{3}{2}\frac{1}{\delta_c} \tag{5.28}$$

and equating

$$k_m = D(\partial c/\partial y)_0 \frac{1}{c_\infty} \tag{5.29}$$

one obtains the local Sherwood number

$$Sh_x = \frac{k_m x}{D} = \left(\frac{13}{280}\right)^{1/2} \left(\frac{14}{13}\right)^{1/3} Re_x^{1/2} Sc^{1/3} = 0.331\, Re_x^{1/2} Sc^{1/3} \tag{5.30}$$

---

[5] There is a similar difference between the thickness δ_c as defined by the boundary conditions Eq. (5.25 a) and the polynomial Eq. (5.25) and the thickness of the Nernstian diffusion layer $\delta_N = c_\infty \left(\frac{\partial c}{\partial y}\right)_0^{-1}$ as between δ_w and δ_{Pr} which both define the extension of the velocity boundary layer

5.4 Mass Transport by Convective Diffusion

integrating Sh_x from zero to $x = L$, the length of the plate, and dividing by L leads to Eq. (5.20 a) for the mean Sherwood number:

$$\overline{Sh} = \frac{1}{L}\int_0^l Sh_x dx = 0.662 \, Re_L^{0.5} Sc^{1/3} \tag{5.31}$$

5.4.5
Channel Flow and Mass Transfer to Electrodes of Parallel Plate Cells for Free and Forced Convection

Free and forced convection through an interelectrodic gap between two planar electrodes is the most often encountered case in electrochemical engineering practice. Hydrometallurgical electrorefining and electrowinning of metals like copper, zinc and lead as well as electrochemical gas evolution at vertical electrodes are typical cases for electrolysis cells with free convection being caused by density differences of the electrolyte.

5.4.5.1
Free Convection at Isolated Planar Electrodes and between Two Vertical Electrodes

At the anode of an electrolysis cell for copper refining the dissolution of copper increases the $CuSO_4$-concentration and the electrolyte density close to the anode surface, whereas $CuSO_4$-depletion and a corresponding decrease of the electrolyte density has to be taken into account at the cathode (Fig. 5.8 a) due to cathodic copper deposition. This causes a downward flow along the anode and an upward flow along the cathode, Fig. 5.8 b, both flows improving the mass transfer condi-

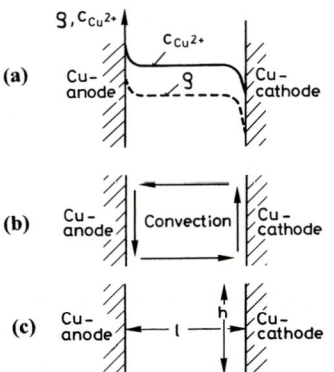

Fig. 5.8. a–c. Schematic of development of natural convection driven by density differences in the electrolyte induced by cathodic deposition of copper from copper sulfate solutions: **a** concentration and density profiles of Cu^{2+}; **b** convection pattern; **c** defining electrode height, h, and cell width, l

tions at the respective electrode, which is of particular importance at the cathode for preventing dendritic growth of the refined cathodically deposited copper.

Because of the relative far distance of cathode and anode in copper refining cells (6–10 cm), there is little interference between the two electrolyte flows. Each flow along an electrode in such cells can be treated as flow along a separate planar electrode. At the leading edge of the electrode the flow starts as laminar flow converting to turbulent flow after reaching a critical distance from the edge, which is determined by the critical values of Gr=10^9 and GrSc of approx. $5 \cdot 10^{12}$.

The dimensionless correlation for laminar free convection reads

$$Sh = 0.66 (ScGr)^{0.25} \tag{5.32}$$

and for free, turbulent flow it is given by:

$$Sh = 0.31 (ScGr)^{0.28} \tag{5.33}$$

For laminar as well as for turbulent flow the characteristic length L used in Gr is equal to the height h of the electrode or the distance from the lower (cathode) or upper (anode), respectively, edge, Fig. 5.8 c.

These correlations apply also for electrodes at which gases are evolved or at which additionally to the main reaction electrochemical gas evolution occurs as a side reaction (for instance for Zn-electrowinning this would be cathodic hydrogen evolution). Usually the gas bubbles formed are small compared to the extension of the hydrodynamic boundary layer so that the gas/electrolyte emulsion can be approximately considered as a homogeneous phase of density $\rho \cdot (1 - \varepsilon)$ with ε meaning the volume fraction of the gas in this emulsion.

If the gap between a counter electrode and the respective electrode or a diaphragm contrary to the initially assumed case is very narrow compared to the height of the electrode then the developing free convection is significantly influenced by the distance between diaphragm and electrode. The dimensionless correlation for this case reads

$$Sh = 0.19 (ScGr)^{1/3} \tag{5.34}$$

where the characteristic length in Gr is no longer the height or length but the width, l, of the interelectrodic gap.

5.4.5.2
Convective Mass Transfer for Parallel Plate Cells with Forced Convection: Planar Plate Cells

Unidirectional forced convection through the narrow gap between two parallel plates is realised in some types of electrolyzers particularly developed for organo-electrosyntheses for instance the bipolar gap electrolyser (Fig. 5.9 a), the Swiss roll cell (Fig. 5.9 b) or the circular capillary gap cell (Fig. 5.9 c). Fluid dynamics of such cells was dealt with in Sect. 5.3. For laminar flow conditions ve-

5.4 Mass Transport by Convective Diffusion

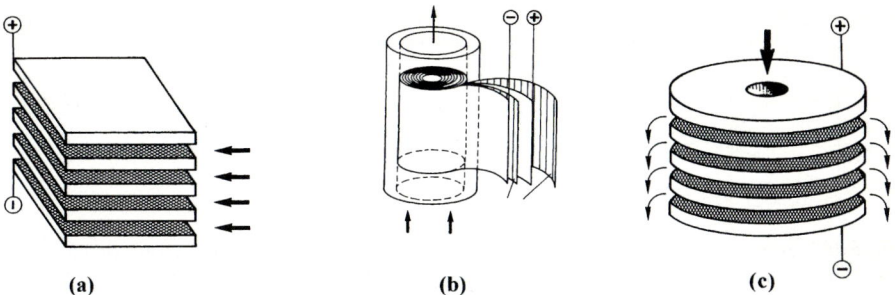

Fig. 5.9a–d. Particular electrolyzer configurations which are operated with forced electrolyte flow: **a** bipolar gap electrolyzer; **b** Swiss roll cell; **c** circular capillary gap cell; **d** circular gap cell

locity boundary layers develop at the leading edge of both electrodes, both growing in thickness until they meet in the middle of the gap. Then further downstream the closed parabolic velocity profile, Eq. (5.35), does not change any further:

$$w(y) = w_{max}\left(1-(y/b)^2\right) \tag{5.35}$$

Figure 5.10 a depicts schematically the solution of the differential equation for mass transport under mass transfer limited condition ($i = i_{lim}$).

$$0 = D\left(\partial^2 c/\partial y^2\right) + w\left(\partial c/\partial x\right) \tag{5.36}$$

This differential equation is solved under observation of the velocity distribution given in Eq. (5.35) and keeping to conditions where $\delta_N < b$ that means main-

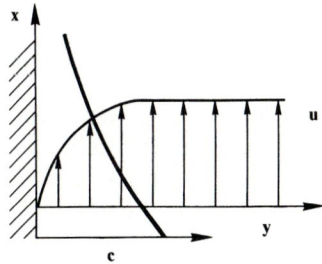

Fig. 5.10. a Schematic presentation of concentration depletion along the electrode in the flowing electrolyte as a solution of Eq. (5.36) describing the superposition of convective mass transfer along the flow direction and diffusive mass transport perpendicularly to the electrolyte flow towards the electrode.

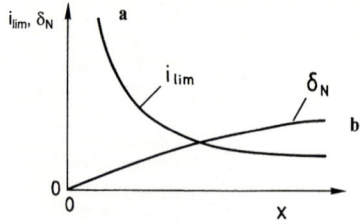

Fig. 5.10.b Schematic presentation of decrease of mass transfer limited current density and extension of Nernst diffusion layer thickness with increasing distance from the entrance or leading edge

taining the extension of the Nernst diffusion layer below the middle of the gap. Figure 5.10 b demonstrates how steeply the mass transfer limited local current density declines at the leading edge of the electrode.

The mean mass transfer limited current density in the gap of width $d = 2b$ for laminar flow (Re<3000) of an electrode of length L equals

$$\bar{i}_{lim} = 2.54\, D^{2/3}\, L^{-1/3} (2b)^{-1/3} (w_{max})^{1/3} c_\infty v_e F \qquad (5.37\ a)$$

and the mean Sherwood number is:

$$\overline{Sh} = 2.54\, Re^{1/3} Sc^{1/3} (2b/L)^{1/3} \qquad (5.37\ b)$$

Local values of Sh and i_{lim} are obtained from these expressions by differentiation with respect to L.

For turbulent forced convection (Re>3000) the correlation reads

$$Sh = 0.023\, Re^{0.8} Sc^{1/3} \qquad (5.38\ a)$$

and does not vary spatially that means it is constant and independent of the distance from the entrance.

Mass transfer limited current densities obtained from Eq. (5.38 a) amount to

$$i_{lim} = 0.023\, D^{2/3} (2b)^{-0.2} (w_{max})^{0.8} c_\infty v_e F \qquad (5.38\ b)$$

Although real parallel plate cells have a finite width, the electrodes were treated above mathematically as being of infinite bredth so that the side walls were not assumed to influence viscous flow and velocity or concentration profiles. For the case of cells made of two concentric cylinders with axial flow the treatment of the parallel plate cell applies therefore equally well. Instead of the electrode distance b the equivalent hydraulic diameter is introduced ($d_{hydro} = 4(R_1^2 - R_2^2)/(R_1 + R_2)$) into the corresponding equations for the parallel plate cell.

But a different correlation is expected for flow through quadratic ducts, which sometimes are used for particular experimental investigations since in such

5.4 Mass Transport by Convective Diffusion

ducts cell gap and electrode width are identical. For quadratic ducts the correlation for the local Sherwood number for laminar flow reads

$$Sh_{local} = 0.849\, Re^{1/3}\, Sc^{1/3} (2b/L)^{1/3} \qquad (5.39\ a)$$

For turbulent flow the correlation is

$$Sh = 0.0115\, Re^{0.9}\, Sc^{1/3} \qquad (5.39\ b)$$

5.4.5.3
Mass Transfer in Circular Capillary Gap Cells

For the circular capillary gap cell depicted schematically in Fig. 5.9 c the particular flow characteristics – declining linear velocity for outward flow $w_r = w_{r_0} \dfrac{r_0}{r}$ – have their consequence on mass transfer: the mass transfer rate decreases according to $r^{-2/3}$.

Calculating local mass transfer at radius r in the circular capillary gap cell of width 2 b for laminar flow using the usual boundary layer concept one obtains:

$$k_{m,r} = 0.415 \left(\frac{D}{rb}\right)^{2/3} \dot{V}^{1/3} \qquad (5.40\ a)$$

where $\dot{V}$ is the volumetric flow rate of the electrolyte entering through the central bore of the cell.

Transformation into the adimensional correlation leads to:

$$Sh_r = 1.66 (b/r)^{2/3}\, Re^{1/3}\, Sc^{1/3} \qquad (5.40\ b)$$

By integration along r from the inlet, r_0, to the outlet of the gap, R, one obtains for the mean Sherwood number for laminar flow:

$$\overline{Sh} = 1.25 (b/R)^{2/3} \left(\frac{1-(r_0/R)^{1/3}}{1-(r_0/R)^2} \right) Re^{1/3}\, Sc^{1/3} \qquad (5.40\ c)$$

The Reynolds number for the concentric capillary gap cell is defined by

$$Re = \frac{\dot{V}}{2 b v} \qquad (5.40\ d)$$

Laminar conditions are prevailing for $Re < 10^5$.

For turbulent flow (Re $>4\times10^5$) one obtains experimentally

$$\overline{Sh} = \text{const}\left(\frac{2b}{R}\right)^{1/2}\left(\frac{4b^2}{R^2-r_i^2}\right)^{0.8} Re^{0.8} Sc^{1/3} \qquad (5.40\ e)$$

5.4.6
Convective Mass Transfer Toward Rotating Electrodes

5.4.6.1
Rotating Cylinder

For rotating cylinder electrodes, which are sometimes used for electrode kinetic investigations, the correlation for turbulent flow condition reads

$$Sh = 0.079 Re^{0.7} Sc^{0.356} \qquad (5.41)$$

holding for $10^3 < Re < 10^5$ and $800 < Sc < 11{,}500$. The Reynolds number is defined with r = radius of the cylinder as the characteristic geometric length (Re = $r^2\omega/\nu$).

5.4.6.2
Rotating Disc Electrode

The rotating disc electrode – sometimes surrounded by a ring electrode – is one of the most important and most frequently used devices for electrode kinetic measurements, since it allows to establish well defined and accurately calculable mass transfer conditions with equal mass transfer rates everywhere on the disc surface – compare Fig. 5.21 b.

Figure 5.11 demonstrates the axial and radial flow pattern at the disc electrodes, which ejects the electrolyte radially and sucks in an axial stream of electrolyte towards the rotating disc electrode which is usually embedded into the flat face of a rotating cylindrical rod[6] made of electrically insulating material in order to elimi-

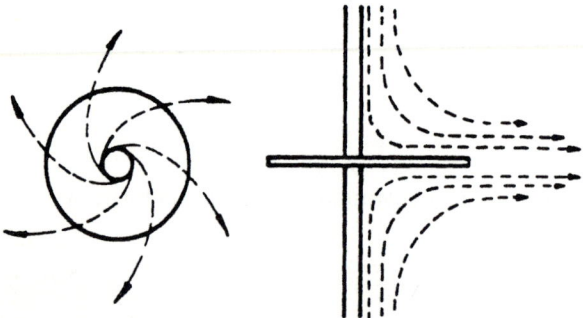

Fig. 5.11. Radial and axial flow pattern in front of a rotating disc electrode

nate the influence of flow discontinuities of the outer edge of the electrode. If a double faced circular disc is attached to a central shaft, the flow pattern becomes symmetrical with respect to the middle plane of the rotating disc and resembles from both sides that of the rotating disc. For laminar flow conditions Levich calculated for the rotating disc the mass transport properties in adimensional form

$$\overline{Sh} = 0.6 \, Re^{1/2} \, Sc^{1/3}$$
$$\left(10^2 < Re = 10^4 \text{ to } 10^5\right)$$
(5.42 a)

and for turbulent flow ($Re > 10^6$):

$$\overline{Sh} = 0.01 \, Re^{0.87} \, Sc^{1/3}$$
(5.42 b)

The correlations hold for the double-faced disc as well. The Reynolds number of the rotating disc is defined by

$$Re = \frac{r^2 \omega}{\nu} \quad \text{with } \omega = 2\pi f; \text{ speed of rotation}$$
(5.42 c)

Table 5.1 collects mass-transfer correlations of the most important types of electrodes and cells.

5.4.7
Mass Transfer at Gas Evolving Electrodes

Mass transfer at gas evolving electrodes can be remarkably enhanced, provided gas evolution is intense enough to influence the flow of the electrolyte along the electrode. As radii of electrochemically evolved gas bubbles are usually relative small (5–50 μm), bubbles can perturb concentration boundary layers very effectively thereby enhancing mass transfer and compressing Nernst-diffusion layers.

As the number of the bubbles evolved per unit surface and unit time increases with increasing current density, the growth of the bubbles, their detachment from the electrode surface and the changed density of the electrolyte/bubble emulsion begin to enhance mass transfer.

Bubble induced mass transfer is superimposed on macroconvective mass transfer and governs convective mass transfer more and more the higher the volumetric rate of gas evolution that means the higher the current density of the gas evolving reaction.

The effective mass transfer can be modelled by vectorial addition of normal convective and bubble induced mass transfer:

$$k_m = \left(k_{m, \, macro}^2 + k_{m, \, bubble}^2\right)^{0.5}$$
(5.43)

---

[6] This device was invented and explored scientifically and practically by the Russian physicochemist Levich.

Table 5.1. Dimensionless correlations involving mass transfer

System	Correlation[a]	Characteristic length and velocity used for definition of dimensionless quantities	Validity limits[a]
Forced convection along parallel plate cell	1) $Sh = 2.54\,(ReSc)^{1/3}(d/L)^{1/3}$ 2) $Sh = 0.025\,Re^{0.8}\,Sc^{1/3}$	electrode gap and length	1) $Re \leq 3000$, laminar flow 2) $Re > 3000$, turbulent flow
Free convection along vertical plate	$Sh = 0.66\,(ScGr)^{1/4}$	height of electrode	$500 < Sc < 80{,}000$ $10^4 < Gr < 10^9$ $5 \cdot 10^6 < ScGr < 5 \ast 10^{12}$
laminar flow	$Sh = 0.31\,(ScGr)^{0.28}$	height of electrode	$4 \cdot 10^{13} < ScGr < 10^{15}$
turbulent flow	$Sh = 0.19\,(ScGr)^{1/3}$	height of electrode	$10^8 < ScGr < 1.4 \ast 10^{12}$ $2.1 \cdot 10^3 < Sc < 5.2 \ast 10^4$
Free convection between vertical electrode and diaphragm	$Sh = 0.0225\,(ScGr)^{0.85}(d/h)^2$	e = electrode-diaphragm distance; h = height	$2 \cdot 10^6 < ScGr < 2 \ast 10^8$ $0.005 < e/h < 5.48\,(ScGr)^{-0.3}$
Forced flow along plate	$Sh = 0.66\,Re^{1/2}Sc^{1/3}$	height of electrode in flow direction, linear flow velocity	$Re < 3 \cdot 10^5$ to 10^6
Axial flow through gap between two coaxial cylinders		equivalent diameter, flow velocity	
laminar	$Sh = 1.62\,(ReSc)^{1/3}\left(\dfrac{d_e}{L}\right)^{1/3}$	d_e = (gap width) × 2	$300 < Sc < 3000$ $300 < Re < 2100$
turbulent	$Sh = 0.023\,Re^{0.8}Sc^{1/3}$		$2100 < Re < 30{,}000$
Rotating cylinder	$Sh = 0.079\,Sc^{0.356}Re^{0.7}$	diameter of cylinder, circumferential velocity	$1000 < Re < 100{,}000$ $835 < Sc < 11{,}500$
Rotating cylinder within coaxial stationary hollow cylinder	$Sc = 0.079\,Re^{0.7}Sc^{0.356}(d_2/d_1)^{0.7}$	diameter of counterelectrode, circumferential velocity	$1000 < Re < 53{,}000$ $0.22 < d_2/d_1 < 0.68$

[a] Sh = Sherwood number = $j_1\,L/Dc_{x=\infty}$; Gr = Grashof number = $gL^3(\rho_0 - \rho_e)/\rho_0\,v^2$; Re = Reynolds number = wd/v; Sc = Schmidt number = v/D; d = electrode gap; d_1 = diameter of stationary working electrode; d_2 = diameter of rotating counterelectrode; d_e = equivalent diameter = $(d_2 - d_1)$; e = diaphragm-electrode distance; h = height of electrode; L = electrode length

5.4 Mass Transport by Convective Diffusion

The bubble induced mass transfer coefficient, $k_{m,\,bubble}$ is influenced by many parameters, the most important of which is the volumetric flow rate of gas evolution or current density respectively.

5.4.7.1
Calculating $k_{m,\,bubble}$ According to the Penetration Model or Model of Periodic Boundary Layer Renewal

If one assumes that the most important effect of mass transfer enhancement at gas evolving electrodes consists of the creation of a wake flow following bubble detachment, by which the electrolyte of the boundary layer is replenished quasiperiodically, one obtains due to this non-steady-state assumption:

$$Sh = \frac{2.76}{C}(Re\,Sc)^{0.5}(1-\Theta)^{0.5} \qquad (5.44)$$

with Θ the mean degree of gas bubble coverage of the electrode.

The dimensionless groups used in Eq. (5.44) are

$$Sh = \frac{k_{m,\,bubble} \cdot d_b}{D} \qquad (5.44\text{ a})$$

with d_b the mean bubble diameter of approximately 50 μm and

$$Re = \frac{\dot{V}_{surface}\, d_b}{\nu} \qquad (5.44\text{ b})$$

$\dot{V}_{surface}$ is the volumetric flow rate of gas evolution per unit electrode surface in cm s^{-1}; the factor C in Eq. (5.44) is a factor introduced to account for the form of the bubbles. C = 1.59 for hemispheres and C = 2 for spheres.

5.4.7.2
Calculating Bubble-Enhanced Mass Transfer According to Flow Model

Another theoretical approach, the microconvective model, accounts for the lateral movement of the electrolyte in the vicinity of bubbles caused by their growth. The resulting equation is functionally similar and numerically close to Eq. (5.44) for low values of the degree of bubble shielding, Θ:

$$Sh = 1.652(Re\,Sc)^{0.5}\left[\Theta^{0.5}(1-\Theta^{0.5})\right]^{0.5}(1+0.29Sc^{1/6}+0.0047Sc^{0.053}) \qquad (5.44\text{ c})$$

For most practical cases it can be approximated by

$$Sh = 0.93(Re\,Sc)^{0.5} \qquad (5.44\text{ d})$$

5.4.8
Mass Transfer in Three-Dimensional Electrodes

Since the rate of conversion in electrochemical cells is proportional to the electrode surface area, there exists the incentive for designing electrochemical cells with high electrode area to volume ratio as actualised in so called "three-dimensional electrodes" as porous electrodes, packed bed electrodes, fluidised bed electrodes and comparable designs.

Mass transfer and potential and current density distribution must be handled simultaneously in the theoretical treatment of such electrodes. First of all one has to visualize that the potential difference between electrolyte and electrode, the "local electrode potential", is changing throughout the bed. The so called "effective" bed depth defines the depth, up to which for a given electrode reaction and overpotential fully established mass transport limited current can still be established. Only if the real electrode depth does not exceed the effective electrode depth, along which spatially equalised mass transfer limited current density prevails, it is sensible to apply data and correlations for mass transfer throughout the entire electrode (for details see below.)

The kinetic characteristics of a fully utilised three dimensional electrode discussed in Chapter 7 are illustrated in Figs. 7.9 a and b, which compares polarisation curves for silver deposition from 10^{-3} mol l^{-1} Ag$^+$ solutions for channel flow and within a packed bed electrode contained in the same channel, both being operated at identical linear flow velocities (as referred to the empty channel).

The current voltage curve measured at the planar electrode reveals that silver deposition is mass transfer controlled. For the packed bed electrode one observes diffusion limited current densities (related to the electrodes' cross section), which are 100 times greater than that for the planar electrode due to the larger actual surface and due to the enhanced mass transfer in the packed bed. As it was experienced for channel flow in the correlation Sh vs Re, one expects a Re$^{1/3}$ dependence if flow in the bed between the particles is essentially laminar but the exponential m is expected to increase from 0.333 towards 1 above a critical Reynolds number as the flow in the bed begins to become turbulent:

$$Sh = C\, Re^m\, Sc^{1/3} \qquad (5.45)$$

Table 5.2 summarises the data obtained experimentally for mass transfer correlations in packed and fluidised beds.

Table 5.2. Parameters C and m obtained for mass transfer correlation. Sh = C Re^{m}Sc$^{1/3}$ for packed and fluidized bed electrodes

Type of electrode	C	m	Range	
Packed bed	0.50	0.61	50<Re<5000	[1]
	0.32	0.66	100<Re<10,000	[2]
Fluidized bed	5.7	0.22	Re : 1–30	[3]
	1.52	0.50	Re : 2–7	[4]
	0.29	0.75	100<Re<5000	[2]
additional factor: $(1-\varepsilon)^{-0.44}$	0.763	0.56	260<Re<340	[5]

The adimensional numbers Sh and Re take account of the particle diameter d_p and the void fraction $(1-\varepsilon)$ of the bed:

$$Sh = \frac{k_m \varepsilon d_p}{(1-\varepsilon)D} \qquad (5.45\ a)$$

$$Re = \frac{w d_p}{(1-\varepsilon)v} \quad \text{w is the flow velocity in the empty volume of the bed} \qquad (5.45\ b)$$

5.4.9
Summary

The quantitative description of mass transfer towards electrodes is one of the most important prerequisites of electrochemical process engineering.

Almost for any practical cell design and flow condition we are today able to describe mass transfer by dimensionless correlations, which are either theoretically derived (for laminar flow) or which rest on reliable experimental determination – see Table 5.1. The correlations obtained for bed electrodes (see Table 5.2) and multiphase flow, however, are still somewhat questionable and need still further experimentation and confirmation.

It must be stressed that mass transfer in real electrolyzers, in particular in big parallel plate cells, is usually not as well defined as in laboratory devices. Furthermore, in big cells mass transfer conditions may be pronouncedly unevenly distributed across the surface of an electrode of an extension of, e.g., say 1×1 m².

5.5
Heat Transport

5.5.1
Chilton–Colburn Analogy of Mass and Heat Transfer

The characteristic adimensional quantity of mass transfer by convective diffusion is the Sherwood number. The Nusselt number characterises in analogous way the heat transfer by convective heat conduction. Equation (5.46) defines the Nusselt number with α the so called heat transfer coefficient and λ the heat conductivity.

$$Nu = \frac{\alpha \cdot l}{\lambda} \qquad (5.46)$$

The characteristic ratio of the physical data of viscous flow and heat conduction of a fluid are given by its Prandtl number Pr:

$$Pr = \frac{v}{a} \qquad (5.47)$$

With ν the kinematic viscosity and a the so called temperature conductivity $a = \lambda/\rho\, c_p$

For a given system, mainly determined by a particular, streaming fluid, its flow velocity and the special geometry of the considered flow device, heat and mass transfer can very often be described by empirically analogous adimensional relations:

$$Sh = C \cdot Re^\alpha \cdot Sc^\beta \cdot \left(\frac{1}{L}\right)^\gamma \text{ and } Nu = C Re^\alpha \cdot Pr^\beta \left(\frac{1}{L}\right)^\gamma \tag{5.48}$$

The existence of these analogous correlations are known as the Chilton-Colburn analogies and have proved useful for instance for pipe flow under turbulent conditions, for flow through packed beds or flow around rotating cylinders in liquids whose physical properties are comparable to that of water. For e.g. mercury with much higher heat conductivity but comparable diffusivity the Chilton–Colburn analogy would not be expected to hold.

The analogy is applicable whenever the concentration and temperature boundary layers do not extend beyond but are rather less extended than the velocity boundary layer. In such cases one may anticipate that the adimensional correlation of heat transfer and mass transfer contains the same numerical factors and the same exponents of Re or Gr on one hand and of the Schmidt-number for mass transfer and the Prandtl number for heat transfer on the other hand. Thus for instance for turbulent flow between two plates one obtains the adimensional correlation for heat transfer:

$$\overline{Nu} = 0.023 \, Re^{0.8} \, Pr^{1/3} \tag{5.49 a}$$

compared to the correlation for mass transfer:

$$\overline{Sh} = 0.023 \, Re^{0.8} \, Sc^{1/3} \tag{5.49 b}$$

5.5.2
General Description of Heat Generation and Heat Transfer in Electrolyzers and Fuel Cells

Heat transfer in electrolyzers and fuel cells is generally brought about by convective heat conduction, i.e. by superposition of heat conduction close to a stationary wall or phase boundary, and heat transport by convection across longer distances further away from the wall. Heat is generated in electrochemical devices mainly by dissipation of Joule's heat and almost always (but to a minor part) by energy dissipation due to overvoltage. The heat generation per unit volume and unit time $1/V \cdot dQ/dt$ by dissipation of Joule's heat caused by current density i in an electrolyte of conductivity κ amounts to:

$$\frac{1}{V}\left(\frac{dQ}{dt}\right)_{diss,\,i} = i^2 \kappa^{-1} \tag{5.50}$$

5.5 Heat Transport

The heat generation per unit volume and unit time at an electrode working with overpotential η at current density i reads:

$$\frac{1}{V}\left(\frac{dQ}{dt}\right)_{diss,\eta} = i\eta \, a_e \tag{5.51}$$

a_e is the electrode surface to volume ratio of the electrolyzer which in parallel plate electrolyzers is usually the inverse of the electrode distance. Heat transfer per unit volume and unit time from the electrode to the bulk of the electrolyte is described by the general heat-transfer equation:

$$\frac{1}{V}\left(\frac{dQ}{dt}\right)_{transf} = \alpha\left(T_{electrode} - T_{bulk}\right)a_e \tag{5.52}$$

with α equal to the heat transfer coefficient (electrode→bulk electrolyte).

Heat transfer across a wall – which might by the electrolyzer frame – from the electrolyte to the surrounding is described by formulating Eq. (5.53) for unit volume of the electrolyzer:

$$\frac{1}{V}\left(\frac{dQ}{dt}\right)_{transf,\,out} = a_{cell} k_h \left(T_{electrol} - T_{surround}\right) \tag{5.53}$$

a_{cell} is the ratio of outer surface of the electrolyzer to its volume and k_h is the heat transfer coefficient of the whole heat exchanging system that means the cell. If α_1, the internal heat transfer coefficient (electrolyte→wall), and α_2, the external heat transfer coefficient (wall→environment), are known, then k_h can be calculated with d_w, the thickness, and λ_w the coefficient of heat conduction of the wall material:

$$\frac{1}{k_h} = \frac{1}{\alpha_1} + \frac{1}{\alpha_2} + \frac{d_w}{\lambda_w} \tag{5.54}$$

5.5.2.1
Heat Balance and Steady-State Temperature of Cells

The steady state condition defining the cell temperature T_{bulk} is given by the equation

$$\sum\left(\frac{dQ}{dt}\right)_{diss} = \sum\left(\frac{dQ}{dt}\right)_{transf.} \tag{5.55}$$

Since Prandtl numbers of electrolytes are of the order of magnitude of 1 compared to typical Schmidt numbers of approximately 1000, heat transfer is much more efficient than mass transfer and local overheating and spatially uneven temperature distribution is not a serious problem in electrolysis cells. That is different in fuel cells, which are not operated with circulating but stagnant electrolytes and out of which heat can only be exported by the heat contents of the working gases. However, overheating of the whole cell can become a problem also for electrolyzers, if electrolyte pumping is too sluggish. Neglecting for almost adiabatic cells heat removal by evaporationand heat transfer to the surrounding, the simple balance

$$\dot{v} \cdot \rho \cdot c_p (T_{out} - T_{in}) = I(U_{cell} - U_{th}) \tag{5.56}$$

with U_{th} equalling the thermoneutral cell voltage, compare Eq. (3.17), allows one to determine the overtemperature $(T_{out} - T_{in})$ under steady state condition, which can be reduced by increasing the volumetric flow rate $\dot{v}$.

5.6
Ionic Charge and Mass Transport in Electrolytes

The ionic conductivity, κ, of the electrolyte together with the cell geometry and the current density, i, determines the ohmic voltage drop in the interelectrodic gap.

For a parallel plate electrolyzer with electrode distance d and equally distributed current density i the ohmic-voltage drop amounts to

$$\Delta U_\Omega = i \frac{d}{\kappa} \tag{5.57}$$

The electrolytic conductivity of dilute and to certain extent also concentrated electrolytes can to a good approximation be attributed to virtually independent single ion movement in an electric field or electric potential gradient respectively. In the absence of diffusion or concentration gradients and fluid flow the migrative flux of the ionic species, i, with charge, z, and ionic equivalent conductivity, λ_i, is given by

$$\dot{n}_{i,\, mig.} = \frac{c_i \lambda_i}{F} \operatorname{grad} \varphi = \frac{z_i F D c_i}{RT} \operatorname{grad} \varphi \tag{5.58}$$

considering the combined action of fluid flow, diffusive mass transfer and ionic migration leads to the general equation for the migrative flux of species, i

$$\dot{n}_i = w c_i + D_i \operatorname{grad} c_i + \frac{z_i F D_i c_i}{RT} \operatorname{grad} \varphi \tag{5.59}$$

In fast flowing electrolytes the first term overrules the two others.

5.6.1
Strong Electrolytes

For dilute aqueous solutions of strong electrolytes the total conductivity of the electrolyte is the sum of partial conductivities of the different ionic species i with an equivalent conductivity λ_i. If the charge of the i-th species is z_i and its concentration is c_i one obtains

$$\kappa = \sum |z_i|\, c_i\, \lambda_i \tag{5.60}$$

The equivalent conductivities λ_i are weakly dependent on the concentration and charges of all ionic species, provided the ionic strength $I = 1/2\, \Sigma\, c_i z_i^2$ does not surmount 10^{-2} moles dm^{-3} according to

$$\Lambda(I) = \Lambda_0 - \left[\frac{2.801\cdot 10^6 |z^+\cdot z^-| q}{\eta(\varepsilon T)^{3/2}(1+\sqrt{q})}\Lambda_0 + \frac{41.25(|z^+|+|z^-|)}{\eta(\varepsilon T)^{1/2}}\right]\sqrt{I} \tag{5.61}$$

For low concentrations Onsager's theory predicts the ionic conductivity correction

$$q = \frac{|z^+ z^-|}{|z^+|+|z^-|} \cdot \frac{\lambda_0^+ + \lambda_0^-}{|z^+|\lambda_0^+ + |z^-|\lambda_0^-} \tag{5.62}$$

which for 1:1 electrolytes (for instance NaCl) at 298 K reads

$$\Lambda(c) = \Lambda_0 - (0.229\Lambda_0 + 60.32)\sqrt{c} \tag{5.63}$$

Table 2.1 collects limiting values Λ_0 at infinite dilution of equivalent conductivities of some anions and cations for aqueous solutions at 298 K. The table shows the exceptionally high ionic conductivities of protons and hydroxyl ions due to the well known Grotthus mechanism of H$^+$ and OH$^-$ migration.

5.7
Temperature Dependence of Electrolyte Conductivities

The Onsager equation fails for higher concentrations and experimental determination and mathematical, rather than physical modelling of the dependence of κ on concentration and temperature is the rule.

To give an example, the conductivity of KCl solutions is modelled according to a power series in concentration c and temperature t:

$$\kappa = \kappa(c) = A + Bc + C\cdot c^2 \tag{5.64}$$

The parameters A, B and C as functions of the temperature are then tentatively written as polynomials of third order.

A simplified approximation for the temperature dependence of the conductivity of aqueous electrolytes on temperature assumes: $\dfrac{1}{\kappa}\cdot\dfrac{d\kappa}{dT} = 0.02\,/\,\text{K}$.

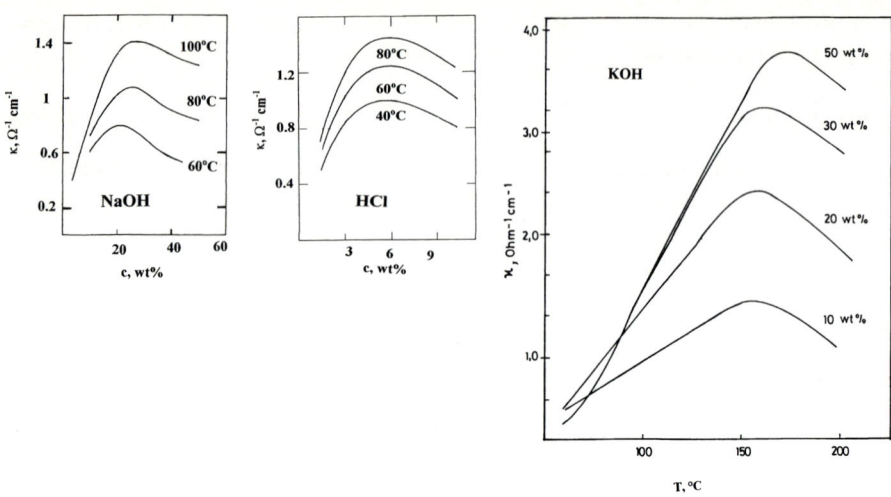

Fig. 5.12a,b,c. Concentration and temperature dependence of the specific conductivity of highly concentrated electrolytes: **a** NaOH; **b** HCl; **c** KOH

For aqueous KCl solutions such a presentation reads

$$A = \alpha_0 + \alpha_1\left(\frac{t}{100}\right) + \alpha_2\left(\frac{t}{100}\right)^2 + \alpha_3\left(\frac{t}{100}\right)^3 \tag{5.65}$$

$$B = \beta_0 + \beta_1\left(\frac{t}{100}\right) + \beta_2\left(\frac{t}{100}\right) \tag{5.66}$$

$$C = \gamma_0 + \gamma_1\left(\frac{t}{100}\right) + \gamma_2\left(\frac{t}{100}\right)^2 \tag{5.67}$$

with t in degree Celsius and

$\alpha_0 = +0.0671477;\quad \beta_0 = -0.02683625;\quad \gamma_0 = +0.062169544$
$\alpha_1 = -0.27901096;\quad \beta_1 = +0.59002202;\quad \gamma_1 = -0.11169075$
$\alpha_2 = +0.21209696;\quad \beta_2 = -0.28912264;\quad \gamma_2 = +0.06978237$
$\alpha_3 = +0.0008966$

As shown in Figs. 5.12 a and 5.12 b, for aqueous NaOH and HCl solution the conductivity of aqueous electrolyte solutions plotted versus their concentration passes through a maximum, provided the solubility is high enough to attain sufficiently high concentrations. Figure 5.12 c shows for KOH solutions of various concentrations, that there exists also a conductivity maximum with temperature.

The maximum value of κ increases with increasing temperatures and often shifts to higher concentrations with higher temperatures. Figure 5.12 c shows that for concentrated aqueous KOH solutions a conductivity maximum is reached at approximately 150 °C. Although the maximum value of κ is clearly determined by the concentration, the temperature at which it is reached is almost independent of the chosen concentration.

Such highly concentrated electrolyte solutions are relatively good conductors with κ of the order of 1 ohm^{-1} cm^{-1} so that for an electrode gap of 1 cm, which is a relatively high value, and a current density of 0.1 Acm^{-2} or 1 Acm^{-2} one calculates ohmic voltage drops of the order of 0.1 V or 1 V respectively. The conductivities of highly concentrated mixtures of strong electrolytes often show a negative deviation from additivity.

5.8 Molten-Salt Electrolytes

Since viscosities, densities and other physical data like ionic mobilities and surface tensions of molten salts (ionic melts) and highly concentrated aqueous electrolytes are comparable, their electric conductivities are comparable (Table 5.3), too, although they are measured often at very different temperatures. But not every salt, which in water is a strong electrolyte, forms ionic melts of high conductivity. As shown in Table 5.4, which compares the ionic conductivity of several metal halides at their respective melting points, three of them: $ZnCl_2$, $AlCl_3$ and $HgCl_2$ are poor conductors, molten $AlCl_3$ and $HgCl_2$ being almost electrical insulators.

Molten electrolytes play a decisive roll in electrowinning of the expressedly non-noble metals Al and Mg where molten kryolite, Na_3AlF_6 and $MgCl_2$ serve as electrolytes. Molten alkali cyanides are used for deposition of coatings of noble

Table 5.3. Comparison of specific data and specific and equivalent conductivity of NaCl in aqueous solution and in molten NaCl

Electrolyte	Viscosity η, m Pa s	Diffusivity D, cm^{2}s^{-1}	Surface tension δ, mNm^{-1}	Density ρ, gcm^{-3}	κ Ω^{-1}cm^{-1}	Λ Ω^{-1}cm^{2}mol^{-1}
aqueous 25 °C, 5 mol l^{-1} NaCl	8.95	3·10^{-5}	72	1	2.13	42
molten NaCl 850 °C	12.5	Na$^+$: 1.53·10^{-4} Cl$^-$: 0.83·10^{-4}	111.8	1.539	3.58	117

Table 5.4. Specific conductivity of different molten metal chlorides at their respective melting points

Salt	LiCl	NaCl	InCl$_3$	ZnCl$_2$	AlCl$_3$	HgCl$_2$
κ, Ω^{-1} cm^{-1}	5.67	3.58	0.42	0.0082	5.6·10^{-7}	8·10^{-5}

metals e.g. Pt on non-noble substrates like titanium. In electrochemical electricity generation by molten carbonate fuel cells, either the molten binary Li/K-carbonate eutectic (62mol% Li_2CO_3+38 mol% K_2CO_3) or the Li/Na carbonate eutectic (52mol% Li_2CO_3+48 mol% Na_2CO_3) are used as electrolyte.

5.9
Segregation in Stagnant Electrolytes of Binary Molten Carbonates in Fuel Cells

In molten-carbonate fuel cells the electrolyte is immobilised in a porous matrix, so that spatial concentration differences cannot be equalised by effective fluid flow but only by less effective back-diffusion. Ionic mass and charge transport is carried out exclusively by migration of carbonate anions, which are cathodically produced by reduction of oxygen

$$\frac{1}{2}O_2 + CO_2 + 2e^- \rightarrow CO_3^{2-} \tag{5.68}$$

and anodically consumed by oxidation of hydrogen which leads to the release of carbon dioxide

$$H_2 + CO_3^{2-} \rightarrow H_2O + CO_2 + 2e^- \tag{5.69}$$

It is known, that in general in binary mixtures of salts with a common anion the two different cations exhibit different ionic mobilities.

Against the steadily moving stream of carbonate anions which is directed from the cathode to the anode the cations are migrating – but according to their different mobilities with different velocities. Under steady state conditions the relatively faster cation would accumulate in the catholyte. Steady state conditions for the cations are defined by zero velocities of the summed convective, migrative and diffusive mass transfer for each of the two cations :

$$0 = \dot{n}_{(total)} = \dot{n}_{conv} + \dot{n}_{mig} + \dot{n}_{diff} \tag{5.70}$$

With t_i equalling the so-called intrinsic transfer number of cation i, defined against the background of the anion common to both salts and x_i equalling the mole fraction of the salt $(M_i)_2CO_3$ in the binary mixtures one obtains for the three transport terms

$$\dot{n}_{conv} = -\frac{i}{2}\frac{2}{F} \cdot x_i ; \quad \dot{n}_{mig} = \frac{i}{F} \cdot t_i ; \quad \dot{n}_{diff} = -\frac{\partial c_i}{\partial y} \cdot D_i \tag{5.71}$$

where y is the distance from the outer anode surface (compare Fig. 5.14) and calculates the steady state balance

$$0 = -\frac{i}{F}(x_i - t_i) - \frac{\partial c_i}{\partial y} \cdot D \tag{5.72}$$

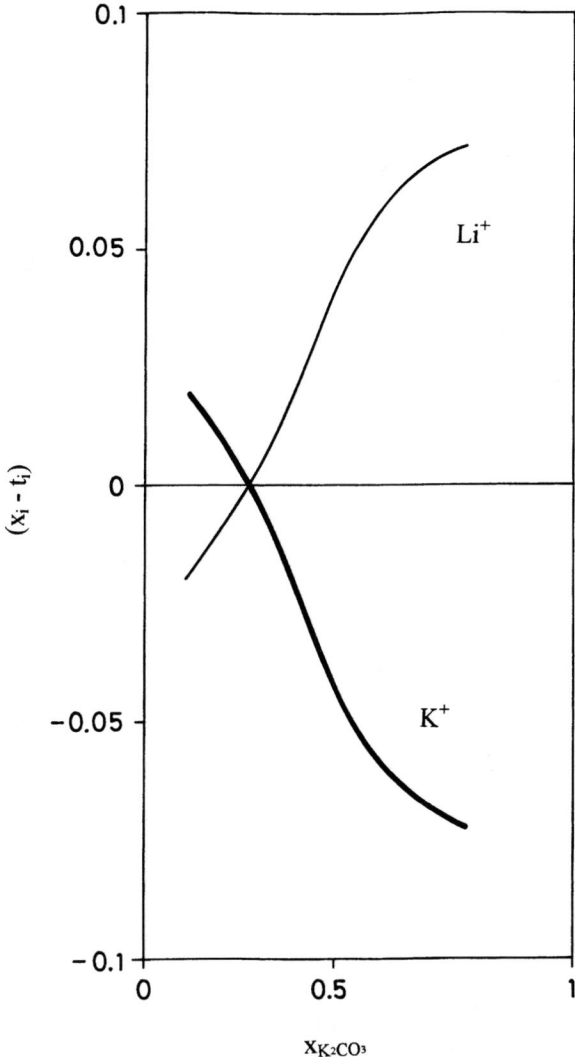

Fig. 5.13. Concentration dependence of the quantity {x_i(mole fraction) – t_i (intrinsic transfer number)} for Li_2CO_3/K_2CO_3-melts [6], [7]

With the volumetric concentration of the cation, i, c_i, which is given to a good approximation by $c_i = x_i \rho / \overline{M}$ with ρ equal to the density of the melt and $\overline{M}$ mean molar weight of the molten binary mixtures of composition x_i one arrives at

$$\frac{dx_i}{dy} = -\frac{i}{F}(x_i - t_i) \cdot \frac{\overline{M}}{\rho} \cdot \frac{1}{D_i} \tag{5.73}$$

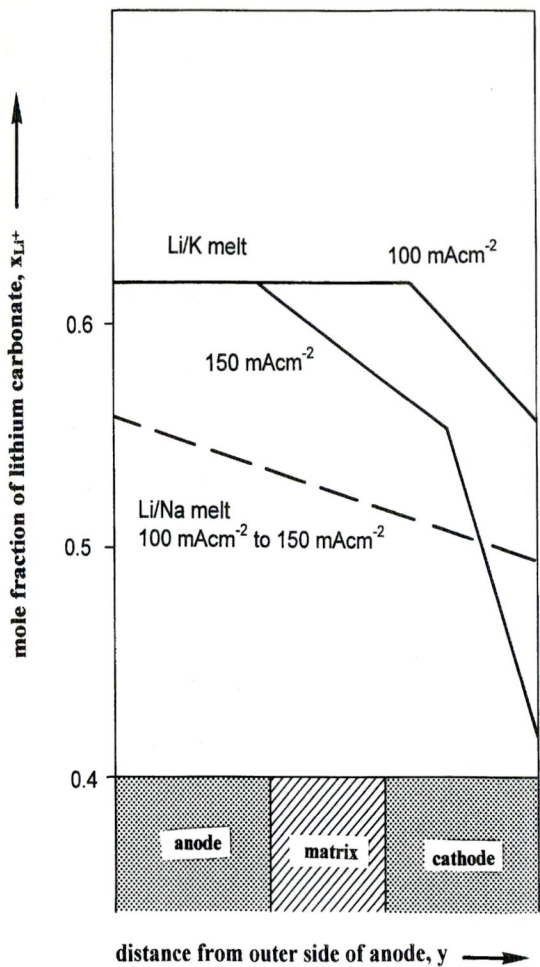

Fig. 5.14 Segregation of binary electrolytes demonstrated by steady state concentration distribution in molten carbonate fuel cells operated with: **a** Li_2CO_3/K_2CO_3 (62/38 mol/mol); **b** Li_2CO_3/Na_2CO_3 (48/52 mol/mol)

where y is the distance of the point of consideration in the cell from the outer surface of the fuel cell anode, t_i, as the intrinsic transfer number is defined by the intrinsic mobilities, b_i and b_j of the two cations against the background of the carbonate anion and by the mole fractions x_i and x_j of the two salts in the binary mixture:

$$t_i = \frac{x_i b_i}{x_i b_i + x_j b_j} = \frac{x_i b_i}{x_i(b_i - b_j) + b_j} \tag{5.74}$$

According to Eq. (5.74) the sign of the gradient dx_i/dy depends on the sign of the quantity $(x_i - t_i)$ and in its magnitude depends also on the current density, i. For $x_{K+} > 0.32$ the quantity $(x - t)$ becomes negative for potassium and $\partial x_{K+}/\partial y$ is positive for Li_2CO_3/K_2CO_3 melts as is demonstrated in Fig. 5.14. The potassium concentration is expected to increase in going from the anode to the cathode, whereas the concentration of lithium decreases correspondingly. Figure 5.14 shows that indeed x_{Li+} becomes lower in the cathode than in the matrix and the anode, after a steady concentration distribution has been established after several hours of cell operation. Figure 5.14 shows also the steady state concentration distribution of lithium established in a molten carbonate fuel cell with lithium/sodium carbonate melts. Evidently the segregation effect is much weaker, so that the Li/Na carbonate melt is at some advantage, as segregation might lead to irregularities in cell operation.

5.10
Current-Density Distribution in Cells and Electrochemical Devices

Equation (5.75) describing the cell voltage of an electrochemical cell as function of current density, i

$$U_{cell} = E_0 + \eta_a - \eta_c + i \cdot d \cdot \kappa_{elyte}^{-1} \tag{5.75}$$

may be used under the condition of equal current density distribution of both electrodes which prevails if the electrodes are parallel to each other and their surfaces match the cross section of the cell. This geometry may be a relatively good approximation for, e.g., metal electrowinning or refining cells but is rather an exception than the rule with other processes and even in copper electrorefining cells current densities at the rim of the electrodes can be twofold enhanced over the current density at the main part of the electrode surface, where an even current density distribution prevails. In technical cells used for cathodic and anodic gas evolution corrugated, extended metal sheets or louvered electrodes (see Chap. 10, Industrial Electrodes) are used which impose pronouncedly uneven current distributions. It is not the aim of this chapter to calculate explicitly current density distributions for particular electrode/diaphragm/counter electrode geometries as the arithmetic problems in current density calculations are very complex and demanding[7], but to outline the fundamentals and general implications of e.g. the adimensional treatment of the problem.

The distribution of current density along the electrode is of great industrial importance since it may cause in electroplating a local variation of the thickness of the deposited metal, a non-uniform corrosion of electrodes in industrial electrolysis cells, a poor energy efficiency in primary and secondary batteries, or a decreased space time yield with three-dimensional electrodes.

---

7 The interested reader may refer to B. Steffen and I. Rousar, Electrochim. Acta, *40* (1995) 379.

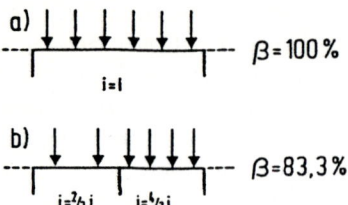

Fig. 5.15 a,b. Schematic presentation of: **a** even; **b** uneven current-density distribution with equal total current or mean current density respectively

Furthermore, a non-uniform current-density distribution may decrease the current efficiency of electrolytic processes. This can easily be understood by the schematic of Fig. 5.15 showing two patterns of current distribution represented by the density of the current lines on the two halves of a planar electrode. Assumed that current I flows to the electrode and the average current density may match the limiting current density i_l of the considered electrode reaction.

If the current-density distribution is uniform (Fig. 5.15 a) a current efficiency of 100% is achieved. If the current is distributed on the electrode in such a way, that the local current density is 2/3 i_l on the left half of the electrode and 4/3 i_l on the right (see Fig. 5.15 b), a mean current efficiency of only 83% is obtained because on the right half of the electrode that part of the current density which exceeds the limiting current density is consumed by an electrochemical side reaction for instance reduction or oxidation of the solvent. This example illustrates the practical relevance of the current distribution.

The main factors which influence the current distribution are:
(1) geometry of the cell system;
(2) conductivity of the electrolyte and the electrodes;
(3) charge transfer overpotentials and slope of the current–voltage curve;
(4) concentration overpotentials which are mainly controlled by mass-transport processes.

Depending on which of these factors is dominant three types of current distribution can be distinguished, as characterised in Table 5.5

Since the electrode potential is strictly correlated to the current density by the electrode kinetic rate equation, an uneven current distribution is always accompanied by an uneven electrode potential distribution.

In homogeneous bulk phases without space charge the fundamental equations describing potential distributions, is the Laplace equation which defines the divergence of the electrical field strength as zero in charge-free space.

$$\Delta \varphi = \partial^2 \varphi / dx^2 + \partial^2 \varphi / \partial y^2 + \partial^2 \varphi / \partial z^2 = 0 \tag{5.76}$$

and Ohm's Law

$$\partial \varphi / \partial x + \partial \varphi / \partial y + \partial \varphi / \partial z = \vec{i} \kappa^{-1} \tag{5.77}$$

5.11 Primary Current-Density Distribution

Table 5.5. Types of current distribution

Type of current distribution	Assumptions	Parameters	Examples
primary	absence of overpotential	geometry (cell, electrodes)	electropolishing
secondary	charge transfer overpotential concentration variations near the electrode	geometry, charge transfer overpotential, conductivity of electrolyte and electrodes	production cells fuel cells
ternary	charge transfer and concentration overpotential	geometry, conductivity, charge transfer and concentration overpotential	cells for deposition from effluents

or

$$\text{div } i = -\kappa \Delta \varphi = -\kappa \left(\frac{\partial^2 \varphi}{dx^2} + \frac{\partial^2 \varphi}{\partial y^2} + \frac{\partial^2 \varphi}{\partial z^2} \right) \quad (5.78)$$

If one deals with particulate or porous electrodes, whose inner surface acts as source or sink of ionic currents in Eq. (5.77) div i is not zero. The boundary conditions at the phase boundary electrolyte/electrode and electrolyte/insulating wall are most important for solving Eq. (5.76):

φ = const at the surface of an electrode (5.79)

and

$\partial \varphi / \partial n = 0$ in the electrolyte at an insulating wall; (5.80)

n is the space vector in the direction perpendicular to the electrode surface of an electrode of any arbitrary shape or an insulating wall respectively.

Frequently even for relatively simple geometries the Laplace equation is not easy to solve and for more complicated geometries highly sophisticated numerical procedures, for instance finite elements, have to be applied to perform the necessary integrations. Today standard computer programs, which solve the problem, are commercially available [8].

5.11
Primary Current-Density Distribution

Neglecting charge-transfer kinetics and mass-transfer limitations leads immediately to the so called primary current-density distribution by solving the Laplace equation. In general solving the Laplace equation leads to the construction of equipotential surfaces, whose gradient in space according to Ohm's law serves to calculate current density distributions in space. Of the simpler cell geometries the rectangular box cell and the cylindrical coaxial gap cell have some practical

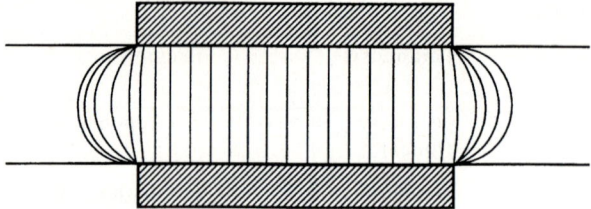

Fig. 5.16. Electrode configurations with infinite current densities at the edge according to the primary current density distribution: two parallel plate electrodes mounted flush with surrounding insulating plate

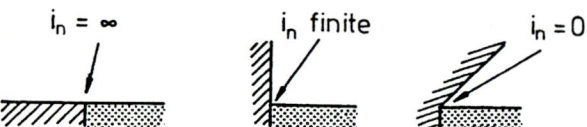

Fig. 5.17. Primary current density near the electrode edge for various angles between the electrode and the surrounding insulating wall

relevance and both obviously have a perfectly uniform primary current density distribution on each electrode.

As soon as the side walls of the rectangular box cell are removed and both parallel electrodes are embedded flush into insulating walls of infinite extension (Fig. 5.16), the primary current density distribution becomes non-uniform with infinite current density at the edges and steadily decreasing but more and more uniform distribution of current densities in going towards the centre of the electrodes. The primary current density distribution becomes more uniform, the narrower the gap between the electrodes is. A circular disk electrode surrounded by a hemispherical counter electrode very far away has also infinite current density at its rim, according to Eq. (5.74):

$$i_r = \frac{0.5i}{\left(1-(r/R)^2\right)^{0.5}} \tag{5.81}$$

Infinite current densities are, of course, not realistic in cells and in any electrochemical device, because electrode kinetic and mass transfer limitations decrease the currents to finite values and help to equalise current density distributions.

That the primary current distribution on electrodes mounted flush into an insulating wall yields an infinite current density at the electrode edge is restricted to the case, where the insulator and the electrode form an angle of 180° or greater. In Fig. 5.17 the edge current density is given for three typical cases with respect to the angle formed between the electrode surface and the insulating wall. If the insulating wall and the electrode form a right angle, then a finite edge current density results. If the angle is less than 90° the current density in the corner

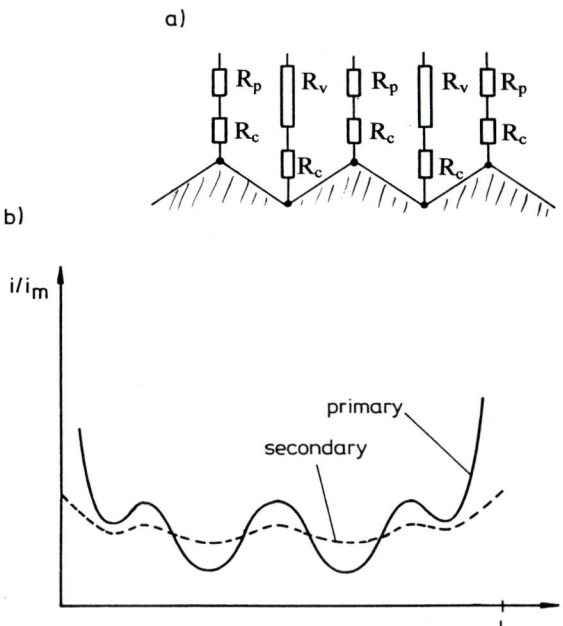

Fig. 5.18. a Electrode with serrated surface profile. **b** Schematic of current density distribution of serrated electrode becoming more equal due to charge transfer resistance

is zero. The primary current distribution may be uniform, if the electrodes completely cover two opposite walls of the cell. This situation is very well approximated in many industrial cells, such as plate and frame cells.

5.12
Secondary Current Density Distribution

In secondary current distributions the influence of an additional charge transfer overpotential is considered. In general, the secondary current distribution is more uniform then the primary one. This can be illustrated by regarding the charge transfer overpotential in terms of the so called polarisation resistance, R_c, which is given in Ω cm^2 by the slope of the current voltage curve

$$R_c = \frac{d\eta}{di} \tag{5.82}$$

Looking at an electrode as shown schematically in Fig. 5.18 a with a serrated surface profile shows clearly, that the electrolyte resistance to the peaks (R_p) is smaller than that to the valleys (R_v). If overpotentials are neglected, according to Kirchhoff's law the ratio of peak to valley current density is given as

$$\frac{i_p}{i_v} = \frac{R_v}{R_p} > 1 \tag{5.83}$$

If an additional polarisation resistance R_c is taken into account in series to the electrolyte resistance, then the current density ratio is modified to

$$\frac{i'_p}{i'_v} = \frac{R_v + R_c}{R_p + R_c} < \frac{R_v}{R_p} \tag{5.84}$$

resulting in a more uniform current density distribution as also shown schematically in Fig. 5.18.

5.13
Secondary Current Density Distribution and "Throwing Power" in Electrodeposition and Electrocoating

Equalisation of current densities on electrode surfaces of any shape and surface structure is the aim in electroplating objects of many different forms and patterns with protective corrosion resistant or decorative metal layers. An almost constant thickness of such coatings irrespective of the form of the object can only be brought about by overruling the primary uneven current distribution by the secondary current density distribution, which is mainly determined by the electrode kinetics of the galvanic process. This means, that by appropriate chemical means like complex formation of the electrodeposited metal, or the use of surfactants which adsorb at the surface of the object, galvanic deposition of the metal has to be decelerated kinetically. If almost everywhere on an arbitrarily formed object the galvanic coating has the same thickness, the process is said to possess a high throwing power. In practice the cells of Haring and Blum (Fig. 5.19 a) on the one hand and Hunt (Fig. 5.19 b) on the other, are used to determine throwing powers of galvanic bathes and galvanic coating processes.

The Haring and Blum cell consists of two parallel plate cathodes and a corrugated plate anode between and parallel to them, which is placed in an asymmetric position, so that its distance to one of the cathodes is at the most half of the distance to the second cathode or less. In the Hunt cell the cathode has a given angle of inclination versus the upright anode, so that the anode–cathode distance changes continuously from top to bottom. In the Haring and Blum cell the difference of coating thicknesses at both cathodes or the ratio of the two partial current densities for both cathodes is a measure of the throwing power, as both partial currents should almost match at high throwing power – whereas in the Hunt cell the change of the thickness of the deposited metal from top to bottom is the indicator of the throwing power.

Treating the division of the total current, I, flowing through the Haring and Blum cell into the partial currents I_1 and I_2 flowing to the cathodes C_1 and C_2 for the case of a linear current voltage correlation one obtains

$$I = I_1 + I_2 \tag{5.85}$$

$$U_{cell} = E_{anode} - E_{cathode,\,1} + i_1 L_1 / \kappa = E_{anode} - E_{cathode,\,2} + i_2 L_2 / \kappa \tag{5.86}$$

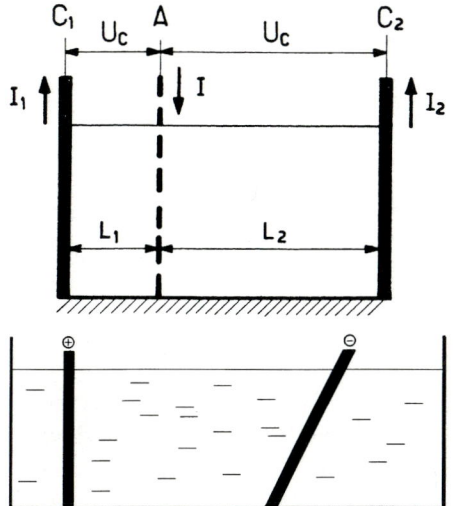

Fig. 5.19. a,b. Experimental cells for determining the throwing power of galvanic baths. **a** Haring and Blum cell allows to measure throwing power by comparing currents of two cells with different electrode gap. **b** Hunt cell can only distinguish plating thickness at different deposition sites

The potential of the two cathodes is given by

$$E_{cathode,\,1} = E_0 + \eta_1 \tag{5.87 a}$$

$$E_{cathode,\,2} = E_0 + \eta_2 \tag{5.87 b}$$

with E_0 the equilibrium potential and $\eta_{1,2}$ the respective charge-transfer overpotentials at both chemically identical cathodes. From Eqs. (5.86) and (5.87 a,b) and using the postulated linearity of the current–voltage curve, Eq. (5.87 c) for low overpotentials:

$$\frac{\eta_2 - \eta_1}{i_2 - i_1} = \frac{d\eta}{di} \tag{5.87 c}$$

one arrives at

$$\frac{i_2}{i_1} = \frac{L_1 + \kappa\,(d\eta/di)}{L_2 + \kappa\,(d\eta/di)} \tag{5.88}$$

It is obvious, that the higher the electrolyte conductivity and the steeper the slope $d\eta/di$, that is the higher the polarisation resistance of the electrode process the closer the ratio of the two current densities comes to unity.

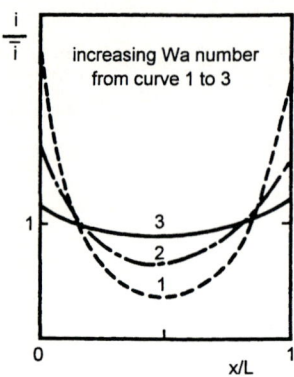

Fig. 5.20. Secondary current density distribution at pair of planar electrodes depicted in Fig. 5.16. Increasing Wagner numbers are effecting the current distribution to become more even

5.14
The Wagner Number

Dimensional analysis of the problem of current density distributions in the Haring–Blum cell and also in any other electrochemical device shows that the Wagner number, Wa, the dimensionless ratio of the surface specific polarisation resistance $d\eta/di$ and the surface related ionic resistivity $(1/\kappa)$ L is the relevant adimensional quantity determining current density distributions:

$$Wa = \frac{(d\eta/di)}{(1/\kappa)L} \qquad (5.89)$$

The higher the Wagner number, the more uniform is the current-density distribution. Returning to the problem of the current density distributions on two parallel planar plate electrodes being inserted flush in an insulating wall having the distance L, the secondary current distribution reduces at the edge from infinite, the value obtained from the primary current density distribution, to the value obtained from current voltage correlation by inserting the applied overpotential and as shown in Fig. 5.20 increasing Wagner numbers equalise the current density distributions more and more. The following rules are applicable.

The current distribution is the more uniform:
(1) the higher the slope of the polarisation curve, $d\eta/di$, (the polarisation resistance);
(2) the higher the conductivity of the solution;
(3) the smaller the characteristic length of the system.

Some interesting conclusions may be drawn in detail with respect to industrial applications.

(a) The current distribution depends on the composition of the electrolyte system. Adding supporting electrolyte (increasing the solution conductivity) or use of additives acting as inhibitors (increasing the charge transfer resistance) lead to a more uniform current distribution.
(b) Many industrial electrochemical processes, for instance water electrolysis, take place under charge transfer control with exponential current voltage correlations at catalytically activated electrodes. In this situation an increase of the mean current density causes a less uniform current distribution, because the charge transfer resistance decreases with current density.
(c) The current distribution depends not only on the geometry, but also on the size of the system. This implies that for scale-up from the laboratory to the industrial plant the consequences of an enlarged size to current distribution have to be considered carefully in terms of the Wagner number.

5.15
Tertiary Current Distribution

Calculating tertiary current-density distributions means taking also into account additional current density limitations due to mass transfer. The most extreme case occurs if the current density at the electrodes is limited everywhere by mass transfer. For a planar electrode along which the electrolyte is flowing the resulting limiting current density distribution is shown schematically in Fig. 5.21 a.

For a serrated electrode surface profile two Fig. 5.18 different limiting cases of mass transfer control have to be discussed. In the case of the so-called macroprofile extension of the diffusion layer thickness is assumed to be small as compared to the electrode roughness. Then the diffusion resistance enhances the polarisation resistance resulting in a more uniform current distribution than the secondary one.

The so called microprofile is given if the electrode roughness is small compared to the diffusion layer thickness. Then the peaks are better accessible to diffusive

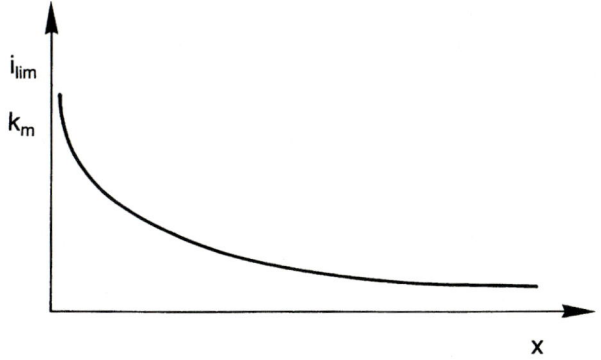

Fig. 5.21a. Current density distribution under mass transfer control on a planar electrode with laminar parallel electrolyte flow

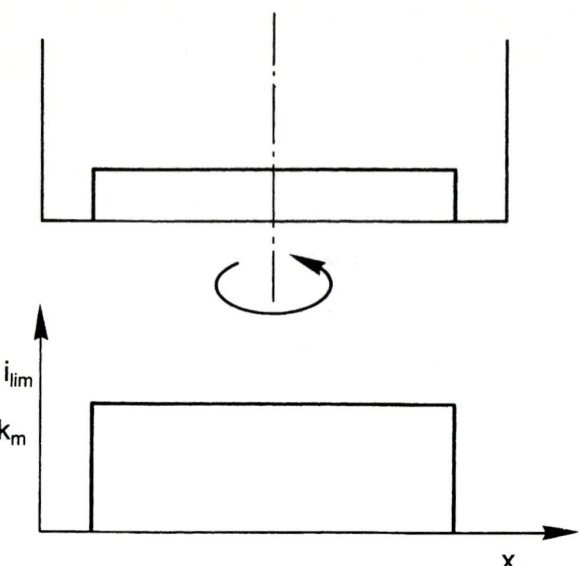

Fig. 5.21b. Current-density distribution under mass transfer control at rotating disc electrode

mass transfer than the valleys. This means that the highest current densities are obtained at the peaks. Some practical implications of this situation should be briefly discussed.

a) Let us assume that at the electrode a diffusion controlled anodic dissolution takes place. Then the peaks are dissolved with a higher rate than the other parts of the electrode and the surface roughness is diminished. This technique is practically applied in electropolishing.
b) If a diffusion limited metal deposition takes place then the highest deposition rates are given at the peaks. This means that the surface roughness will be enhanced. In an extreme case a dense metal deposition is no longer possible and dendrites or even metal powders, which detach from the surface are obtained. This effect is used in electrochemical metal powder production.
c) Electroplating aims at uniform metal deposition, keeping off mass transport limitations. Therefore plating baths contain usually the electrodeposited metal salt in highest possible concentrations, keeping the applied current densities at a small fraction of mass transport limitations. In high current density metal deposition as for instance high-speed zinc coating of steel wire or strip, mass transfer has to be enhanced correspondingly for instance by vigorous stirring or high linear velocities of the substrate moving through the plating bath. The extreme case of completely even current density distribution as a consequence of tertiary distribution is realized at the rotating disc operated at mass transfer limited current densities. This is demonstrated schematically in Fig. 5.21 b.

References

1. F. Colquhoun-Lee, J. Stepanek, Chem. Ind. (London) (*1974*), 108
2. G.H. Hughmark, AIChE J. *18* (1972) 1020
3. J.C. Chu, J. Kalil, W.A. Wetteroth, Chem. Ing. Progr., *49* (1953) 141
4. D.J. Pickett, J. Appl. Electrochem. *5* (1975) 101
5. Ramdonglert, J.P. Conderc, H. Angelino, Trans. Inst. Chem. Eng. *53* (1975) 175
6. C. Yang, R. Tagaki, K. Kawamura, I. Okada, Electrochim. Acta, *32* (1987) 1607
7. P.-H. Chou, I. Okada, Internal Cation Mobilities in the molten binary system $(Li,Na)_2CO_3$, in H. Wendt ed., MS 5. Proc. of the Vth Int. Symp. on Molten Salt Chemistry and Technology, Molten Sal Forum, *5–6* (1998) 83
8. B. Steffen, I. Rousar, Electrochim. Acta, *40* (1995) 379

Further Reading

Vol. 6, Electrodics: Transport, in: Comprehensive Treatise of Electrochemistry, E. Yeager, J. O'M Bockris, B.E. Conway, S. Sarangaponi Eds., Plenum Press New York and London, 1983

W. Vielstich, Zeitschr. f. Elektrochemie, *57*, 646, (1953)

J.T. Davis, Turbulence Phenomena, Academic Press, New York and London 1972

E.R. Eckert, Heat and mass transfer, 2nd ed., Mc Graw Hill, New York, 1959

CHAPTER 6

Electrochemical Reaction Engineering

6.1
Introductory Remarks

Electrochemical reaction engineering deals with modelling, calculation and prediction of the rate of electrochemical conversion or production in real reactors, that means the calculation of space time yields, selectivities and energy efficiencies of electrolyzers. Such predictions are based on one hand on microkinetic investigations of the relevant electrochemical reactions that means on information extracted from current voltage curves and preparative laboratory scale electrolyses and on the other hand on calculated spatial current density and potential distributions in electrolyzers along or across the electrodes or for instance in porous electrodes.

6.2
Microkinetic Models

Microkinetic models are generally based on the measurement of current-voltage curves at close to real operation condition.

Repeating the most important ones, the Butler Volmer equation at Eq. (6.1) must be mentioned for cases where applied current densities are much lower than mass transfer limited current densities:

$$i = i_0 \left(\exp\left\{ \frac{\alpha_a \eta F}{RT} \right\} - \exp\left\{ -\frac{\alpha_c \eta F}{RT} \right\} \right) \text{ with } i_0 = i_0^0 \Pi c_i^{\eta_i} \tag{6.1}$$

The current voltage curve for mass-transfer controlled metal deposition ($M^{z+} + ze^- \rightarrow M$) is also repeated by Eq. (6.2):

$$i = i_1 \left(1 - \exp\left\{ -\frac{v_e F |\eta|}{RT} \right\} \right) \tag{6.2}$$

with i_l, the mass transport limited current density. The current–voltage curve for mass-transfer controlled oxidation or reduction of a dissolved reactant producing a dissolved reaction product $\left(A+v_e e^- \rightleftarrows B\right)$, neglecting differences of the diffusion coefficients D_{ox} and D_{red}, is described by Eq. (6.3):

$$E = E^0 + \frac{RT}{v_e F} \ln\{(i_1 - i)/i\} \quad (6.3)$$

or in a form which also includes irreversible charge transfer

$$E - E_{1/2} = m \, \log\{(i_1 - i)/i\} \quad (6.3\,a)$$

where m is the slope of the mass transfer corrected semilogarithmic current voltage curve. The slope m at 25 °C usually equals 120 mV for irreversible charge transfer and 60 mV for reversible 1e-charge transfer.

Such equations based on a theoretical model of the electrode process should always be preferred if it is assured that they describe the current voltage curve of a well defined electrode reaction across concentration and current density ranges which are relevant to the process conditions. If such theoretically based equations fail to describe the current–voltage correlation, one would prefer experimentally measured current–voltage curves. The cell voltage current density relation at any point in the electrolysis cell is obtained by composing the cell voltage from equilibrium potential, anodic and cathodic overpotentials and ohmic potential drop, Eq. (6.4):

$$U_c = E_0 + \eta_a - \eta_c + IR \quad (6.4)$$

Writing Eq. (6.4) in a more general form Eq. (6.4 a) would allow for uneven current distribution

$$U_c = E_0 + \left(\eta_a - \eta_c + \Delta u_\Omega\right)_{local} \quad (6.4\,a)$$

6.3 Mode of Operation

In reaction engineering continuous and batch operations are distinguished. Batch operation of electrolyses means performance of an electrolysis process in a closed system without any in- and outflows of electrolyte (Fig. 6.1 a). A continuously operated electrolyzer is supplied with a steady flow of fresh electrolyte while a steady flow of depleted electrolyte is leaving the electrolyzer so that the amount of electrolyte in the electrolysis reactor does not change with time. (Fig. 6.1 b).

Typical continuously operated electrolyzers are kept under steady operation over long time periods extending over many hours, days and even weeks or months. Continuously operated are "large" processes like chloralkali electrolysis plants, water electrolyzers and metal electrowinning and refining plants, for instance copper refining or aluminium, magnesium and zinc electrowinning. Processes for the electrowinning of noble metals, for electrosynthesis of precious or-

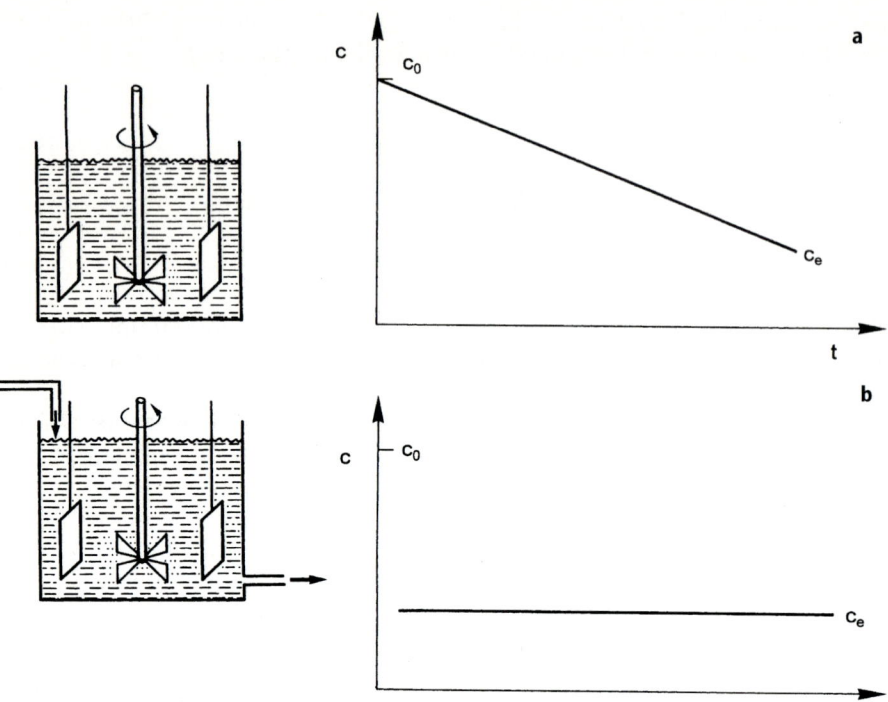

Fig. 6.1. a Schematic of batch operated electrolysis with linear reactant concentration decay with time at constant current. **b** Continuously operated electrolysis cell with time independent reactant concentrations at inlet, c_0, and exit, c_e

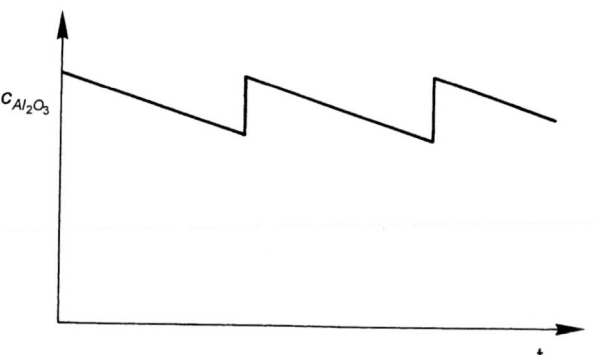

Fig. 6.2. Schematic of interrupted reactant depletion with time in semi-batch operated aluminium electrowinning cells

ganic compounds and other "small" products like perchloric acid production are performed in batch electrolyzers. Aluminium electrowinning in Hall–Heroult cells is run as semibatch reactors because aluminium oxide is added at times after depletion of the Al_2O_3 content in the cell to a critical level, Fig. (6.2).

6.4
Electrical Control of Cells

Contrary to experimental electrochemical investigation in laboratory cells for which the potential control of the working electrode is an indispensable prerequisite, industrial electrolyses are almost always executed with current (and this means current density) control because it is technically much easier to control currents than electrode potentials. Furthermore due to uneven current density distributions in technical electrolyzers the electrode potential is not determined unequivocally by the applied current. In technical electrolyzers the applied current rules cell voltage and a local distribution of electrode potentials.

6.5
Macrokinetic Models

Based on the microkinetic description of the electrode reaction by current–voltage curves of cathode and anode processes the performance of an electrochemical reactor is predicted by calculating the spatial current-density distribution which is determined not only by the applied cell current or the cell voltage but which is also determined by the spatial concentration distribution of the electroactive species as their concentration together with the spatial distribution of mass transfer determines the detailed form of the current–voltage curve at any point of an electrode. Local mass-transfer limited current densities in particular are mainly determined by the flow pattern in the electrolyzer and the current density distributions at the electrode.

6.5.1
Stirred-Batch Tank Reactor

A stirred-batch tank reactor has a high mixing rate compared to the rate of chemical conversion. Therefore a spatially uniform concentration of any species in the electrolyte is usually established in stirred tank reactors. The electrochemical conversion of substance R according to $R + v_e e^- \rightarrow S$ with current density, i, at the electrode of area A changes the concentration of the reactant R in a batch electrolyzer with time according to Eq. (6.5):

$$dc_R/dt = -iA/v_e F V_R \qquad (6.5)$$

Introducing the quantity $a_e = A/V_R$, the volume specific electrode area, one obtains

$$dc_R/dt = -ia_e/v_e F \qquad (6.5\text{ a})$$

Therefore an imposed current I changes the concentration with time-independent rate until at sufficiently low concentration the imposed current density equals and further exceeds the mass transfer-limited current density i_l. If i exceeds i_l, the current efficiency ϕ^e begins to drop below unity because that part of the current density which exceeds i_l is consumed by a parasitic electrochemical reaction, for instance by electrochemical conversion of the solvent (e.g. O_2 or H_2 evolution from water). Therefore at time $t > t_1$ the concentration changes according to Eq. (6.6):

$$dc_R/dt = -i_l a_e / v_e F = -k_m c_R a_e \qquad (6.6)$$

with the solution for $i > i_l$ and time t:

$$c_R = c_{R,0} \exp\{-k_m a_e (t - t_1)\} \qquad (6.6\,a)$$

One obtains the mass-transfer limited, time dependent, current efficiency for mass transfer limited electrochemical conversion:

$$(\Phi^e)_1 = i_l / i = k_m c_R v_e F / i = k_m c_{R,0} v_e F \exp\{-k_m c_{R,0} v_e F\} / \text{ if } i_l < i \qquad (6.7)$$

Figure 6.2 depicts the depletion of reactant R with time according to Eqs. (6.5) at constant current density. Referring to aluminium electrowinning, the aluminium oxide concentration in the cell is replenished before the condition $i = i_l$ at $t = t_1$ becomes valid.

6.5.2
Continuously Stirred Tank Reactor

Continuously stirred tank reactors (CSTR) are fed continuously with a stream of reactants and electrolyte of volumetric flow rate $\dot{V}$. Since the electrolyte flow at the reactor outlet is the same as at the inlet and does not change with time the electrolyte inventory or electrolyte volume V_R does not change with time. Intensive stirring ascertains spatially equal concentrations in the electrolyzer volume V_R.

The CSTR is a good approximation for electrolyzers which are intrinsically well stirred by for instance vigorous electrolytic gas evolution or by well developed thermal or density driven free convection. Gas-stirred cathode chambers of chloralkali electrolyzers of the diaphragm or membrane type and copper refining cells which are well stirred by concentration-difference induced free convection can be modelled according to CSTRs. Figure 6.1 b shows that according to the general material balance which is given in Eqs. (6.8 a) and (6.8 b) the concentration c_e of the electroactive species in the reactor and at its exit (subscript e) is substantially lower than the concentration c_0 in the electrolyte stream which enters the reactor with the volumetric flow rate $\dot{V}$.

6.5 Macrokinetic Models

The material balance of a chemical continuously stirred reactor reads for steady state conditions:

$$\dot{V}(c_e - c_0) = V_R r_V \quad (6.8\text{ a})$$

introducing the residence time $\tau = V_R/\dot{V}$ one obtains

$$\tau^{-1}(c_e - c_0) = r_V \quad (6.8\text{ b})$$

where r_V is the conversion rate per unit time and volume with which the chemical species under consideration is consumed ($r < 0$) or produced ($r > 0$). For an electrochemical CSTR the balance is given by Eq. (6.8 a)

$$\dot{V}(c_e - c_0) = I\phi^e / \nu_e F = ia_e \phi^e V_R / \nu_e F \quad (6.8\text{ c})$$

which may be written also as

$$c_e - c_0 = \pm i a_e \phi^e / \nu_e F \quad (6.9)$$

minus holds for starting materials and plus for products. The space–time yield, ρ, is defined by $\dot{V}(c_e - c_0)_{\text{prod}} / V_R$ and is given by Eq. (6.10):

$$\rho = i a_e \phi^e / \nu_e F \qquad \text{(in mol per volume and time)} \quad (6.10)$$

As long as the applied current density, i, is lower than the mass transfer limited current density i_l, the current efficiency ϕ_e is at the highest possible level approaching unity in the absence of parasitic reactions. According to Eq. (6.8 a) the condition $i < i_l$ can always be fulfilled by adjusting the flow rate $\dot{V}$ to sufficiently high values thereby increasing the concentration c_e.

6.5.3
Plug-Flow Reactor (PFR)

In ideal plug flow reactors (PFR) each volume increment of reactants or of the electrolyte which passes through the reactor has equal linear velocity w. Therefore complete absence of backmixing is postulated though perpendicular to the flow direction complete mixing with negligible concentration gradients is assumed which is equivalent to the assumption of equal residence time for each volume increment of the electrolyte, which passes through the electrolyzer.

The model of the PFR is a good approximation for chloralkali electrolysis cells of the amalgam technology or electrolyzers with channel flow as for instance the chlorate electrolysis cell and the model may also be a good approximation for flow-through electrodes as packed grids, tissues, and packed-bed electrodes.

Figure 6.3 a describes the principle of a channel flow reactor with uniform velocity across the cross section of the channel – the literal meaning of "plug flow"

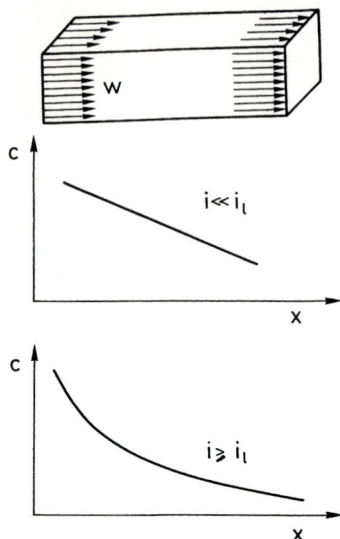

Fig. 6.3a Piston flow in tube electrolyzer: **b** establishing space independent almost constant current density if i<<i_l with linear concentration decay; **c** establishing exponentially decaying current density if i >i_l

or "piston flow". Figure 6.3 b explains that flow-through cells can be operated along their electrodes at uniform current density i – which can be realised if i<<i_l. Such electrolyzers would show typically a linearly decreasing concentration of the electroactive species, R, with increasing electrode and cell length.

An exponential decrease of the concentration, however, is predicted if the cell is operated at or above limiting current density, i_l, as shown in Fig. 6.3 c.

Under steady-state conditions the mass balance along the differential length dx of the reactor reads

$$w(dc/dx) = r_v \tag{6.11 a}$$

Equations (6.11 b and c) translate this general mass balance along the differential length dx into the mass balance of an electrochemical rectangular plug flow channel reactor of electrode width b and interelectrodic gap h for steady state conditions dc/dt=0 and using $h^{-1}=a_e$, the volume specific electrode area of a parallel plate of the cell:

$$\dot{V}(dc/dx) = bhw(dc/dx) = -b \cdot i\Phi^e / v_e F \tag{6.11 b}$$

$$w(dc/dx) = h^{-1}\Phi^e \frac{1}{v_e F} = \frac{a_e i\Phi^e}{v_e F} \tag{6.11 c}$$

6.5.3.1
Plug Flow Electrolyzer with Uniform Current Density

Introducing a_e into Eq. (6.11 b) and solving this equation for the reactor length L leads to the concentration difference c_0-c_e between reactor inlet and outlet for assumed uniform current-density distribution (Fig. 6.4 b)

$$(c_0 - c_e) = bLha_e i\phi_e / v_e F\dot{V} = \tau a_e i\phi^e / v_e F \tag{6.12}$$

Operation at uniform current density formally refers to zero-order reactions in terms of conventional chemical kinetics. Therefore the expression for the space time yield is the same as in Eq. (6.10) for the continuously-stirred tank reactor with $i>i_1$.

6.5.3.2
PFR Operated at Mass Transfer Limited and Higher Current Density

If the cell voltage is kept high enough so that at the working electrode everywhere an electrode potential is established which provides the condition of mass transfer limited current density, then according to Eq. (6.13) the electrochemical reaction becomes first order with respect to the converted species R:

$$i_1 = k_m c_R v_e F \tag{6.13}$$

Introducing the volume specific electrode surface a_e yields for the volume specific reaction rate under mass-transfer-limited current density $r_v = a_e k_m c$ so that Eq. (6.11 a) reads $w(dc/dx) = -k_m a_e c$ from which by multiplying with the cross section of the cell lumen bh and introducing $a_e = li^{-1}$ Eq. (6.14) is obtained:

$$\dot{V}(dc/dx) = -k_m bc \tag{6.14}$$

At the reactor outlet (x=L) c_e is calculated:

$$c_e = c_0 \exp\{-k_m bL/\dot{v}\} = c_0 \exp\{-k_m a_e \tau\} \tag{6.15}$$

with $\tau = Lhb/\dot{V}$ and $a_e = h^{-1}$
The space time yield in this case reads

$$\rho = \tau^{-1} c_0 \left(1 - \exp\{-k_m a_e \tau\}\right) \tag{6.16}$$

Figure 6.3 c depicts the exponential decrease of the concentration with flow length which is for instance typical for a packed-bed electrode in which poisonous metal ions are extracted from effluents of the galvanic industry by cathodic deposition under mass-transfer-limited current densities.

6.5.4
Cell Cascades

It is sometimes easier or more profitable to use several continuously-stirred tank reactors in series, a reactor cascade, instead of either one large stirred tank reactor or a plug flow reactor. Figure 6.4 demonstrates the principle of a reactor cascade in which the concentration of the reactant is decreased in several steps. For an electrolysis under mass-transfer-limited current density the advantage of a reactor cascade over a single stirred tank is evident. For a given degree of conversion of the reactant R in one stirred tank reactor or electrolyzer which works at a low concentration level c_e and with the low mass transfer limited current density $i_1 = k_m c_e v_e F$ a relatively large electrode area is necessary for achieving the concentration difference $c_0 - c_e$. In contrast the stepwise depletion of c_R in a cascade allows to reduce the total electrode area significantly as only in the last reactor of the cascade the lowest current density is applied, to achieve the concentration difference $c_{n-1} - c_e$ whereas in the electrolyzers upstream the electrolysis can be performed much more efficiently at higher current densities. A cascade of several continuous stirred electrolyzers may also serve as a mathematical model depicting a continuously operated chanel electrolyzer by subdividing the truly continuous reactor into many small subunits, as differential continuously-stirred tank reactors. Any numerically treated homogeneous plug-flow-reactor model performs such subdivision mathematically in going from differentials dc/dt to differences $\Delta c/\Delta x$.

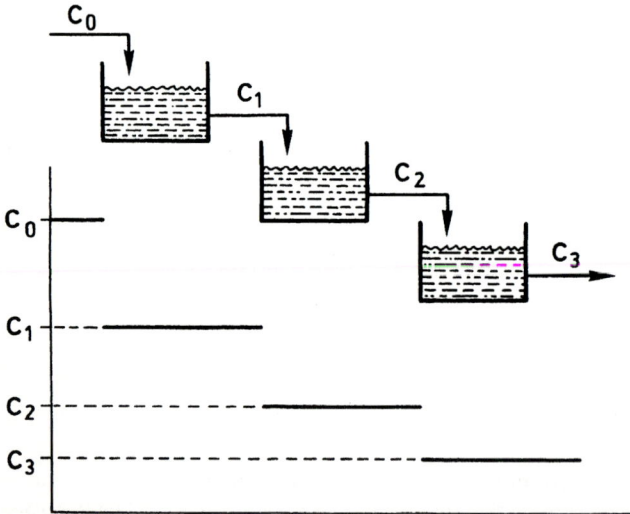

Fig. 6.4. Schematic of a cascade of backmix reactors decreasing the reactant concentration in steps

6.5 Macrokinetic Models

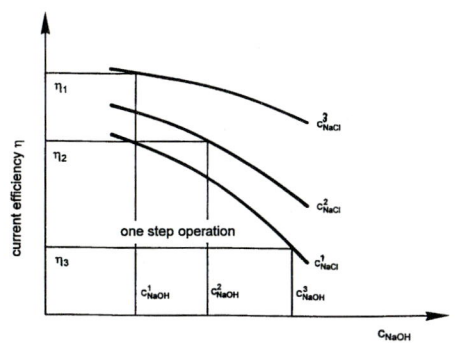

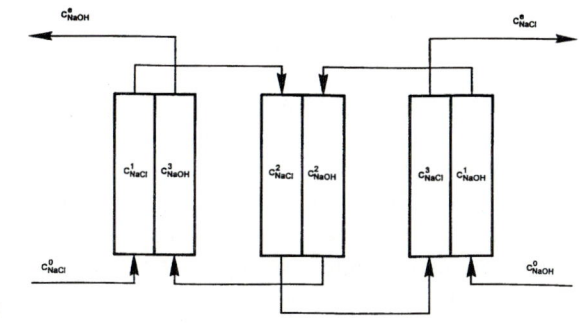

Fig. 6.5a,b. Schematic of three membrane chloroalkali-electrolysis cells with cascade-operated cathodes: **a** current efficiency vs. concentration of generated caustic-soda solution for three different brine-depletion rates (c_0-c_e) at constant c_0; **b** schematic of applied electrolyte flows

An example, but today only "ad usum delfini", as the technology is no longer necessary and outdated, for using electrolyzer cascades is the proposal to run membrane cells for chloralkali electrolysis in cascade-arrangement. At the beginning of the history of this process only membranes for which the current efficiency of caustic soda generation decreased drastically with the concentration of sodium hydroxide were available. At 10 wt% NaOH the current efficiency was almost 100% but at 30 wt% it dropped to less than 80% (Fig. 6.5 a). Therefore it was sensible to run the cathode compartments of three electrolyzers – which by themselves establish each one a continuous stirred tank reactor – in series forming a cascade. At the same time the anode compartments were supplied with lye in series and in counterflow. The current efficiency decreases also with the degree of conversion of the anolyte (lye). Accounting for this effect shows that counterflow cascading as in Fig. 6.6 b give optimal overall current efficiencies.

This technology, however, is today obsolete as largely improved membranes assure today close to 100% current efficiency at caustic-soda concentration of 30 wt% or even higher.

6.5.5
Extended Modelling of Electrolyzers

Hitherto the application of reactor theory to electrolyzers confined to the discussion of two different operation modes:
(I) mass transfer limited or higher current densities $i \geq i_l$
(II) uniform current density $i \ll i_l$

which are the two extremes between which the operation mode of electrolyzers are established.

Mass-transfer-limited current densities are only used in cases where species dissolved in relatively low concentration $c < 10^{-2}$ molar are to be converted or electrochemically removed from a process stream to the highest possible degree achieving lowest possible effluent concentrations. Examples are waste-water treatment of effluents containing noxious substances as for instance Cu^{2+} or CrO_4^{2-} ions.

Almost uniform current density is usually established if the electroactive species is highly concentrated ($c > 1$ molar) and only a small fraction is converted in the electrolyzer. Chloralkali electrolyses according to the amalgam, the diaphragm or the membrane process are typical examples. In these processes the anolyte is depleted or converted by 10–30% at the most. Very often, however, neither mass-transfer-controlled current densities nor negligible depletion of the depolarizer prevail. In such cases the current density in electrolyzers which would be described according to the plug-flow reactor model decreases along the electrode almost perfectly exponentially as under purely mass transfer control.

Referring to Eq. (6.3) and (6.4) in calculating the overpotentials and from them current densities or vice versa for a given cell voltage, spatially changing depolarizer concentrations and accordingly spatial current-density distributions induced by concentration changes have to be accounted for in modelling of such electrolyzers. Since a decrease in the ohmic potential drop IR due to decreased current densities at the given cell voltage U_c would allow to establish higher overpotentials η_a and η_c the concentration effect is partially compensated resulting in a more uniform current density distribution than with mass transfer controlled current density.

Another effect, not dealt with here, is the influence of electrochemically evolved gases on the effective electrolyte conductivity and on the term IR and ΔU_Ω which affects current density distributions very strongly (see Chap. 7). With gas evolving electrodes e.g. in water electrolyzers or chloralkali electrolyzers with vertical electrodes this effect augments towards the top of the electrode as the evolved gas bubbles ascend and accumulate in the upper part of the cell. This leads to decreasing current densities at the upper part of the electrolyzer. In industrial electrolyzers very often such quite complicated situations prevail and only a detailed numerical simulation based on a more elaborated physical model of the electrolyte flow allows for a good and suitable description of the operation parameters of the cell [1].

6.5.6
Residence-Time Distribution

Residence time distributions are an indicator to what extent a given electrolyzers can be modelled by the aforementioned idealised reactor types. Ideal continuous stirred tank and plug flow reactors exhibit distinctly different residence time distributions. For an ideal plug-flow reactor of length, L, all volume increments are residing in the reactor for the same time, which equals $t = L/w = V_R/\dot{V}$. Real residence time distributions deviating from this idealised assumption can be measured by any method which is able to measure the time lag between related concentration changes of a dissolved inert marker substance at the inlet and the outlet of the reactor. Most often either a concentration pulse of the marker is induced at the entrance by injecting within the shortest possible time a certain amount of the marker into the reactant or electrolyte stream (pulse method) or a quick change of the marker concentration (step method) is applied mostly from zero to a time independent value. Figures 6.6 a and b depict both measuring methods together with the concentration response at the reactor outlet. The pulse method defines the differential residence time distribution E(t) as determined by the marker concentration $c_M(t)$ measured at time t after injection of the amount m_M of the marker at the reactor inlet:

$$E(t) = (1/\tau) V_R c_M(t)/m_M \tag{6.17}$$

The step method defines the integral residence time distribution F(t):

$$F(t) = c_M(t)/c_M(t=\infty) = \int_0^t E(t)dt \tag{6.18}$$

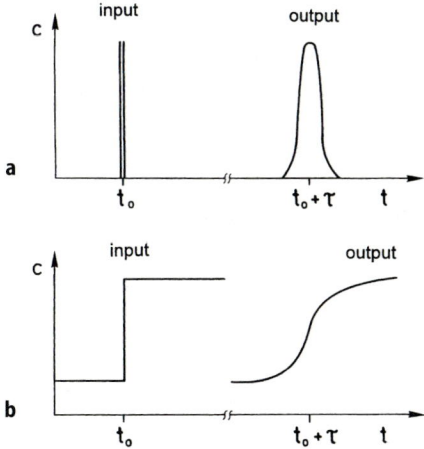

Fig. 6.6a,b. Schematic of procedure to measure residence-time distributions in tube reactor: **a** pulsed addition of marker for differential residence-time distribution; **b** displacement method for measuring integral residence-time distributions

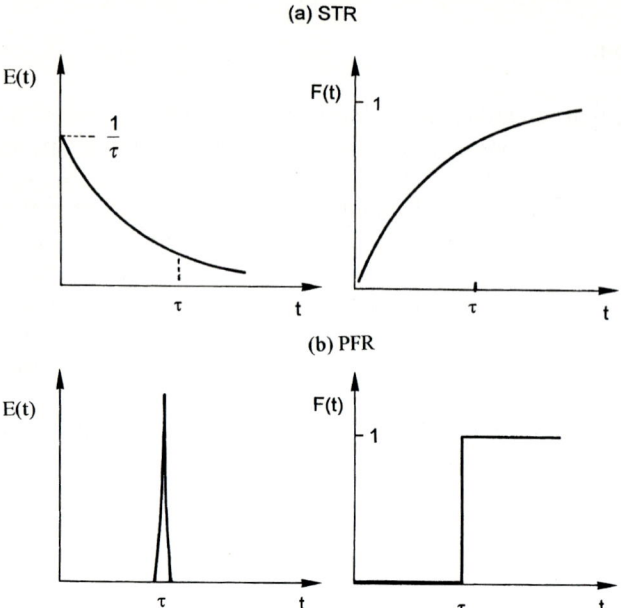

Fig. 6.7a–d. Differential E(t) and integralF (t) residence-time distributions of: **a,b** stirred-tank reactor; **c,d** plug-flow reactor or electrolyzer

For an ideal plug-flow reactor (Fig. 6.7 a,b) both types of residence-time distributions read

$$E(t) = \frac{1}{\tau}\delta(t) \tag{6.19}$$

with

$$\begin{aligned}&\delta = 0 \text{ for } t < \tau \\ &\delta = 1 \text{ for } t = \tau \\ &\delta = 0 \text{ for } t > \tau\end{aligned} \tag{6 a}$$

(Dirac's delta function)

$$\begin{aligned}&F(t) = 0 \text{ for } t < \tau \\ &F(t) = 1 \text{ for } t \geq \tau\end{aligned} \tag{6.20}$$

(step function)

For the ideal stirred tank reactor (Fig. 6.7 c,d) both functions read

$$E(t) = 1/\tau \, \exp\{-t/\tau\} \tag{6.21}$$

6.5 Macrokinetic Models

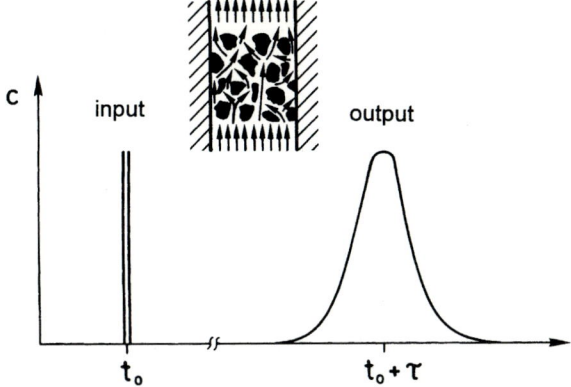

Fig. 6.8. Schematic presentation of a typical packed-bed residence time distribution as induced by frequent diversion of partial streams

and

$$F(t) = 1 - \exp\{-t/\tau\} \tag{6.22}$$

The electrochemical generation of a marker substance at the reactor inlet and the electrochemical measurement of the marker concentration – for instance by measurement of mass transfer limited current densities for electrochemical reconversion of the marker – at the reactor exit is an elegant experimental means for determining residence time distributions of electrolyzers. Every real reactor and electrolyzer shows a residence time distribution which deviates from the ideal behaviour. Such deviations are due to non-ideal flow patterns. As an example a packed bed electrolyzer will be discussed.

In packed bed electrolyzers the fluid flow is ever and ever subdivided and diverted by the many partial streams of the electrolyte flowing around the particles and flowing through the bed's voids. The resulting typical relatively broad residence time distribution of a packed bed is schematically depicted in Fig. 6.8.

The phenomenon of a broadened, Gaussian-like residence time distributions is formally noted as "dispersion" or backmixing which is described in a formalism similar to that of molecular diffusion Eq. (6.23) with the axial dispersion coefficient D_{ax} describing axial back mixing. Such experimentally determined dispersion coefficients are by several orders of magnitude larger than the molecular diffusion coefficient (10^{-3} to 10^{-2} cm^{2}s^{-1} compared to $D_{mol} \approx 10^{-5}$ cm^2 s^{-1} in liquids). The one-dimensional, axial, dispersion of the marker is described by Eq. (6.23):

$$dc_M/dt = w(dc_M/dx) - D_{ax} d^2c/dx^2 \tag{6.23}$$

For axial dispersion an appropriate adimensional quantity, the axial Peclet-number P_{ax}, is defined for adimensional correlations relating the axial Peclet number, Pe_{ax}, to Re and Sc. (d_p is the particle diameter and w the linear velocity in the empty reactor)

$$Pe_{ax} = wd_p/D_{ax}$$

In the 0.1<Re <10 range Pe_{ax} varies between 1.5 and 2.5 for packed beds. For particle diameters 0.25–0.7 mm which are very often used in packed-bed electrodes, axial dispersion coefficients had been measured of the order of $10^{-2} cm^2 s^{-1}$. Theoretical considerations based on probability theory predict under turbulent flow conditions for packed beds composed of balls or beads limiting value of $wd_p/D_{ax} \varepsilon$ of 2, which seems to be in reasonable agreement with experimental data (ε is the void fraction of the packed bed). The discussed complicated flow pattern in packed beds results in a reduced residence time of a smaller part of the electrolyte and a correspondingly longer residence time of other parts of the flowing electrolyte. The consequence is evident: In those parts which leave the electrolyzer with shorter than the anticipated, mean, residence time, the depletion of for instance a noxious metal ion by cathodic reduction has not proceeded as far as it had been predicted for ideal plug flow and in those parts which reside longer in the electrolyzer than mean residence time, τ, the depletion had proceeded beyond the postulated limit. If the electrolysis is performed at mass transfer limited current densities residence time distributions which deviate sizeably from the ideal lead to reduced space time yields and an enhanced mean exit concentration of the reactant R, compared to predicted values calculated for an ideal plug flow reactor.

Equation (6.24), gives the general expression for the mean exit concentration for non-ideal channel or bed electrodes:

$$\bar{c} = \int_0^\infty E(t)c(t)dt \qquad (6.24)$$

If the residence time distribution is at variance to that of an ideal plug-flow reactor, only reactors with uniform current density i yield the anticipated exit concentration c_e.

6.5.7
The Selectivity Problem of Consecutive Reactions in Batch Reactors

Consecutive electrochemical reactions in batch, plug-flow and continuously stirred reactors

$$R \xrightarrow{k_1} S \xrightarrow{k_2} T \qquad (6.25)$$

constitute a well known selectivity and yield problem in chemical reaction engineering, which is also of concern in electrochemical reactors. In particular organoelectrosynthesis reactions very often proceed in consecutive steps and

6.5 Macrokinetic Models

would produce a mixture of the products S and T instead of only one product if the desired substance is the intermediate. An example is the mediated and the direct anodic oxidation of toluenes to benzaldehydes, Eq. (6.26):

$$ArCH_3 + H_2O \rightarrow ArCHO + 4H^+ + 4e^- \qquad (6.26\ a)$$
$$\ \text{R} \qquad\qquad\qquad \text{T}$$

which proceeds in two distinct oxidation steps. The first is the formation of benzyl alcohols or their ethers. If the reaction is executed in alcoholic solvents according to Eq. (6.26 b) benzylalcohol alkylethers are formed:

$$ArCH_3 + 2HOR \rightarrow ArCH_2OR + 2H^+ + 2e^- \qquad (6.26\ b)$$
$$\ \text{R} \qquad\qquad\qquad \text{S}$$

The second step is the oxidation of benzylalcohols or benzylethers to benzaldehydes or dialkylacetals, Eq. (6.27):

$$ArCH_2OR + HOR \rightarrow ArCH(OR)_2 + 2H^+ + 2e^- \qquad (6.27)$$
$$\ \text{S} \qquad\qquad \text{T}$$

But the oxidation of toluenes does not stop with the formation of benzaldehydes as they can be further converted to less well defined products and benzoic acid or its esters, Eq. (6.28):

$$ArCH(OR)_2 + yHOR \rightarrow yArCOOR + \text{polym. oxid. products} + xH^+ + xe^-$$
$$\ \text{T} \qquad\qquad\qquad \text{acid}$$
$$(6.28)$$

If polymer formation can be neglected, the selectivity with respect to S is defined by

$$S_S = \frac{n_S}{n_S + n_T + n_{acid}} \qquad (6.29)$$

Figure 6.9 depicts schematically the evolution of the concentrations c_R, c_S and c_T with time for a batch experiment in order to demonstrate that the selectivity S_S of the formation of the intermediate S is diminishing with time (it is unity at $t = 0$).

The concentration of S passes through a maximum which is of especial interest, as halting the conversion of R at the time of the concentration maximum of S would facilitate the working-up procedure for recovery of S, provided benzyl alcohols would be the desired product. The height of the concentration maximum of S is directly related to the ratio k_1/k_2 of the rate coefficients of the two consecutive reactions. The higher this ratio, the higher is $c_{S,max}$ and the higher the fractional conversion of R at which $c_{S,max}$ is obtained, the reaction stopped and the separation procedure performed. Also the selectivity obtained at $c_{S,max}$ increases with this rate ratio.

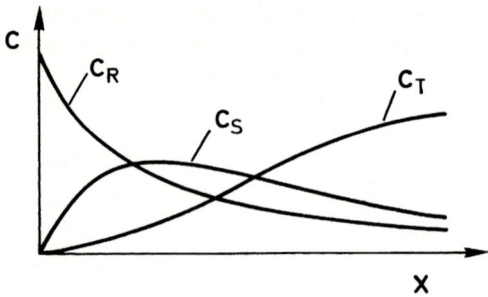

Fig. 6.9. Schematic presentation of the evolution of the concentrations with time of reactant (R), intermediate product (S) and stable end product (T) due to consecutive reaction steps (R → S) and (S → T)

For electrochemical reactions the analogue of homogeneous rate coefficients k_1 and k_2 at a given electrode potential E are exponential functions of the difference between the working potential E and half wave potentials of the substrates R, S and T and the slope, m, of the mass transfer corrected, semilogarithmic current voltage curve. At a given electrode potential E these relations read

$$\log k_1 \approx \pm(E - E_{1/2,R})/m_R \qquad (6.30\ a)$$

$$\log k_2 \approx \pm(E - E_{1/2,S})/m_S \qquad (6.30\ b)$$

$$\log k_3 \approx \pm(E - E_{1/2,T})/m_T \qquad (6.30\ c)$$

where m_R, m_S and m_T are the slopes of the semilogarithmic mass-transfer-corrected–current voltage curves $\left(\lg\dfrac{i_l - i}{i_l}\ vs\ E\right)$ of the substances R, S and T. The positive sign holds for anodic and the negative for cathodic reactions. For stirred-batch electrolyzers performing reaction sequence, Eq. (6.25) electrochemically, the rates of conversion of R, S and T per unit electrode area and the condition $i_{R,S,T} \ll i_{lim,R,S,T}$ at a given electrode potential, E, would be given by

$$\dot{n}_R = k_m c_R 10^{\{(E-E_{1/2,\ R})/m_R\}} \qquad (6.31\ a)$$

$$\dot{n}_S = k_m c_S 10^{\{(E-E_{1/2,\ S})/m_S\}} \qquad (6.31\ b)$$

$$\dot{n}_T = k_m c_T 10^{\{(E-E_{1/2,\ T})/m_T\}} \qquad (6.31\ c)$$

6.5 Macrokinetic Models

with the partial current densities i_R and i_S for conversion of R and S reading

$$i_R = k_m c_R V_{e,R} F \cdot 10^{\{(E-E_{1/2,\,R})/m_R\}} \qquad (6.32\ a)$$

$$i_S = k_m c_S V_{e,S} F \cdot 10^{\{(E-E_{1/2,\,S})/m_S\}} \qquad (6.32\ b)$$

Both partial currents sum up to the total current density i, provided the partial current i_T by which the aldehyde is converted and parasitic currents can still be neglected because the aldehyde is oxidised at much higher potential than the toluene, R, and the benzyl alcohol, S.

$$i = i_R + i_S \qquad (6.33)$$

For cases where the condition $i_{R,S,T} \ll i_{lim,R,S,T}$ does not hold but mass-transfer-limited current densities of these species become comparable to the respective partial current densities i_R and i_S Eqs. (6.32 a) and (6.32 b) have to be exchanged by the complete expressions Eqs. (6.34 a) and (6.34 b):

$$i_R = \frac{k_m c_R V_{e,R} F}{1 + 10^{\{(E-E_{1/2,\,R})/m_R\}}} \qquad (6.34\ a)$$

$$i_S = \frac{k_m c_S V_{e,R} F}{1 + 10^{\{(E-E_{1/2,\,S})/m_S\}}} \qquad (6.34\ b)$$

Table 6.1 collects characteristic electrochemical data for the anodic oxidation of differently substituted toluene, benzylalcohol-ether and benzaldehyde-dimethylacetat. The half-wave potential of the ether is approximately 60 mV more anodic than that of the toluene from which with a slope of the semilogarithmic current voltage curve of approximately 80 mV per decade a rate ratio, k_1/k_2, of approximately 5 is obtained at E equalling $E_{1/2}$ of the toluene. Figure 6.10 demonstrates for the anodic oxidation of chlorotoluene the development of the concentrations of the toluene, the ether and the acetal. From this experimental plot a rate ratio, k_1/k_2, of approximately 3, which differs from but comes close to the predicted value, can be obtained from the maximal concentration of the chlorobenzylmethyl ether. This plot shows also the generation of polymers as the dotted line representing the sum of the concentrations of toluene, alcohol and aldehyde decreases steadily with time and this decrease enhances after $c_{s,max}$ is reached. The situation depicted in Fig 6.10 for a batch reactor is transferable to plug flow reactors. As is common place for homogeneous chemical reactions also for electrolytic reactions the continuous stirred tank reactor is less advantageous for obtaining high selectivities for an intermediate, S, than the batch and the plug flow reactor because it operates steadily at reduced concentration levels of R and enhanced concentration levels of S. Also for the continuous stirred tank reactor the concentration of the intermediate, S, passes

Table 6.1. Anodic voltammetric data of different toluenes, benzylalcohol methylethers and benzaldehyde dimethylacetals in 0.005 mol dm^{-3} molar ethanol solutions

Depolarizer	$E_{1/2}$/mV vs NHE	V_e	dE/dlg[$(i_l-i)/i$] mV
p-Methoxytoluene	1366	4.1	66
p-Methoxybenzyl-ethylether	1305	1.8	66
p-Methoxybenzyl diethyl acetal	1643	6.8	≈120
Toluene	1750	3.9	69
Benzylethylether	1790	2	104
Benbzaldehyde diethyl acetal	1880	6	74
p-Chlorotoluene	1790	3.9	43
p-Chlorobenzyl-ethyl ether	1790	1.9	120
p-Chlorobenzaldehyde diethyl acetal	1870	4.1	120

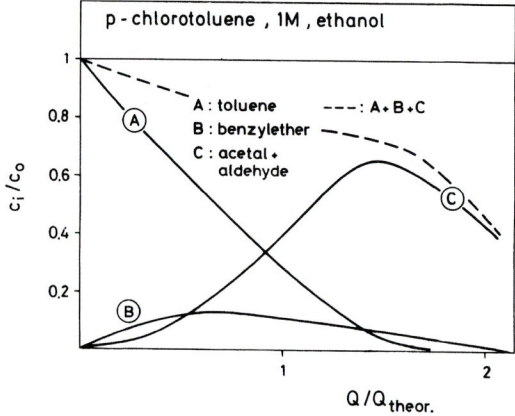

Fig. 6.10. Development of the concentrations of p-chloro-toluene, p-chloro-benzylalcohol methylester and p-chloro-toluene dimethylacetal in a batch reactor. The ordinate represents relative charge consumed $Q/Q_{th} = 1$ refers to 4 mols of electricity per mol of toluene

through a maximum as residence time and conversion increases, but the concentration maximum is lower than obtained in batch and plug flow reactors.

6.6
Coupling of Electrochemical and Chemical Reactors

If the electrochemically produced species is intended to be consumed in a slow chemical consecutive reaction very often the volume of the electrolyzer is not sufficient for the performance of this slow, homogeneous reaction because the chemical reaction is too sluggish. An additional reactor volume has to be added to the volume of the electrolyzer in order to drive the chemical reaction almost to completion.

6.6 Coupling of Electrochemical and Chemical Reactors

One of the commercially more important processes, the chlorate process, comprises a relatively slow homogeneous chemical reaction – hypochlorite synproportionation – Eq. (6.38).

Hypochlorite formation, Eq. (6.37), from anodically produced chlorine, Eq. (6.35), with cathodically generated hydroxyl ions, Eq. (6.36), is a relatively fast reaction.

$$Cl^- \rightarrow 1/2\,Cl_2 + e^- \qquad \text{(anode)} \qquad (6.35)$$

$$H_2O + e^- \rightarrow 1/2\,H_2 + OH^- \qquad \text{(cathode)} \qquad (6.36)$$

$$Cl_2 + 2OH^- \rightarrow ClO^- + Cl^- \qquad \text{(fast)} \qquad (6.37)$$

$$ClO^- + 2HClO \rightarrow ClO_3^- + 2HCl \qquad \text{(slow)} \qquad (6.38)$$

It is the quantity space time yield ρ of reaction Eq. (6.38) times reactor volume V_R, which has to be matched with the anodic production rate of chlorine at sufficiently low steady state-hypochlorite concentrations. Too high hypochlorite concentrations would favour undesired anodic hypochlorite oxidation, associated with oxygen evolution and detrimental current efficiency losses – Eq. (6.39):

$$6ClO^- + 3H_2O \rightarrow 2ClO_3^- + 4Cl^- + 6H^+ + 3O_2 + 6e^- \qquad (6.39)$$

Figure 6.11 shows that the electrolyte is circulating in a loop which comprises the electrolyzer and an additional reactor volume. The electrolyzer is essentially a bipolar stack of flat electrodes forming a set of flow channels. Due to turbulent flow of the electrolyte the chlorine-saturated anolyte and the basic catholyte are rapidly mixed and hypochlorite is formed. The greater part of the loop serves as reaction vessel which at high circulation rate of the electrolyte can be modelled as a continuous stirred tank reactor. At steady state the chlorine-generation rate at the electrode surface A_e matches the six-fold production rate r_{v,ClO_3^-} of chlorate times in the reactor volume V_R according to Eq. (6.40):

$$I\phi^e = i\phi^e a_e V_R / F - 6 r_{v,ClO_3^-} V_R \qquad (6.40)$$

The concentration of chlorate in the loop – which depends on the current, I, and the volumetric flow rate, $\dot{v}$, of the brine containing chlorate which leaves the reactor and of added fresh brine is determined by Eq. (6.41)

$$\dot{V} c_{e,ClO_3^-} = I\phi^e/6F = i\phi^e a_e V_R / 6F \qquad (6.41)$$

A similar but still more complex situation arises in mediated oxidation of organic substrates, Eqs. (6.42) and (6.43), which often are only little soluble in aqueous electrolytes

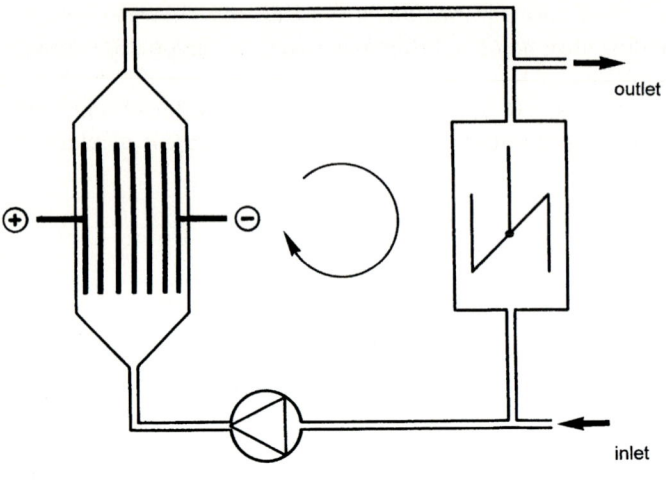

Fig. 6.11. Schematic of a chlorate production loop. The recirculated electrolyte passes through the electrolyzer and a reactor which is to good approximation a stirred-tank reactor

$$M^{z+} \rightarrow M^{(z+1)+} + e^- \quad \text{generation of oxidising mediator} \quad (6.42)$$

$$nM^{(z+1)+} + R \rightarrow S + nM^{z+} \quad \text{(chemical conversion)} \quad (6.43)$$

Such sparingly soluble substances, R, are usually reacted in form of a dispersion or emulsion in an aqueous electrolyte so that the mass transfer of the substrate, R, from the surface of the dispersed droplets into the aqueous electrolyte, where the reaction Eq. (6.43) proceeds according to homogeneous kinetics has some influence on the total rate. Very often mass transfer from the emulsion determines the macrokinetics of the chemical reaction of the little soluble organic substrate with the mediator $M^{(z+1)+}$, which follows the electrochemical generation of the redox-reactant.

6.7
Electrolyzer Design and Chemical Yield Losses Due To Parasitic Chemical Reactions

With the exception of electrochemical metal deposition $\left(M^{z+} + z\,e^- \rightleftarrows M\right)$ and redox reactions of simple redox couples $\left(A^{n+} + e^- \rightleftarrows A^{(n-1)+}\right)$, charge transfer at electrodes is generally accompanied by purely chemical, most often solvolytic reactions of the reduced or oxidised depolarizer. Solvolytic reactions are occurring as a consequence of base and acid production by cathodic reduction or anodic oxidation of organic and inorganic substrates or of the solvent. Examples are given in Eqs. (6.44) and (6.45) for cathodic generation of base R^- as a consequence of cathodic olefin hydrogenation and of protons H^+ as a consequence of anodic olefin oxidation:

$$R_1R_2C = CR_3R_4 + 2HR + 2e^- \rightarrow R_1R_2CH - CHR_3R_4 + 2R^- \qquad (6.44)$$

$$R_1R_2C = CR_3R_4 + 2HOR_5 \rightarrow R_5OR_1R_2C - C - R_3R_4OR_5 + 2H^+ + 2e^- \qquad (6.45)$$

Strong acids and bases usually catalyse numerous chemical reactions – solvolysis reactions in particular. For instance olefins undergo base-catalysed solvent addition at the C=C double bond. Such reactions may give rise to substantial material losses as the chemically converted material – even if it is recovered from the exit stream it often cannot be recirculated, because it is inert to the purposely performed electrochemical conversion.

Poor design of the electrolyzer may cause dead-water or backflow conditions, with sluggish mass transfer of the produced acid or base from the electrode surfaces into the bulk of the solution, where they would be neutralised. Under these conditions solvolysis-induced material losses may become intolerably high. It is well known that solvolysis of acrylonitrile producing cyanoethanol, Eq. (6.46), may capture and destroy non-negligible quantities of this valuable substrate in the Monsanto–Baizer process which consists of electrohydrodimerization of acrylonitrile to adipodinitrile, Eq. (6.47):

$$CH_2 = CHCN + H_2O \rightarrow HOCH_2 - CH_2CN \qquad \left(OH^- \text{-catalysis}\right) \qquad (6.46)$$

$$2CH_2 = CHCN + 2e^- + 2H_2O \rightarrow (CH_2 - CH_2CN)_2 + 2OH^- \qquad (6.47)$$

This reaction impairs the mass yields in improperly designed electrolyzers and necessitates also additional efforts for removal of the undesired product from the electrolyzer effluent stream. Indeed the back mix behaviour and residence time distribution of the chosen reactor, which can be largely influenced by flow velocity and flow pattern, bears seizeably on the result and success of this electrosynthesis.

6.8
Performance Criteria of Electrochemical Reactors

Performance criteria of electrochemical reactors should be chosen, which relate as closely as possible to performance criteria of chemical reactors. In certain points, however, the very nature of electrolyzers, in which the consumption of electric charge and mass is coupled by Faraday's law necessitates the adoption of especially defined criteria as for instance current efficiency or specific electric energy consumption.

6.8.1
Fractional Conversion, X

If a reactant A undergoes an electrochemical reaction

$$A + v_e e^- \rightarrow v_p P \tag{6.48}$$

the fractional conversion is defined as

$$X_A = 1 - \frac{n_A}{n_{A,0}} \tag{6.49}$$

where $n_{A,0}$ is the starting molar amount and n_A is final molar amount of substrate. In the case of multiple reactants indices, e.g. X_A, may be used to sign the key reactant.

6.8.2
Relative Amount of Charge-Q_r

$$Q_r = Q/Q_{th} \tag{6.50}$$

Relative amount of charge is the actual charge Q passed, related to the theoretical amount of charge Q_{th}, which is needed for a complete conversion of the substrate assuming a current efficiency of 1. In the case of multiple reactants indices, e.g. $Q_{r,A}$, may be used to sign the key reactant.

6.8.3
Overall Conversion Related Yield Θ_p

Θ_p - overall conversion related yield

$$\Theta_p = \frac{n_p}{v_p(n_{A0} - n_A)} \tag{6.51}$$

Θ_p' - point conversion related yield

$$\Theta_p' = \frac{v_A r_p}{v_p r_A} \tag{6.52}$$

with r_i the actual rates of production of p and consumption of A.

6.8.4
Current Efficiency Φ^e

The classical definition of current efficiency by Faraday's law is:

$$\Phi^e = \frac{mv_e F}{MIt} \quad (6.53)$$

where m is mass of product, M its molar weight, and v_e is the stoichiometric number of electrons of the reaction. Because the number of electrons involved in an electrochemical reaction represents a stoichiometric coefficient the symbol v_e is used according to the ECRE convention recommending v_i for stoichiometric coefficients. The current efficiency bears a strong analogy to any other kind of yield, e.g. to the conversion related yield.

$$\text{conversion related yield} = \frac{\text{stoichiometric amount of product}}{\text{converted amount of reactant}} \quad (6.54)$$

The current efficiency is a "yield of charge" and one distinguishes the following definitions.
Overall current efficiency Φ_P^e

$$\Phi_P^e = n_p v_e F / v_p Q \quad (6.55)$$

with

$$Q = \int_0^t I \, dt \quad (6\text{ e})$$

and point current efficiency $\Phi^{e'}$

$$\Phi^{e'} = \frac{dN_P}{dt} v_e \cdot F/I \quad (6.56)$$

Overall selectivity S_{P_i} with respect to i-th product

$$S_{P_i} = \frac{n_{P_i}/v_i}{\sum n_{P_j}/v_j} \quad (6.57\text{ a})$$

Point selectivity - S'_{P_i} is given by the respective differential expression

$$S'_{P_i} = \frac{(dn_{P_i}/dt) \cdot (1/v_i)}{\sum (dn_{P_j}/dt) \cdot (1/v_j)} \quad (6.58)$$

6.8.5
Parameters for Energy Considerations

The energy costs of an electrochemical process are closely correlated to the energy efficiency. For characterising this quantity the following definitions are recommended.

Energy yield γ_G

$$\gamma_G = \frac{\Delta G \Phi^e}{U_c v_e F} = \frac{U_0}{U_c} \Phi^e \tag{6.59}$$

Enthalpy related (thermal) energy yield- γ_H

$$\gamma_H = \frac{\Delta H \Phi^e}{U_c v_e F} = \frac{U_{th}}{U_c} \Phi^e \tag{6.60}$$

$$U_{th} = \frac{\Delta H}{v_e F} \tag{6.61}$$

where U_{th} is the thermoneutral cell voltage.

Specific electric-energy consumption-E_s (J kg^{-1}, J m^{-3}) – this quantity usually is mass related for solid or liquid products

$$E_{S,m} = \frac{U_c v_e F}{M \Phi^e} \left(J mol^{-1} \right) \tag{6.62}$$

For gas evolution the specific energy consumption might also be related to standard temperature – pressure (STP) volumes.

References

1. L. Mertens, Kittertwig, Electrochim. Aeta 40 (1995) 387

Further Reading

K. Scott, F. Goodrich, Electrochemical Reaction Engineering, Academic Press, London 1991.

CHAPTER 7

Electrochemical Engineering of Porous Electrodes and Disperse Multiphase Electrolyte Systems

7.1 Introduction

Industrial electrochemistry has often to deal with a situation where either the electrolyte is not homogeneous but contains a second solid, liquid or gaseous dispersed phase or the electrode possesses a particulate structure comprising the electrolyte in pores, channels or holes. Considering the phase structure of such more involved multiphase systems the following classification can be given.

1) Porous, particulate or dispersed electrodes as for instance gas diffusion electrodes in fuel cells, porous coatings on gas evolving electrodes, battery electrodes and packed and fluidized bed electrodes. They are mainly used in cases, where low current densities demand a significant increase of true electrode surface beyond the geometrical cell cross section in order to achieve acceptable space time yields. Using porous or particulate electrodes means, that the inner surface of the porous matrix is used, which can be greater than the cell cross section by several orders of magnitude. Such electrodes are often denominated "three-dimensional electrodes." The different types of porous or particulate electrodes are mainly distinguished by pore dimension and porosity.
2) Electrolyzers operated with non-soluble electrolysis products contained and dispersed in the electrolyte and interelectrodic gap as in the form of bubbles in electrochemical gas evolution or in the form of powders in cathodic metal powder production.
3) Electrolyzers converting dispersed gaseous reactants as in electrochemical gas absorption and other processes converting gaseous substances.
4) Electrolyzers converting dispersed, liquid or solid reactants or scarcely water-soluble reactants which are dissolved in a dispersed second liquid phase of low solubility in the aqueous electrolyte.

For modelling porous or particulate electrodes the spatial current density distribution caused by ohmic-voltage losses in electrolyte pores and slow diffusive mass transfer through the stagnant or slowly moving electrolyte phase in the pores is of highest concern.

For electrolytes containing a second, dispersed solid, gaseous or liquid phase predictions concerning their electrolytic conductivity and the quantitative treatment of mass transfer from the dispersed into the homogeneous electrolyte phase are of higher interest.

7.2
Three-Dimensional Electrodes

7.2.1
General Considerations

Flat massive electrodes are too expensive and not appropriate for performing particular electrochemical processes which due to their very nature can only be performed at low current densities. This might be the case, for instance, because they are operated at low reactant concentration or because it is necessary to increase relative rates of first order vs second or higher order consecutive reactions- which can only be executed with unusually low current densities – e.g. with current densities which are lower than 0.01 A cm^{-2}. For such purposes porous electrodes are the electrode type of choice.

Porous electrodes have an internal surface which, depending on pore diameter, electrode porosity and thickness, may exceed their geometric surface by several orders of magnitude. Therefore very low current densities of the order of 1 mA cm^{-2} or less prevailing at their internal surface may add to usual current densities of 0.1–1 A cm^{-2} at their outer surface so that electrolysis processes may be performed with nominal current densities of usual magnitude and reasonable space time yields respectively although the intrinsically applied current density is very low. Often such electrodes are called three-dimensional as their inner surface extends also in the third dimension. Three dimensional electrode structures of different porosity, thickness and pore geometry are applied for[1].

(i) Fuel cell electrodes which are composed of nanoporous catalyst particles (pore diameter 1 nm>d_p>10 nm) which constitute a microporous electrode structure (micropore dimensions 0.1–10 µm) (Fig. 7.1 a) and in the form of nanoporous, catalytically active, coatings for gas evolving electrodes. (Fig. 7.1 b)

(ii) Microporous electrodes (d_p > 0.5 µm) which are used most often in batteries. Their thickness is usually of the order of several millimetres and pore diameters are typically 10–100 µm (Fig. 7.1 c)

(iii) Packed and fluidized bed electrodes which are composed of particles with diameters of from millimetres to centimeters. Packed-bed and fluidized-bed electrodes are used for the electrochemical conversion of homogeneously dissolved species at low concentration levels (c < 0.01 mol dm^{-3}). Such electrode-beds measure centimeters to fractions of meters in thickness or bed depth respectively (Fig. 7.1 d).

---

1 Different from the frequently used words micro, meso- and macropores – which we assume misleading though historically sanctioned we distinguish only nanopores from micropores whose pore diameters are in the nano- and micrometer range respectively.

7.2 Three-Dimensional Electrodes

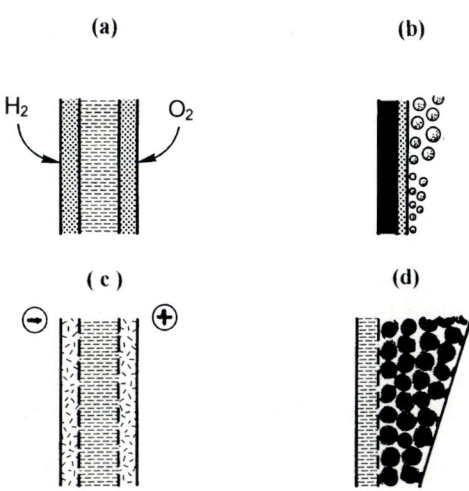

Fig. 7.1 a–d. Schematic presentation of three-dimensional electrodes. Nanoporous: **a** fuel cell; **b** gas evolving electrodes with catalytic coating; **c** schematic of microporous battery electrode; **d** packed bed electrode, whose particles measure in millimetres up to centimetres

7.2.2
Fundamental Equations

Current density distributions in three-dimensional electrodes are governed by ohmic potential-losses and mass transfer induced local concentration differences which both change the local electrode potential. Concentration distributions in three dimensional electrodes and in singular pores of porous electrode particles are determined by competition of electrochemical depolarizer consumption in the porous matrix expressed as volume related electrochemical reaction rate r_v and diffusive and/or convective mass transport into or out of the porous matrix.

Equations (7.1)–(7.3) constitute the fundamental set of differential equations describing potential distribution in the electrolyte (index 1) and electrode matrix (index 2) according to Ohm's law and accounting for the volume related electrochemical reaction rate.

$$i_1 = \kappa_1 \operatorname{grad} \varphi_1 \quad \text{(electrolyte)} \tag{7.1 a}$$

$$i_2 = \kappa_2 \operatorname{grad} \varphi_2 \quad \text{(electrode)} \tag{7.1 b}$$

If electrode potential shifts which are induced by concentration changes can be neglected, then Eqs. (7.1 a) and (7.1 b) by accounting for (7.2 a) result in Eq. (7.1 c):

$$\operatorname{div} \operatorname{grad}(\varphi_1 - \varphi_2) = \operatorname{div} \operatorname{grad} \eta = (1/\kappa_1 + 1/\kappa_2) \operatorname{div} i \tag{7.1 c}$$

Coupling the divergence of current densities in the flooded pores, (i_1), and the electrode matrix, (i_2), to the volume specific electrochemical conversion rate r_v leads to Eqs. (7.2 a) and (7.2 b)

$$\text{div } i_1 = -\text{div } i_2 = r_v v_e F = a_e i_s \tag{7.2 a}$$

$$i_s = i_0 \left[\exp\{\alpha_a F\eta/RT\} - \exp\{-\alpha_c F\eta/RT\} \right] \tag{7.2 b}$$

the quantity i_s is the current density at the internal surface of the porous electrode or wall of the pores resp. described by the Butler Volmer expression, Eq. (7.2 b).

Equation (7.3) accounts for mass transfer limits of the electrochemical conversion at the inner surface of a porous electrode which is given by the local mass transfer coefficient k_m.

$$i_1(x,y,z) = k_m(x,y,z) c(x,y,z) v_e F \tag{7.3}$$

The volume specific mass balance accounting for convective and diffusive mass transfer and electrochemical conversion in a three-dimensional electrode reads

$$dc_i / dt = w \, \text{grad} \, c_i + D \, \text{div} \, \text{grad} \, c_i + r_{v,i} \tag{7.4}$$

For the different types of porous electrodes the following consideration apply.

7.2.2.1
Nanoporous Electrode Particles

Convection does not exist in nanopores. Therefore the term $w \, \text{grad} \, c$ in Eq. (7.4) vanishes and nanopores for the steady state one obtains the equation

$$D_i \, \text{div} \, \text{grad} \, c_i = -r_{v,i} = a_e i_s / v_e F \tag{7.4 a}$$

The low numerical value of the diffusion coefficient of dissolved species ($D \approx 10^{-5}$ cm^2 s^{-1}) limits the reactive parts of the electrode or electrode-particles to several tens to hundred μm at the most and therefore ohmic-potential drops described by Eqs. (7.1 a–c) can very often be neglected so that only Eq. (7.4 a) has to be solved for this type of electrode.

7.2.2.2
Microporous Electrodes

In batteries both solid partners of the energy storing redox-system – for instance $PbO_2/PbSO_4$ – are usually only little soluble in the respective electrolyte. The electrolyte whose components are participating in the charge and discharge reaction are often highly concentrated so that the relative concentration changes

$\Delta c/c$ due to the relatively slow progress of the electrochemical reaction are low although convection is negligible and diffusive mass transfer across distances of several millimetres is relatively slow. Therefore for battery electrodes differential Eqs. (7.1) and (7.2) that means ohmic potential drops in the electrolyte and the electrode matrix are mainly determining the current density distribution in the depth of the electrode to which Eq. (7.4) contributes second order corrections.

7.2.2.3
Packed and Fluidized Bed Electrodes

In packed and fluidized bed electrodes with coarse particles of milimeter up to centimeter size, where low current densities at the electrode particles which are caused by low depolarizer concentrations sum up to high effective bed current densities, the situation is similar as for microporous battery electrodes but because depolarizer concentrations are low and the bed electrode is operated with forced electrolyte flow at close to mass transfer limited current densities, convective mass transfer is essential and results in expressed local differences of the depolarizer concentration and current densities at the inlet and outlet of such electrodes. Therefore the whole set of differential equations (Eqs. 7.1–7.4) must be solved simultaneously for modelling this type of three dimensional electrode.

7.2.3
Gas Consuming Nanoporous Electrodes for Fuel Cells and Nanoporous Catalyst Particles and Layers for Gas Evolving Electrodes

7.2.3.1
Physical Structure of Particulate, Gas Consuming Nanoporous Gas-Diffusion Electrodes

Figure 7.2 a shows the typical structure of a PTFE (Teflon)-bonded Raney-nickel anode in former times used in alkaline fuel cells. The electrode in this case consists of meanwhile outdated, nanoporous Raney-nickel particles of typically 2–5 µm diameter. (The example has been chosen, because it allows to outline the essential features of gas-diffusion electrodes more clearly than any other.) The mean diameter of the nanopores of these particles is approximately 2–5 nm. With porosities of 20% such particles would exhibit a specific inner surface of the order of 50 $m^2 g^{-1}$ or 250–350 $m^2 cm^{-3}$. As nickel is relatively well wetted by caustic potash with a wetting angle of 30° or less, the nanopores are completely filled with electrolyte. At their outer surface the particles are also wetted by the electrolyte and the electrolyte film on the particles' surface constitutes the electrolytic "bridge" coupling the outer and the internal surface of the particles to the electrolyte matrix and counter electrode. Further, hydrophobic PTFE-fibres constitute a gas filled micropore system through which the feed gases are supplied to the electrode interior. After longer use PTFE loses its hydrophobicity

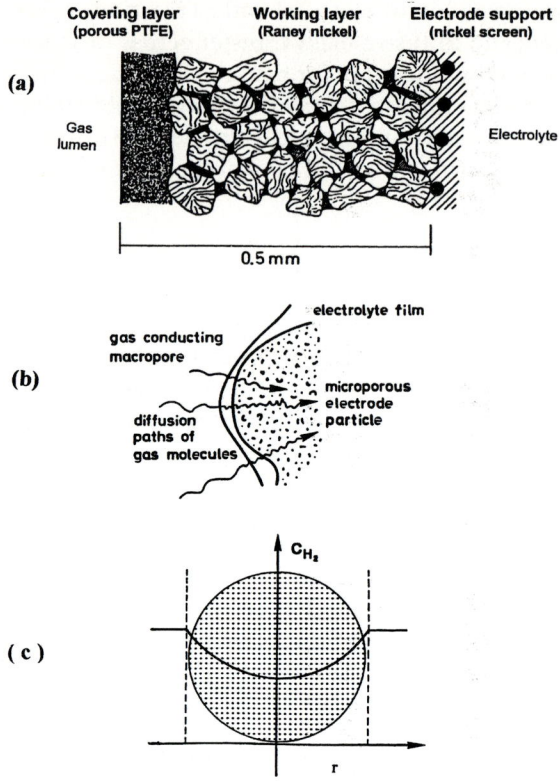

Fig. 7.2 a–c. Fuel cell electrode: **a** typical morphology of fuel cell anode composed of Raney nickel particles; the particles are not drawn to scale but measure approx. 10μm; **b** schematic presentation of dissolution of gaseous reactant into electrolyte film and diffusive transport of dissolved species into the nanoporous electrocatalyst particle on whose inner surface the reactant is electrochemically converted; **c** schematic of concentration distribution in nanoporous particle in fuel cell anode in which the concentration of dissolved hydrogen is diminished due to anodic H_2 oxidation

and nevertheless the coarse pores do not become flooded because of a limited electrolyte supply.

Hydrogen diffuses by fast gas-phase diffusion ($D_{gas} \approx 1$ cm^2 s^{-1}) through the gas-filled micropores towards the electrolyte film covering the electrode particles where it dissolves and diffuses slowly ($D \approx 10^{-5}$ cm^2 s^{-1}) into the nanoporous particle (Fig. 7.2 b). The anodic oxidation of hydrogen at the internal surface of the Raney nickel particles results in progressive depletion of hydrogen the farther the distance to the outer surface of the particle so that at the centre of the particle hydrogen concentration might become much lower than at its outer surface – even nil (Fig. 7.5 c). If the size of the electrode particles is too large so that this mass transfer condition is not properly matched, the catalyst will not be fully utilized as its central parts are not supplied by the depolarizer and therefore cannot contribute to the electrochemical reaction (Compare also Sect. 4.9.2).

7.2.3.2
Physical Structure of Raney Nickel Coatings for Hydrogen Evolving Cathodes

Figure (7.3 a,b) illustrates the structural features of two different types of Raney-nickel-coated hydrogen evolving cathodes which are either fabricated by rolling a mixture composed of smooth and ductile carbonyl nickel particles and very hard Ni/Al: 50/50 wt% particles on a nickel support or by cathodic codeposition of a Ni/Zn-alloy containing more than 50 wt% of zinc. Both coatings are eventually leached in 30 wt% KOH in order to dissolve due to selective corrosion (M + $nH_2O \rightarrow M(OH)_n + n/2\ H_2$) the reactive component (Al or Zn).

Leaching of Ni/Zn-coatings produces under partial shrinking of the nickel matrix the catalytically active nanoporous electrocatalyst. The galvanically deposited layer forms cracks and fissures on shrinking which generates additionally to the nanoporosity and microporosity which is morphologically similar to the microporosity of rolled Raney-nickel layers which resemble to some extent to PTFE-bonded Raney-nickel fuel cell anodes (Fig. 7.2 a). This microporosity is essential for a high degree of utilisation of the active coatings.

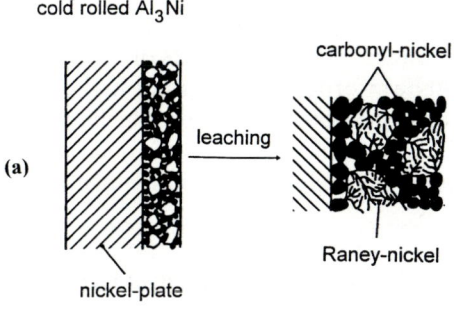

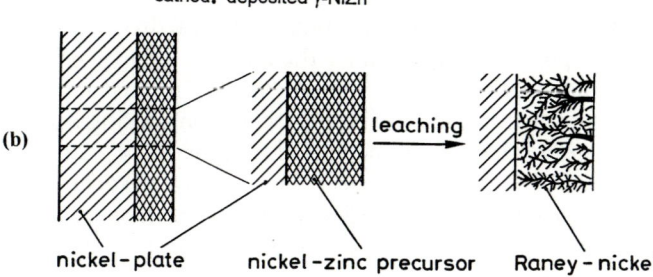

Fig. 7.3 a,b. Schematic of the features of H_2-evolving cathodes catalytically activated by a Raney nickel coating: **a** composite coatings from cold rolled mixture of ductile nickel and Raney nickel particles; **b** smooth Raney nickel coating obtained by cathodic codepositing zinc and nickel followed by alkaline leaching of zinc

As pointed out above for gas consuming fuel cell electrodes, also the nanopores of gas evolving electrodes are completely filled with electrolyte and electrolytic generation of hydrogen or other gases at the internal surface of catalyst particles leads to accumulation of homogeneously dissolved hydrogen or other product gases in the pores which can accumulate to more than one thousand-fold the solubility of hydrogen (or the product gas) at ambient pressure. Under 1 bar of hydrogen the solubility of this gas in 30 wt% KOH amounts to roughly 10^{-4} mol dm^{-3} of hydrogen, so that concentrations of 0.1 mol dm^{-3} of dissolved gases might be envisaged in nanopores. Such high hydrogen concentrations can be achieved since according to the Young–Laplace equation (Eq. 7.5) the equilibrium hydrogen pressure p_{eq} (c_{H_2}) in the narrow pores does not exceed the pressure difference Δp which is necessary to blow the pore free of electrolyte:

$$\Delta p = 2\sigma \cos\Theta / d_p \tag{7.5}$$

(σ is the surface tension at the interface electrolyte/hydrogen amounting to $7.8 \cdot 10^{-2}$ N m^{-1} with 1 mol dm^{-3} KOH solution, and Θ is the wetting angle of this electrolyte on nickel, d_p is the pore diameter with a value of between 1 and 5 nm).

Therefore in a hydrogen producing nanoporous catalyst particle contained in a rolled layer or a galvanically applied catalytic coating there evolves a concentration profile of hydrogen in which c_{H2} increases from c_0, the equilibrium concentration at 0.1 MPa H$_2$ at the pore mouth, towards the deeper parts of the pore or the coating to concentrations which might be by orders of magnitude higher (details are given below). This gives rise to concentration polarisation which renders those parts of the nanoporous electrode, where the concentration polarisation approaches the externally applied overpotential η, electrochemically inactive.

7.2.3.3
Modelling Hydrogen Concentration Profiles and Catalyst Efficiencies for Hydrogen Consuming Fuel Cell Anodes or Other Gas Diffusion Electrodes

Making use of the known dimensions of the small catalyst particles in gas diffusion electrodes whose diameters do not exceed a few micrometers and measure often in tenths of micrometers, allows to neglect ohmic potential drops in the interior of theses particles or pores thereof because the integral $\frac{1}{\kappa}\int i_{pore} dx$ does not exceed a few millivolt under realistic pore current densities. In a first approach the actual spherical form of the particle may be neglected considering a single pore of length L, the latter amounting to half the particle diameter L = 1/2 d_p = r_p. Accounting for the spherical or almost spherical form of the particle would change the result by a numerical factor of the order of one (3/4). The mass transport equation (Eq. 7.4) applied for the pore in one-dimensional form if radial diffusion in so narrow pores are neglected reads

$$D d^2 c / d x^2 = -r_v = i_{w,x} a_e / 2F = i_w / F r_p \tag{7.6}$$

Here $i_{w,x}$ is the actual current density on the wall of the pore.

7.2 Three-Dimensional Electrodes

At the mouth of the pore (x=0) the hydrogen concentration is kept at the equilibrium solubility c_0. The second boundary condition for Eq. (7.6) accounts that at the length x=L which means in the middle of the particle, the gradient $(dc/dx)_L$ is zero.

Describing the current voltage curve for hydrogen oxidation by the Butler Volmer equation (Eq. 7.7):

$$i_{w,x} = i_0 \left[\exp\{\alpha_a F\eta/RT\} - \exp\{-\alpha_e F\eta/RT\} \right] \tag{7.7}$$

and accounting for the concentration polarisation

$$\eta_{c,x} = \frac{RT}{2F} \ln c_x/c_0 \tag{7.8}$$

leads to the corrected Butler–Volmer equation (Eq. 7.9)

$$\begin{aligned} i_s = i_0 \big[&\exp\{\alpha_a F(\eta + RT/2F \ln c/c_0)/RT\} \\ &- \exp\{-(\alpha_c F(\eta + RT/2F \ln c/c_0))/RT\} \big] \end{aligned} \tag{7.9}$$

In order to demonstrate the consequence of Eqs. (7.6) and (7.9) on the concentration of hydrogen in the pore two oversimplified versions of Eq. (7.9) are used:

(a) for high overpotential neglecting the second term in Eq. (7.9) and neglecting the concentration-overpotential against the total overpotential transforms Eq. (7.9) to

$$i_s = i_0 \exp\{\alpha_a \eta F/RT\}^{2)} \tag{7.9 a}$$

(b) for low overpotentials by linearizing the Butler–Volmer equation, setting $\alpha_a + \alpha_c \approx 1$ and neglecting again the concentration overpotential vs the applied overpotential leads to the linear expression

$$i = i_0 F\eta/RT^{2)} \tag{7.9 b}$$

Assuming that hydrogen oxidation is a first order reactions leads to linear dependence of i_0 on c

$$i = i_0^* c \tag{7.10}$$

Introducing Eqs. (7.9 a,b) into Eq. (7.6) leads with $a_e = \dfrac{1}{r_p}$ for a pore to

$$d^2c/dx^2 = i_0^0 c_x \exp\{\alpha_a F\eta/RT\}/(DFr_p) \tag{7.10 a}$$

---

[2] High overpotential would be more appropriate for oxygen reduction and low overpotential would be typical for hydrogen oxidation in gas-diffusion electrodes

and respectively for the linear rate equation

$$d^2c/dx^2 = i_0^0 c_x F\eta/(RTDr_p) \qquad (7.10\text{ b})$$

The equations are of the general adimensional form

$$d^2(c_x/c_0)/d(x/L)^2 = (c_x/c_0)\Phi_{Th}^2 \qquad (7.10\text{ c})$$

L has the meaning of a characteristic length, for instance the pore depth. In reaction engineering modelling the utilisation of porous catalysts leads to the same type of equation with the Thiele modulus Φ_{Th} as the essential dimensionless quantity describing the utilisation of porous particles. By analogy the electrochemical Thiele modulus would read for the two cases

$$\Phi_{Th} = L(i_0 \exp\{\alpha_a \eta F/RT\}/DFr_p c_0)^{1/2} \qquad \text{(high overpotential)} \quad (7.11\text{ a})$$

and

$$\Phi_{Th} = L(i_0 \eta/RTDr_p c_0)^{1/2} \qquad \text{(low polarisation)} \quad (7.11\text{ b})$$

The solution of the differential equation (Eq. 7.10 c) leads to

$$c_x/c_0 = \cosh\{\Phi_{Th}(1-x/L)/\cosh \Phi_{Th}\} \qquad (7.12)$$

Integration of the differential current $di = i_x 2\pi r dx$ leads to the total pore current I_p for high polarisation

$$I_p = 2\pi r i_0 \lambda \exp\{\alpha_a F\eta/RT\} \tanh(L/\Phi_{Th}) \qquad (7.13)$$

with the penetration depth λ equalling

$$\lambda = (DFr_p c_0/i_0)^{1/2} \exp\{-\alpha_a F\eta/2RT\} = L/\Phi_{Th} \qquad (7.13\text{ a})$$

and for low polarisation this pore current would read

$$I_p = 2\pi r_p i_0 \lambda'(F\eta/RT) \tanh\{L/\Phi_{Th}\} \qquad (7.14\text{ a})$$

with the penetration depth λ' reading

$$\lambda' = (RTDr_p c_0/i_0 \eta)^{1/2} = L/\Phi_{Th} \qquad (7.14\text{ b})$$

The degree of utilisation of the pore, q, is defined by the ratio of actual pore current given in Eqs. (7.14 a,b) and pore currents calculated for full utilisation that means for an assumed equal current density at the pore walls in the whole pore.

7.2 Three-Dimensional Electrodes

The degree of utilisation, u, would therefore read

$$u = \frac{\lambda}{L} \tanh \frac{L}{\lambda} \tag{7.15}$$

Introducing the concentration polarisation in Eq. (7.9) leads to the explicit concentration dependence of the two terms in the Butler Volmer equation (Eq. 7.12)

$$i = i_0 \left[(c/c_0)^{\alpha_a/2} \exp\{\alpha_a F\eta/RT\} - (c/c_0)^{-\alpha_c/2} \exp\{-\alpha_c F\eta/RT\} \right] \tag{7.16}$$

An assumed concentration depletion in the depth of the pore by one half would lead to a numerical value of the concentration correction factor $(c/c_0)^{\alpha/2}$ of 0.7 and 1.4 respectively for $\alpha = 0.5$. Neglecting these factors would therefore cause a 30% overestimation of the anodic and 40% underestimation of the cathodic reaction. This is obviously the ultimate limit to which neglect of the concentration polarisation can be reasonably extended.

Applying the complete Butler Volmer equation but neglecting the $(c/c_0)^{\alpha/2}$ correction leads to

$$I_p = 2\pi r_p i_0 \lambda \left(\exp\{\alpha_a F\eta/RT\} - \exp\{-\alpha_c F\eta/RT\} \right) \tanh L/\lambda \tag{7.17}$$

with the penetration depth

$$\lambda = \left(Fr_p Dc_0 / 2i_0 \right)^{1/2} \exp\{-\alpha_a F\eta/2RT\} \tag{7.18}$$

Figure 7.4 depicts the calculated effective utilisation of Raney-nickel particles in dependence of the characteristic data, pore radius and length or depth, exchange current density and applied overpotential and Fig. 7.4 b shows in an experimental measurement of current voltage curves of Raney nickel anodes with equal catalyst loadings that particles which measure more than 10 µm in diameter are no longer fully utilized as evidenced by the strong increase of anodic overpotential.

Historically it is noteworthy, that the mass transfer problem for hydrogen in Raney nickel particles of fuel cell anodes was solved for the first time by Mund (in the early seventies) with the assumed kinetic equation [1]

$$i_x = i_0 \left[c_x / c_0 \right]^{\alpha/2} \exp\{\alpha \eta F/RT\} - \exp\{-(1-\alpha)F\eta/RT\} \tag{7.19}$$

which yields the penetration depths

$$\lambda = \left(v_e Dr_p Fc_0 / 2i_0 \right)^{1/2} \exp\{-\alpha F\eta/RT\} \tag{7.19 b}$$

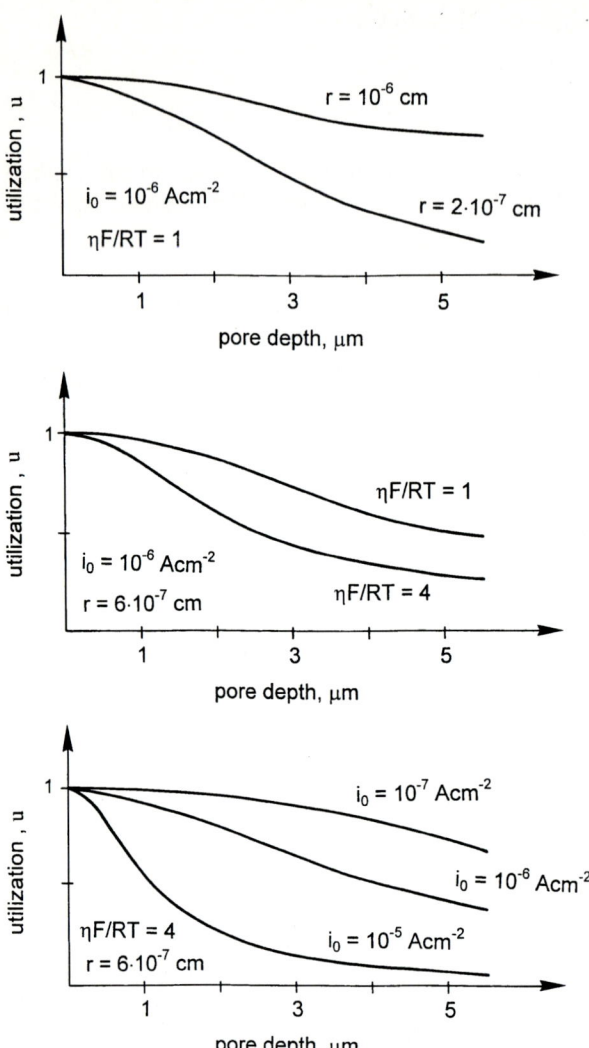

Fig. 7.4 a. Calculated effective utilisation of Raney-nickel H_2-anode particles in dependence on pore radius, length, exchange current density and applied overpotential according to K. Mund and F. v. Sturm Electrochim Acta *20* (1975) 463.

and again the degree of utilisation u

$$u = \lambda/L \tanh L/\lambda \qquad (7.19\ c)$$

Applying the same considerations which were applied to single pores to nanoporous spherical particles that means extending the model from a single pore to a three-dimensional catalyst granule of homogeneous nanoporosity does not change the result substantially. Because of the spherical geometry and corre-

7.2 Three-Dimensional Electrodes

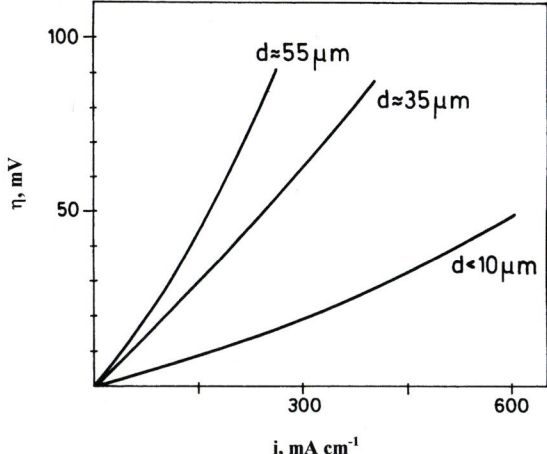

Fig. 7.4 b Demonstration of particle diameter or pore length effect on catalyst utilisation by current potential curves of anodic hydrogen oxidation with anodes composed of Raney-nickel particles of different particle size but equal gravimetric catalyst load

spondingly concentric diffusion, catalyst particles are better utilised than a singular pore. The utilisation for spherical particles is given by

$$u_{particle} = 3/\Phi_{Th} \cdot \left(1/\tanh\Phi_{Th} - 1/\Phi_{Th}\right) \tag{7.20}$$

and the quantities L and r_p in Eqs. (7.11 a,b) are exchanged by d_p and by $2a_e^{-1}$ respectively with a_e being the volume specific internal surface of the particle and d_p the particle diameter. Also the diffusion coefficient D has to be changed accounting for porosity ε and the tortuosity factor f_t by introducing the effective diffusion coefficient, D^*:

$$\Phi_{Th,sph} = d_p\left(i_0 \exp\{\alpha_a F\eta/RT\}/D^* F a_e^{-1} c_0\right)^{1/2} \quad \text{(high overpotential)} \tag{7.20 a}$$

$$\Phi_{Th,sph} = d_p\left(i_0 F\eta/RTD^* a_e^{-1} c_0\right)^{1/2} \quad \text{(low overpotential)} \tag{7.20 b}$$

$$D^* = D\varepsilon f_t \tag{7.21}$$

7.2.3.4
Modelling of Hydrogen Concentration Profiles and Catalyst Efficiencies for Hydrogen Evolving Nanoporous Raney nickel Catalyst Coatings

As in the case of gas consuming nanoporous catalyst particles the following treatment of nanoporous catalyst particles embedded in a catalytically inert microporous matrix, so called composite coatings, and that of cracked and fissured or closed nanoporous catalyst coatings, so called smooth coatings confines mainly to the model of single nanopores. Anticipating the result, that penetration dephth of the current causing evolution of dissolved gases in nanopores of catalytically active coatings of gas evolving electrodes usually is confined to several tens of micrometers, allows to neglect ohmic potential drops in nanopores of gas evolving coatings as well as it was allowed to neglect this effect in nanoporous particles of gas consuming fuel cell electrode particles.

The utilisation of Raney-nickel coatings – smooth or composite – can be limited by concentration polarisation. As in a narrow pore the overpressure Δp, which is given by the Young–Laplace equation (Eq. 7.22) would have to be overcome in order to initiate bubble formation:

$$\Delta p = \frac{2\sigma \cos\Theta}{r_{pore}} \tag{7.22}$$

(Θ=wetting angle, σ=surface tension of the electrolyte)

Δp for hydrogen in aqueous electrolyte solutions according to Eq. (7.21) would exceed 10 MPa in a pore with 3 mm pore diameter, which would mean that an increase of the hydrogen concentration from roughly 10^{-3} to $1\,mol\,dm^{-3}$ would be possible.

Unless hydrogen concentration in the nanopore electrolyte does not exceed the concentration corresponding to the maximal overpressure of Eq. (7.22), the generated and dissolved hydrogen can only leave the pore by diffusion. Therefore accumulation of dissolved hydrogen in the pore at depth, x, reduces there the cathodic overpotential due to concentration polarisation

$$\eta_{eff,x} = \eta_{ext} + RT/2F \ln\left(c_{H_2,x}/c_{H_2,0}\right) \tag{7.23}$$

with $c_{H_2,0}$ the concentration of dissolved hydrogen outside the pore as defined by Henry's law and the working pressure of the electrolyzer.

Figure 7.5 depicts schematically under (a, b, and c) the pore of length L and the increase of hydrogen concentration and decrease of the effective overpotential with increasing pore depth. A quantitative treatment of this situation allows also in this case of hydrogen generation in a nanoporous matrix to define the electrochemical Thiele modulus as relevant dimensionless parameter which determines the utilization of such concentration polarised pore.

The kinetic equation for hydrogen evolution is chosen arbitrarily in a simplified assumption in order to facilitate the numerical mathematics. But the consequence of concentration polarisation on catalyst utilisation can be demonstrated irrespective of the kind of assumed kinetic law provided it implies potential dependent forward and backward rate which match at equilibrium potential. The simplified rate equation is chosen in order to demonstrate the effect of concentration polarisation more clearly than would do the more extended rate equation describing the Volmer–Heyrovski mechanism. The concentration distribution of hydrogen along the pore axis, x, is determined by the current density $i_{s,x}$ at the pore wall at position x from the pore mouth:

7.2 Three-Dimensional Electrodes

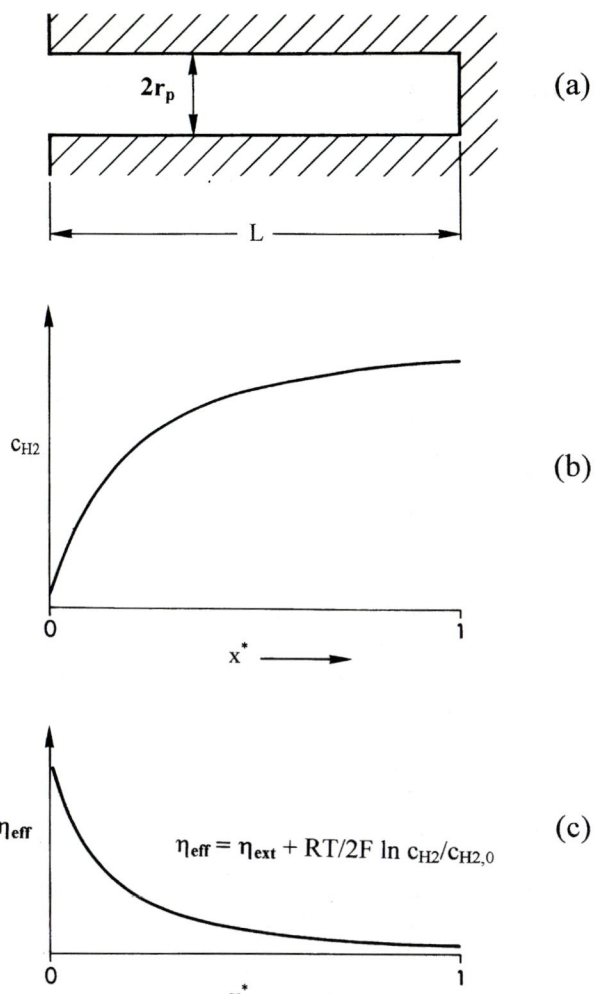

Fig. 7.5. Hydrogen concentration polarisation in H_2-evolving pore: **a** schematic of straight pore; **b** increase of hydrogen concentration with pore depth; **c** decrease of the effective overpotential with increasing pore depth

$$\left(\partial c^2/\partial x^2\right)_x = i_{w,x}/r_p FD \tag{7.24}$$

with r_p the pore radius.

$$\left(\partial c/\partial x\right)_x = i_{p,x}/2DFr_p = \frac{\int\limits_\infty^x i_{w,x}dx}{DFr_p} \tag{7.25}$$

with $i_{p,x}$ the current density in the pore – not at the pore wall – at x. Neglecting ohmic potential drops and applying a simplified rate law

$$i = i_0 \left(\exp\{\alpha_a \eta F/RT\} - \exp\{-\alpha_c \eta F/RT\} \right) \quad (7.26\,a)$$

calculating the current density at the pore wall at the location x one obtains under consideration of the effective overpotential by introducing Eq. (7.22) the local current density

$$i_{w,x} = i_0 \left[(c_x/c_o)^{0.25} \exp\left\{\frac{\alpha_a F \eta_{ext}}{RT}\right\} - (c_x/c_o)^{-0.25} \exp\left\{\frac{-\alpha_c F \eta_{ext}}{RT}\right\} \right] \quad (7.26\,b)$$

By introducing adimensional quantities $c^* = c/c_0$, and $x^* = x/L$ with L equaling the pore length one can extract from Eq. (7.23) the adimensional quantity

$$\Phi_{Th}^0 = L \left(\frac{i_0}{r_p F D c_0} \right)^{1/2} \quad (7.27)$$

which we call the normalised electrochemical Thiele modulus in contrast to the electrochemical Thiele modulus, Eq. (7.27) which is Φ_{Th}^0 times the potential dependent Butler–Volmer term:

$$\Phi_{Th} = \Phi_{Th}^0 \left(\exp\left\{\frac{\alpha_a F \eta}{RT}\right\} - \exp\left\{\frac{-\alpha_c F \eta}{RT}\right\} \right) \quad (7.28)$$

With the reduced hydrogen concentration, c^*, $(c_x/c_0 = c^*)$ and the reduced distance from the pore mouth, x^*, ($x/L = x^*$) and with $\alpha_a = \alpha_c = 0.5$ one obtains from Eq. (7.24) for 20° C numerically the adimensional differential equation:

$$\partial^2 c^*/\partial x^{*2} = \left(\Phi_{Th}^0\right)^2 \left[10^{\eta/120mV} (c^*)^{0.25} - 10^{-\eta/120mV} (c^*)^{-0.25} \right] \quad (7.29)$$

This equation cannot be solved in closed form but only numerically for different overpotentials and different values of the normalized Thiele modulus with the boundary conditions

$$c^*_{x^*=0} = 1, (\partial c^*/\partial x^*)_{x^*=1} = 0$$

and

$$c^*_{x^*=1} \leq \exp\left\{\frac{-\eta F}{2RT}\right\} \quad (7\,b)$$

7.2 Three-Dimensional Electrodes

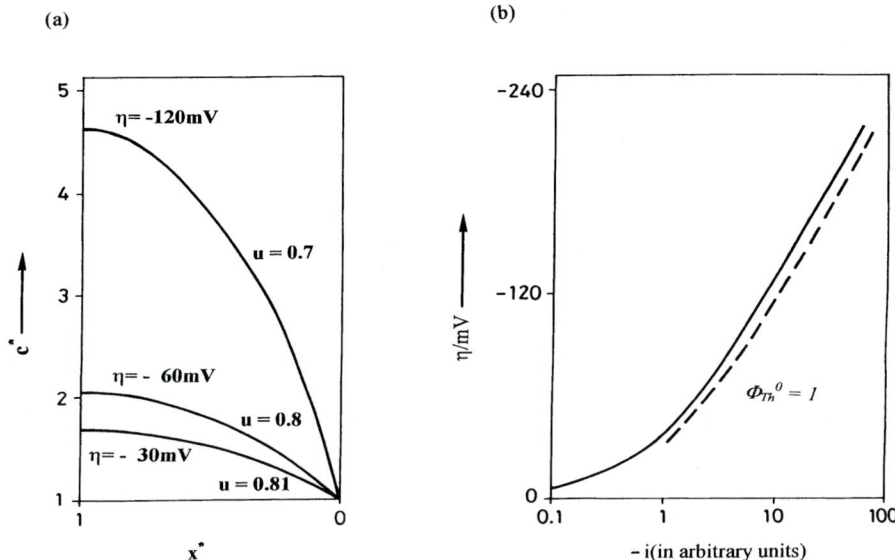

Fig. 7.6 a. Increase of hydrogen concentration with pore depth for three different overpotentials for a normalised Thiele modulus $\Phi_{Th}^0 = 1$. **b** Pore current potential relation calculated therefrom; current density is in arbitrary units. This picture would be representative for, e.g. 1 µm coatings or 2 µm particles of Raney nickel with 2 nm diameter pores and $i_0 = 10^{-4}$ A cm^{-2}. Concentrations are normalised with respect to saturation concentration at the pore mouth ($c_0^* = 1$); x^* is the adimensional pore depth $x^* = x/L$. At $x^* = 0$ is the pore mouth. Dotted line is calculated with assumed 100% utilization

and serves to calculate the pore current density at the pore mouth according to Eq. (7.24) for $x^* = 0$, i.e. it allows recalculating current voltage curves of a single pore for different parameters of Φ_{Th}^0.

Figure 7.6 a shows for $\Phi_{Th}^0 = 1$ the concentration distribution $c^*(r)$ at different overpotential and Figure 7.6 b shows the corresponding pore current density – overpotential correlation. $\Phi_{Th}^0 = 1$ is a value approximately typical of 2 µm Raney nickel particles in composite coatings or smooth Raney nickel coatings of 1 µm thickness with 2 nm pores if i_0 is set to approximately 10^{-4} A cm^{-2} – which is typical for the hydrogen evolution reaction at nickel at 70 °C and if c_0 is $0.3 \cdot 10^{-6}$ mol cm^{-3}. Figure 7.6 a shows that in the interior of the pore of such relatively short pores the concentration of hydrogen is far below the concentration, c^*_{equil}, dictated by the applied overpotential (c^*_{equil} would be 10, 10^2 and 10^4 for η equalling –30, –60 and –120 mV). Therefore the utilisation is not much lower than 1, namely 81, 80 and 70% at these three different overpotentials and the current voltage curve does not deviate much from that calculated with neglect of concentration polarisation (dotted line Fig. 7.6 b).

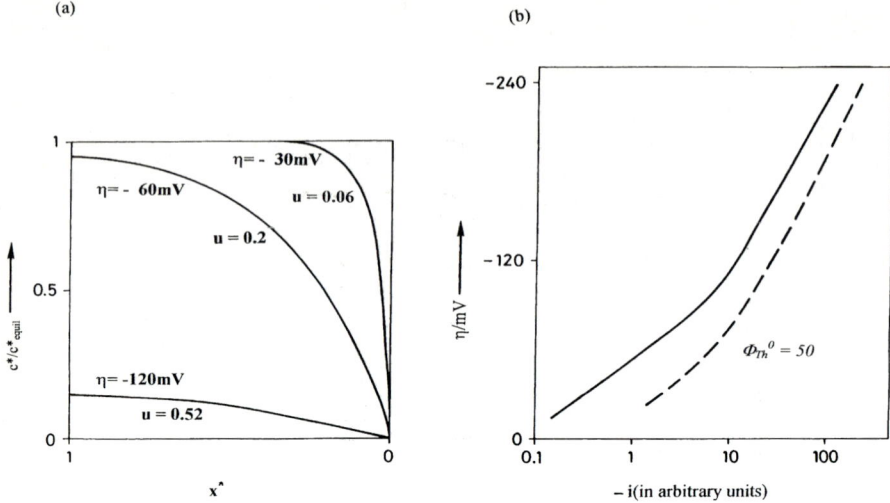

Fig. 7.7 a. Increase of hydrogen concentration with pore depth for three different overpotentials (−30, −60 and −120 mV) for normalised Thiele modulus $\Phi_{Th}^0 = 50$. **b** Pore current potential correlation calculated from these data. This example would be representative for a Raney nickel coating of 50 μm thickness. The dotted line is calculated with assumed full utilization

The utilisation is defined by Eq. (7.30):

$$u = L^{-1} \int_0^L i_{w,x} \, dx / i_{w,0} \tag{7.30}$$

Figure 7.7 a depicts for a normalised electrochemical Thiele modulus of 50 – for instance for a straight 2 nm pore in a smooth Raney nickel coating of 50 μm thickness – the concentration distribution in terms of the relative concentration c^*/c^*_{equil} from the bottom of the pore ($x^* = 1$) to its mouth $x^* = 0$ for three different overpotentials (−30, −60 and −120 mV). c^*_{equil} varies over four orders of magnitude, the concentration c^*_{equil} being determined by the applied overpotential:

$$c^*_{equil} = 10^{(-\eta/30 mV)} \tag{7.31}$$

The difference to Fig. 7.6 a is striking: At −30 mV almost complete equilibrium is established along more than three quarters of the pore depth. Correspondingly the utilisation drops to only 6% at −30 mV, and rises to 20, 52 and 60% at −60, −120 and −240 mV respectively. This creates a typical deviation of the semilogarithmic current voltage curve from the usual straight line with a slope of 120 mV per decade. In Fig. 7.7 b straight segment of lower Tafel slope, which covers almost two orders of magnitude of the current density with a slope of approx. −60 mV dec^{-1} is calculated. For higher values of the normalised Thiele modulus this behaviour is even more expressed, the flatter part of the current voltage curve extending to still higher current densities and overpotentials.

So far these model calculations seem to explain at least partially the observation with smooth, but relatively thick Raney nickel coatings, that semilogarithmic current-voltage plots measured for hydrogen evolution exhibit reduced slopes of approximately −70 mV per decade over at least two orders of magnitudes of current density. The model, however, would fail at overpotentials in excess of −90 to −100 mV, because H_2 concentration would exceed the limit of 1000 bars, hydrogen bubbles would be formed in 2 nm pores, pushing out the electrolyte. Would this block the pore permanently? Certainly not. As soon as the pressure has decreased by, let say, a factor of ten by release of electrolyte and gas, the electrolyte would be sucked back into the pore redissolving the residual gas and filling the pore again. So at higher overpotentials we would expect a periodic H_2 resaturation of the pore-electrolyte, gas bubble formation with subsequent electrolyte and gas ejection and electrolyte replacement resulting in lower than predicted concentration polarisation of the pore on the time average. In Fig. 7.7 b this would result in a moderate extension of the flat part of the current voltage curve to still higher current densities.

7.2.4
Porous Battery Electrodes

Battery electrodes possess relatively coarse pores with pore diameters between 10 and 100 μm and are relatively thick – from 1 to several mm.

The electrochemical reactions in conventional batteries convert solid electrode materials into solid reaction products as given in Eqs. (7.32) and (7.33) for the lead-acid battery and in Eqs. (7.34) and (7.35) for the nickel–cadmium battery respectively:

$$Pb + H_2SO_4 \rightarrow PbSO_4 + 2H^+ + 2e^- \qquad \text{(anode)} \quad (7.32)$$

$$PbO_2 + H_2SO_4 + 2H^+ + 2e^- \rightarrow PbSO_4 + 2H_2O \qquad \text{(cathode)} \quad (7.33)$$

$$Cd + 2NaOH \rightarrow CdO + H_2O + 2Na^+ + 2e^- \qquad \text{(anode)} \quad (7.34)$$

$$2NiOOH + 2Na^+ + 2e^- \rightarrow 2NiO + 2NaOH \qquad \text{(cathode)} \quad (7.35)$$

The electrolyte is involved in these reaction which leads to depletion of sulfuric acid on discharge of the lead acid battery and to accumulation of sodium hydroxide in the cathode and its depletion in the anode during discharge of the nickel–cadmium battery.

But since the electrolyte concentration is high ($c > 5$ mol dm^{-3}) the relative concentration changes $\Delta c/c$ may be neglected and consequently also concentration polarisation in the electrode and electrode pore respectively. Current density distributions in operating battery electrodes are therefore mainly determined by ohmic potential drops.

If, as it is typical for battery electrodes, the conductivity of the electrode matrix which often consists of a metal sponge or a porous electronically relatively well conducting metal oxide (e.g. PbO_2, NiOOH etc.) is by orders of magnitude higher than that of the electrolyte, it may only contribute insignificantly to Ohmic voltage drops. One obtains from Eqs. (7.1 c), (7.2 a) for a narrow pore ($l \gg r_p$) under these conditions the one-dimensional differential equation:

$$d^2\Delta\varphi/dx^2 = d^2\eta/dx^2 = \left(1/\kappa_{elyte}\right)a_e i_s \tag{7.36}$$

Inserting the Butler Volmer equation, which may be applied in linearized form, as overpotentials in batteries are usually low, one obtains for a cylindrical pore with $a_e = 2/r_p$:

$$d^2\eta/dx^2 = 2i_0(\alpha_a + \alpha_c) F\, \eta/\left(\kappa_{elyte} r_p RT\right) \tag{7.37}$$

If the pore length is L the boundary conditions are

$$\eta_{x=0} = \eta_0 \tag{7 c}$$

$$(d\eta/dx)_L = 0 \tag{7 d}$$

wall current density and potential are changing with pore depth as in the case of nanoporous fuel cell electrode particles according to the cosh function:

$$\eta_x = \eta_0 \cosh\{\lambda - 1(L-x)/\cosh L/\lambda\} \tag{7.38}$$

$$i_{w,x} = \{i_0(\alpha_a + \alpha_c)F/RT\}_0 \cosh\{L/(L-x)\}/\cosh L/\lambda \tag{7 e}$$

with the penetration depth λ defined as

$$\lambda = \{\kappa_1 r_p RT/2i_0(\alpha_a + \alpha_c)F\}^{0.5} \tag{7.39}$$

The pore current density at the pore mouth for electrodes of virtually infinite thickness is

$$(i_p)_{x=0} = \frac{2i_0(\alpha_a + \alpha_c)F\lambda}{r_p RT}\eta_0 \tag{7.40}$$

7.2 Three-Dimensional Electrodes

With the slope of the current voltage curves:

$$(di_p/d\eta)_0 = \frac{2i_0(\alpha_a + \alpha_c)F\lambda}{r_p RT} \tag{7.41}$$

$(i_p)_{x=0}$ and $(di_p/d\eta)_0$ differ by the factor $2\lambda/r_p$ from the respective values of a smooth flat electrode.

7.2.5
Packed Bed and Fluidized Bed Electrodes Composed of Coarse Particles

As will be pointed out below in chapter 9, an important economic parameter correlated with the specific investment costs is the spacetime yield ρ of an electrochemical cell which is given by Eq. (7.42):

$$\rho = a_e \frac{i\phi^e M}{v_e F} \tag{7.42}$$

In all cases where solutions with a low depolarizer concentration have to be treated, a low mass-transfer-limited current density results which requires the application of a large volume-specific electrode area in order to keep-space-time yields high. This may be achieved in a tolerably small volume by applying a three-dimensional electrode structure. Typical examples are the packed-bed electrolyzer for purification of metal containing waste water, or the fluidized-bed electrolyzer for metal winning from dilute solutions.

Considering a packed or fluidized bed of conducting solid, compact particles acting as a whole as electrode, different geometrical arrangements of the electrolysis cell may be distinguished. Three of practical relevance are given together with their basic differential equations in Fig. 7.8 and Table 7.1. There are cubic (A), (B) and concentric (C) arrangements. Flow-by systems (A) are characterised by parallel electrolyte and current-flow directions whereas in flow-through systems (B,C) current and electrolyte flow directions are perpendicular to each other. Integration of the second order differential equation systems in Table 7.1 require four boundary conditions which are given in Table 7.1. For a diffusion controlled reaction under limiting current density polarisation where the limiting current density $i_{x,l}$ at point x in the bed prevails (Eq. 7.43).

$$i_{x,l} = k_m v_e F c_x \tag{7.43}$$

the differential equation system in Table 7.1 b can be integrated analytically resulting in the local potential of the electrolyte phase, φ_{elyte} and the particle phase φ_{elode}:

$$\varphi_{x,\,elyte} = \frac{a_e i_l}{\kappa_{elyte}}\left(hx - \frac{x^2}{2}\right) \tag{7.44}$$

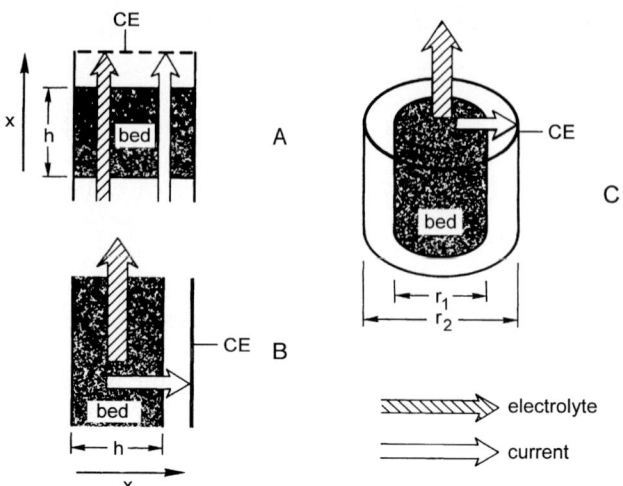

Fig. 7.8. a,b,c Various geometries of porous flow through electrodes. Basic differential equations yielding the current and potential distributions in these three-dimensional electrodes and respective boundary conditions, see Table 7.1

Table 7.1. Basic equations and boundary conditions determining potential distributions in packed bed electrodes

	basic differential equations	definitions	boundary conditions
A	$\dfrac{d^2\varphi_{elyte}}{dx^2} = \dfrac{a_e i}{\kappa_{elyte}}$		$\varphi_{elyte,0}=0$ $\varphi_{elode,h}=\eta_0$
B	$\dfrac{d^2\varphi_{elode}}{dx^2} = \dfrac{a_e i}{\kappa_{elode}}$ $i=i[\eta,x]$	$\eta_x = \varphi_{elode,x} - \varphi_{elyte,x}$ $\kappa_{elyte} = \kappa_{elyte,0}\varepsilon^n$	$\dfrac{d\varphi_{elyte,h}}{dx}=0$ $\dfrac{d\varphi_{elode,0}}{dx}=0$
C	$\dfrac{d^2\varphi_{elyte}}{dr^2} - \dfrac{1}{r}\dfrac{d\varphi_{elyte}}{dr} = -\dfrac{a_e i}{\kappa_{elyte}}$ $\dfrac{d^2\varphi_{elode}}{dr^2} - \dfrac{1}{r}\dfrac{d\varphi_{elode}}{dr} = \dfrac{a_e i}{\kappa_{elode}}$ $i=i[\eta,r]$	$\kappa_{elode} = \kappa_{elode,0}(1-\varepsilon)^n$ $i_{bed} = a_e \displaystyle\int_0^h i\,dx$	$\varphi_{elyte,r2}=0$ $\varphi_{elode,r1}=\eta_0$ $\dfrac{d\varphi_{elyte,r_1}}{dr}=0$ $\dfrac{d\varphi_{elode,r}}{dr}=0$

7.2 Three-Dimensional Electrodes

$$\varphi_{x,\,elode} = \frac{a_e i_l}{\kappa_{elode}}\left(\frac{h^2}{2} - \frac{x^2}{2}\right) \tag{7.45}$$

$$\eta_x = \eta_0 - a_e i_l \left(\frac{h^2}{2\kappa_{elode}} + \frac{hx}{\kappa_{elyte}} - \frac{x^2}{2\kappa_{elode}} + \frac{x^2}{2\kappa_{elyte}}\right) \tag{7.46}$$

x represents the direction from the counter electrode side of the bed electrode towards the contact electrode. The penetration depth, h, of the limiting current density in x direction depends on the ratio between effective particle and solution conductivity and is given as

$$\frac{\kappa_{elode}}{\kappa_{elyte}} \geq 1; \quad h = \left[\frac{2\Delta\eta}{a_e i_l \left(\kappa_{elyte}^{-1} - \left(\kappa_{elode} + \kappa_{elyte}\right)^{-1}\right)}\right]^{0.5} \tag{7.47}$$

$$\frac{\kappa_{elode}}{\kappa_{elyte}} \leq 1; \quad h = \left[\frac{2\Delta\eta}{a_e i_l \left(\kappa_{elode}^{-1} - \left(\kappa_{elode} + \kappa_{elyte}\right)^{-1}\right)}\right]^{0.5} \tag{7.48}$$

$$\frac{\kappa_{elode}}{\kappa_{elyte}} \gg 1; \quad h = \left[\frac{2\kappa_{elyte}\Delta\eta}{a_e i_l}\right]^{0.5} \tag{7.49}$$

Where $\Delta\eta$ is the range of over potential along which the mass transfer limited current density prevails. A comparison of polarisation curves for cathodic silver deposition (10^{-3} mol dm^{-3} solution) at a planar electrode with parallel electrolyte flow and a packed bed electrode consisting of silver beads of 1 mm diameter is given in Fig. 7.9 a and b. The shape of the polarisation curve at the planar electrode is typical for a diffusion controlled reaction with a limiting current density depending on the flow rate. $\Delta\eta$ equals in this case at least 300 mV. The macrokinetic polarisation curve at the bed electrode shows current densities 100 times higher than at the planar electrode. A factor of 20 is due to the increased effective electrode area of the particle bed and a further factor of 5 is due to mass transfer enhancement in the bed electrode. The diffusion limited current density region in the bed electrode is approached at higher potentials than at the planar electrode. This is due to the potential distributions in the bed which may be calculated by Eqs. (7.44), (7.45) or (7.46). Examples are given schematically in Fig. 7.10 a,b for a packed and a fluidized-bed electrode. The local overvoltage is given as the difference between the particle and the solution potential. A limiting current-den-

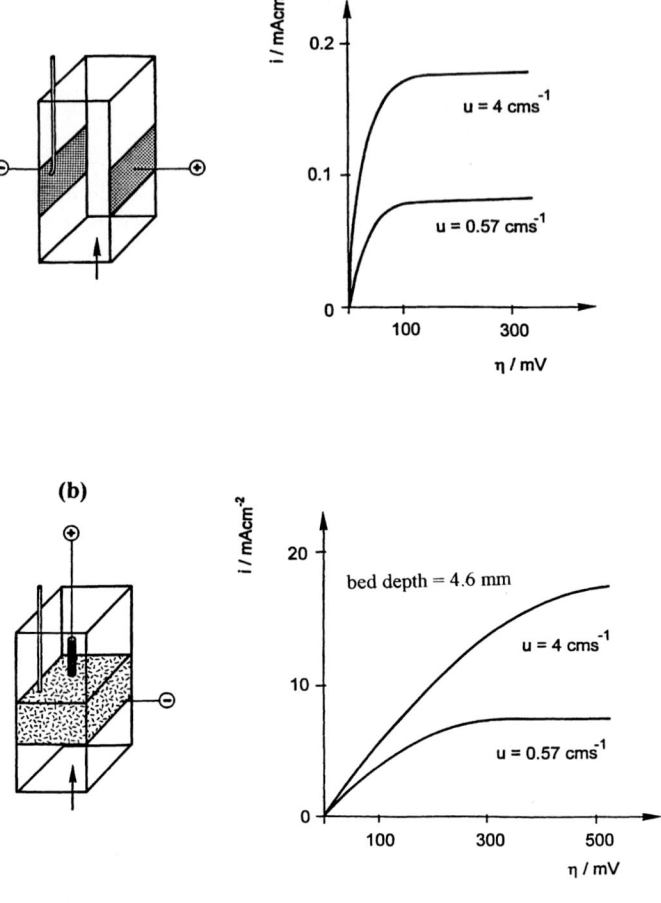

Fig. 7.9 a,b. Current potential curves of silver deposition at: **a** planar; **b** packed bed electrode

sity behaviour of a packed-bed electrode can be only established if the potential at the contact electrode (x = l) is higher than 120 mV, see Fig. 7.9 a. Under this condition the potential at x = 0 representing the bed-electrode overvoltage η_0 is already remarkably larger. This explains the fact that at a bed electrode the potential of the macrokinetic limiting current density region is shifted towards higher values. This effect is the more pronounced the higher the local microkinetic current density, which increases with increasing flow rate and depolarizer concentration. Comparison between Fig. 7.10 a,b shows that the overpotential $\varphi_{elode}-\varphi_{elyte}$ distribution is more uniform if the particle conductivity is of the same order of magnitude as the solution conductivity which is usually the case for the fluidized-bed electrode.

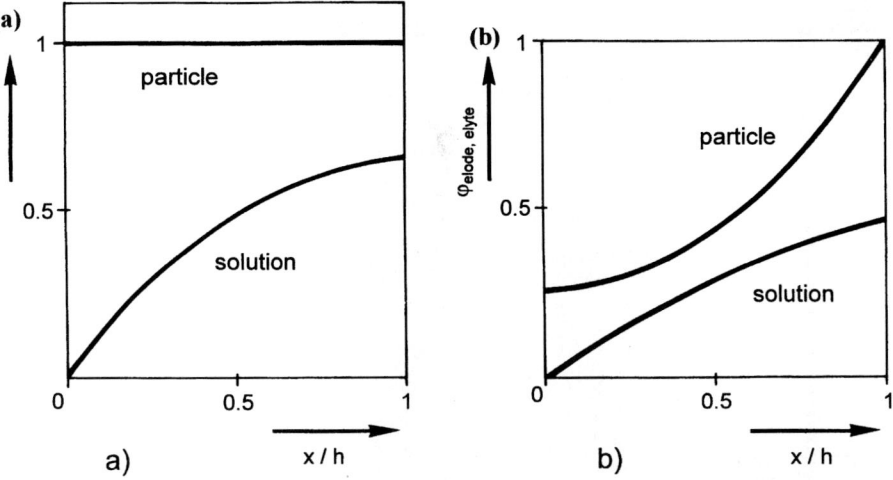

Fig. 7.10 a,b. Schematic presentation of potential distributions in the electrolyte and the particle phase: **a** packed bed; **b** fluidized bed or packed bed with lower bed conductivity respectively

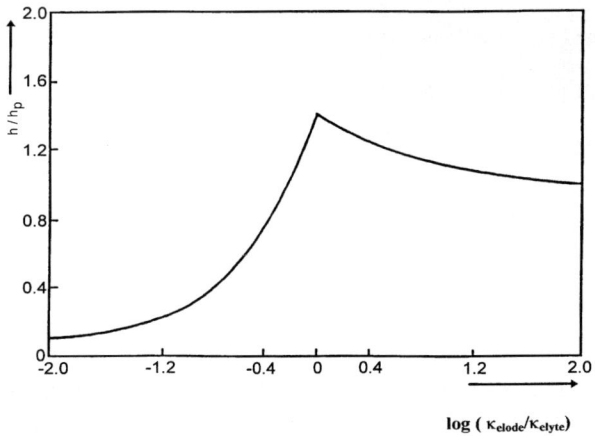

Fig. 7.11. Effective bed height (related to bed height of highly conductive bed) against conductivity ratio in bed electrodes, operated at mass transfer limited current densities

The space time yield of a three-dimensional electrode should be the higher the larger the penetration depth of the limiting current density is. The ratio of the bed depth of a bed with limited electrode conductivity related to the effective bed depth of a packed bed with infinite electrode conductivity, h/h_p, calculated by Eqs. (7.48) and (7.49) is shown in Fig. 7.11 as function of the logarithmic conductivity ratio. A maximum of the penetration depths ratio of 1.44 results if $\kappa_{elode} = \kappa_{elyte}$.

Fig. 7.12. Photography of a packed-bed electrode composed of carbon particles. The bed expands in flow direction (upwards) because the bed depth extends with decreasing concentration of the depolarizer (Ag^+ in this case). The counter electrode is at the left hand side.

The penetration depth of the current into a packed-bed electrode of graphite particles can be visualised by silver deposition from diluted silver-nitrate solutions. In Fig. 7.12 a graphite-packed bed electrode is shown into which silver was deposited from solutions of different concentration. Electrolyte flux in the bed proceeds from the bottom to the top. With decreasing concentration the limiting current density decreases resulting in a higher penetration depth According to Eqs. (7.47) to (7.49). Therefore the bed widens from bottom to top.

7.2.5.1
Fluidized Bed Electrodes

Fluidized bed electrolysis is used today for metal winning from moderately dilute solutions containing only some g dm^{-3} metal e.g. copper II ions and where

7.3 Ionic Conductivity of Electrolytes Containing Dispersed Gas Bubbles in Gas-Evolving Electrolyzers

Electrochemical gas evolution plays an important role in e.g. the chloralkali electrolysis, water electrolysis, and as anodic oxygen evolution in metal winning electrolysis. The main effect of the electrochemically evolved gases is that the effective resistivity of the gas/electrolyte mixture close to the electrode surface is increased by presence of the gas bubbles. This causes an additional IR-drop across the electrolyte gap of the cell resulting in an increased cell voltage and specific energy consumption. Since for large scale electrochemical products like chlorine the total costs are determined up to about 60% by the energy costs this problem is of some economic importance. Several theoretical models have been derived describing the conductivity ratio (effective conductivity κ of the gas/liquid mixture related to the conductivity of the gas-free electrolyte κ_0) as function of the gas voidage ε_g. The different relations available are summarised in

Table 7.2. Theoretical conductivity relations of gas/electrolyte mixtures

Relative conductivity equation	Author
$\dfrac{\kappa}{\kappa_0} = (1-\varepsilon_g)$	–
$\dfrac{\kappa}{\kappa_0} = \dfrac{1-\varepsilon_g}{1+\varepsilon_g/2}$	Maxwell
$\dfrac{\kappa}{\kappa_0} = \dfrac{8(2-\varepsilon_g)(1-\varepsilon_g)}{(4+\varepsilon_g)(4-\varepsilon_g)}$	Tobias
$\dfrac{\kappa}{\kappa_0} = (1-\varepsilon_g)^{3/2}$	Bruggemann
$\dfrac{\kappa}{\kappa_0} = \dfrac{1-0.5\varepsilon_g(3-\varepsilon_g)}{1-3/2\varepsilon_g+0.5\varepsilon_g}$	Prager

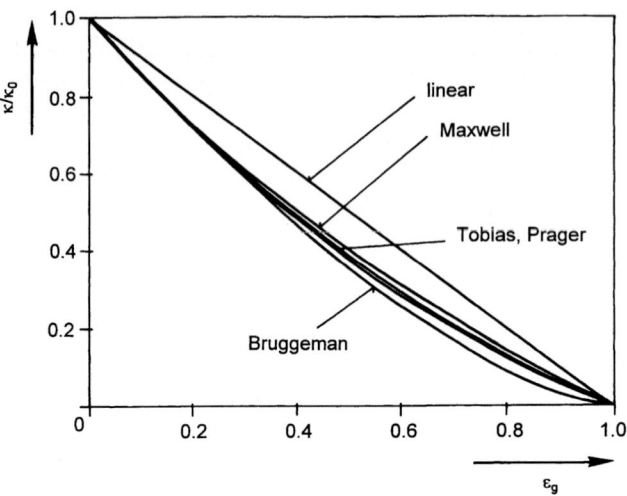

Fig. 7.13. Conductivity ratio κ/κ_0 of gas/liquid emulsions as function of gas voidage according to the different correlations

Table 7.2. A graphic representation of these relations showing the conductivity ratio as function of the gas voidage is given in Fig. 7.13. Apart from the too simple linear model the deviations between the different models are negligible from the engineering point of view within the range of practical gas voidages. The main problem for calculating the electrolyte resistivity as function of the operation parameters of an electrolysis cell is the reliable estimation of the gas voidage under operating conditions in an electrolyzer. A physically reasonable relation for modelling of gas evolving electrolysis cells has been derived by Nicklin for liquid/gas two-phase flow.

$$\varepsilon_g = \left[1 + \frac{w_{0,l} + w_r}{w_{0,g}}\right]^{-1} \tag{7.50}$$

The physical basis of this equation is the idea that the actual gas-bubble rising velocity, w_b, results as a superposition of the empty tube liquid velocity, $w_{0,l}$ the empty-tube gas velocity, $w_{0,g}$, and the relative swarm-rising velocity, w_r, of the gas bubbles.

$$w_b = \frac{w_{0,g}}{\varepsilon_g} = w_{0,l} + w_{0,g} + w_r \tag{7.51}$$

For calculation of the rising velocity of a bubble swarm, w_s, various relations are available. One of them giving reasonable agreement with experimental data in electrolyte solutions is the Marucci equation.

7.3 Ionic Conductivity of Electrolytes Containing...

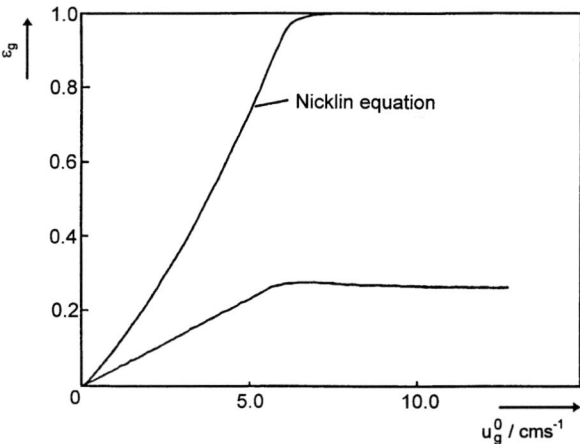

Fig. 7.14. Gas hold-up as function of empty tube gas velocity in water according to the prediction of Nicklin, Eq. (7.50) compared with experimentally obtained results

$$w_s \approx w_r = w_g \frac{(1-\varepsilon_g)^2}{1-\varepsilon^{5/3}} \tag{7.52}$$

w_s is the rising velocity of single gas bubbles which can be calculated be equating buoyant and friction forces.

Considering a steady state bubble column the liquid empty tube velocity is zero and the gas voidage is only a function of the empty-tube gas velocity. For high gas velocities the gas hold-up calculated by the Nicklin equation (7.50) approaches 1 as shown in Fig. 7.14. Contradictory to this predicted behaviour experimental results show a much smaller limiting gas hold-up. The same behaviour which is shown for air in water in Fig. 7.15 is confirmed for other gas/electrolyte combinations of electrochemical importance. The appearance of a limited maximum gas voidage – typically characteristic for each gas/electrolyte combination – may be understood in terms of a coalescence-barrier model. The gas liquid interface is charged due to the formation of an electrochemical double layer. This results in a certain electrostatic repulsion of the gas bubbles. Coalescence of two gas bubbles which is thermodynamically a spontaneous process can only occur if the kinetic energy of the moving gas bubbles is large enough to overcome the electrostatic coalescence barrier. By assuming a mean minimum bubble distance which is related to the strength of the electrostatic repulsion a modification of Nicklins considerations allows the formulation of a new gas-voidage relation.

$$\varepsilon_g = \varepsilon_m \left[\left\{ w_{0,l} \frac{\varepsilon_m - \varepsilon_g}{1-\varepsilon_g} + \varepsilon_m w_r \right\} \left(w_{o,g}\right)^{-1} \right]^{-1} \tag{7.53}$$

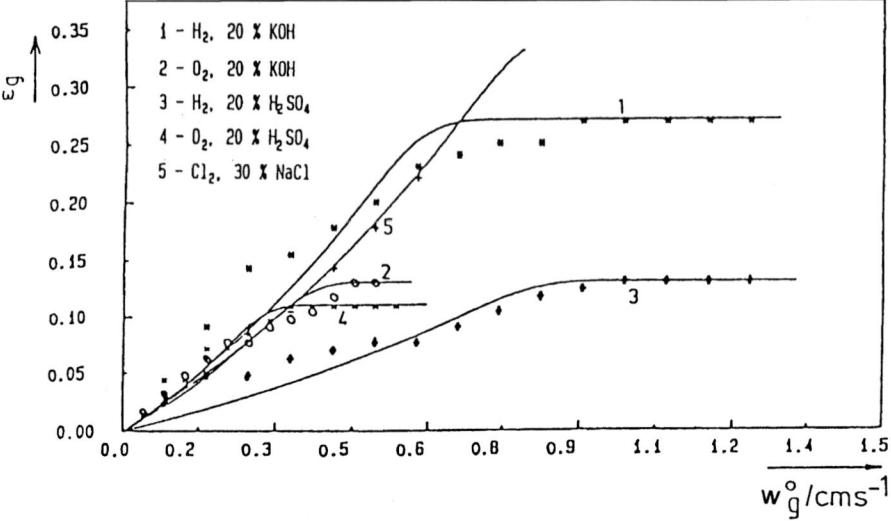

Fig. 7.15. Experimental points and according to Eq. (7.53) modelled gas voidage

The swarm-rising velocity w_r can be calculated by the Marucci equation at Eq. (7.52) but with the reduced gas voidage $\varepsilon_g/\varepsilon_m$ instead of ε_g.

Equation (7.53) represents an implicit equation which can be solved numerically. The full lines (a) and (b) in Fig. 7.15 are calculated with this coalescence barrier model equation. This model contains two empirical parameters for each gas electrolyte combination: the maximum gas voidage ε_m and the rising velocity of a single gas bubble w_s which have to be determined by fitting of experimental data.

An important factor determining the gas-voidage distribution in an electrolysis cell is the hydrodynamic boundary layer along and around the gas evolving electrode which is established by the rising gas bubbles. Flow-velocity distributions in front of various electrodes have been measured by laser-doppler-anemometry. As an example flow velocity distributions in front of a hydrogen and oxygen evolving planar vertical nickel electrode are shown in Fig. 7.16 as function of distance from the counter electrode and electrode height. Close to the electrode the vertical rising velocity component is continuously increasing with height. The flow velocity passes through a maximum and then decreases with increasing distance from the electrode finally resulting in a reverse flow at a larger distance. Under the conditions of these measurements where gas bubble diameters do not exceed tens of micrometers the gas bubble velocity relative to the electrolyte is negligibly small and the gas/electrolyte mixture may be considered as a fluid with a unique flow velocity and averaged density. Due to Bernoulli's equation the decreasing flow velocity with increasing distance from the electrode causes a static pressure gradient directed towards the electrode. This may

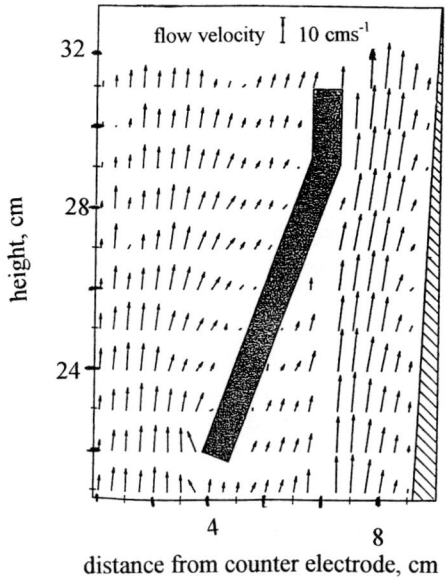

Fig. 7.16. Flow distribution near a gas evolving electrode. The counter electrode is at the left hand side. Superimposed up ward flow velocity of electrolyte · 10 cms^{-1}

explain the observation that in front of such electrodes a relatively thin gas bubble layer is formed which flows along the electrolyte like a "separate phase".

In order to avoid the negative effect of this bubble layer on cell resistance and cell voltage usually open electrode structures are applied like expanded metal sheet electrodes or lantern or Venetian-blind electrodes in order to allow that gas bubbles are diverted and transported behind the electrode. At the upper side of the lamina a flow parallel to the electrode surface is established transporting the gas bubbles behind the electrode where due to an increased gas voidage the bubbles are rapidly transported upwards. This effect limits the IR-drop due to gas evolution in industrial electrolysis cells to several tens of millivolts which is important with respect to energy saving.

7.4
Electrolyzers with Gaseous Reactants

In several industrial applications of electrochemical processes gaseous reactants which are processed in two-phase dispersions have to be converted in the electrolyte. Examples are the electrosynthesis of hexafluoropropylene oxide from gaseous hexafluoropropylene and also electrochemical gas purification. In the latter case waste gas components like chlorine or sulphur dioxide are absorbed in an electrolyte and are converted electrochemically to hydrochloric acid or sulphuric acid. In such cases the overall reaction may be considered as a sequence of an absorption step and a heterogeneous electrochemical conversion. Since under these conditions very often the absorption step is rate limiting the absorption rate has to be studied and its velocity increased by hydrodynamic measures. This non-electrochemical step of gas absorption can be easily studied by electrochemical methods.

A concentration-relaxation method allows to determine mass-transfer velocities in electrochemical reactors converting gaseous reactants. In a stirred agitated vessel a continuous gas stream of volumetric flow rate $\dot{V}_g$ containing oxygen is injected and dispersed by the stirrer. The vessel is equipped with electrodes and at the cathode the oxygen is reduced to water. The electrode is polarised to limiting current density conditions at $t=t_0$. Thereupon the electrolyte which hitherto was saturated with oxygen approaches a new steady state concentration of oxygen which is lowered due to steady electrochemical consumption of oxygen. The oxygen concentration transient being proportional to the current transient measurable at the electrode supplies the desired information of mass transfer rate of oxygen which is transferred from the dispersed gas bubbles into the electrolyte.

Equations (7.54)–(7.56) describe the respective volume specific mass balance during the time of the establishing relaxation of the equilibrium. c_l is the concentration of actually dissolved oxygen whereas c_{g0} and c_{g1} are the actual oxygen concentrations in the gas phase at the inlet (c_{g0}) and the outlet(c_{g1}) of the stirred tank. A_g/V_R is volume-specific gas–liquid interface across which the oxygen is dissolved with mass-transfer rate k_g into the electrolyte.

A_e/V_R is the volume related electrode surface of the electrolyzer and k_e is the mass transfer coefficient which rules the transport of dissolved oxygen towards the cathode surface (electrochemical mass-transfer rate). H is the Henry coefficient for oxygen.

$$\frac{dc_1}{dt} = \left(\frac{dc_1}{dt}\right)_{PT} + \left(\frac{dc_1}{dt}\right)_{EL} \tag{7.54}$$

$$\left(\frac{dc_1}{dt}\right)_{PT} = \frac{A_g k_g}{V_R}\left(\frac{c_{g,e}}{2H} - c_1\right) \tag{7.55}$$

$$\left(\frac{dc_1}{dt}\right)_{EC} = \frac{A_e k_e}{V_R} c_1 \tag{7.56}$$

The indexes $_{PT}$ and $_{EL}$ signify phase transfer and electrode. The unknown gas outlet concentration c_{g1} can be eliminated by the additional mass balance.

$$\left(\frac{dc_1}{dt}\right)_{PT} = \frac{\dot{V}_g(c_{g0} - c_{g1})}{V_R} \tag{7.57}$$

$\dot{V}_g$ is the volumetric flow rate of the gas

7.4 Electrolyzers with Gaseous Reactants

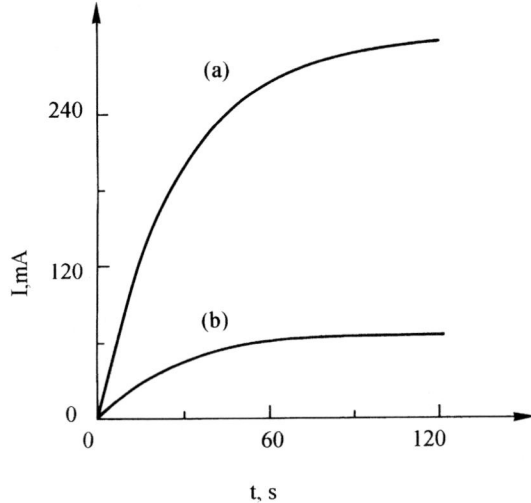

Fig. 7.17. Current transients measuring establishing steady-state oxygen concentrations in an agitated stirred tank reactor for two different oxygen inlet concentrations. The transient allows to determine the quantity ka. **a**: pure oxygen, **b**: air

The solution of the differential Eqs. (7.54)–(7.57) is

$$c_1(t) = c_1(t=0)\exp\{a_1 t\} - \frac{a_2}{a_1}(1 - \exp\{a_1 t\}) \tag{7.58}$$

with

$$a_1 = \frac{A_g k_g + A_e k_e + A_g k_g A_e k_e / 2H\dot{V}_g}{V_R + \frac{A_g k_g V_R}{2H\dot{V}_g}} \tag{7.59}$$

$$a_2 = \frac{A_g k_g c_{g0}}{HV_R \left(1 + \frac{A_g k_g}{2H\dot{V}_g}\right)} \tag{7.60}$$

The electrochemical mass-transfer parameter $A_e k_e$ can easily be determined by a separate experiment measuring the limiting current density of the conversion of any appropriate redox system under the same hydrodynamic conditions applying an inert gas stream. Then the only two unknown parameters of the transient solution of Eq. (7.58) are the volumetric mass-transfer coefficient $A_g k_g$ and the Henry coefficient H of the gas. The latter can also be determined independently. These two parameters can be obtained by a non-linear fitting procedure of experimental transient data. An example of two transients measured with different oxygen inlet concentrations is shown in Fig. 7.17. This is an example illustrating how electrochemical methods may contribute also to investigate non-electrochemical unit operations.

7.5
Electrochemical Liquid/Liquid Systems

Another type of industrial electrochemical reactions is carried out in e.g. an electrolyte containing a dispersed second liquid phase. One step of the PUREX process for reprocessing of spent nuclear fuel consists of the reextraction of dissolved plutonium from an organic extractant phase back into the aqueous phase. The plutonium is dissolved as tributylphosphate (TBP)–complex in the organic solvent in the form of quadrivalent Pu^{4+}. If the Pu^{4+} ions present in low equilibrium concentration in the aqueous phase are reduced to Pu^{3+} the plutonium distribution equilibrium can be shifted towards the aqueous phase – see Eqs. (7.61)–(7.63). This is the principle of the electrochemical plutonium reextraction:

$$\left(Pu^{4+}\right)_{org} \rightleftarrows \left(Pu^{4+}\right)_{aq} \tag{7.61}$$

$$\left(Pu^{4+}\right)_{org} + e^- \rightarrow \left(Pu^{3+}\right)_{aq} \tag{7.62}$$

$$\left[c\left(Pu^{3+}\right)_{aq}/c\left(Pu^{3+}\right)_{org}\right]_{equ} \gg \left[c\left(Pu^{4+}\right)_{aq}/c\left(Pu^{4+}\right)_{org}\right]_{equ} \tag{7.63}$$

Electrochemical mixer-settlers or pulsed sieve-plate columns are equipped with electrodes. At the cathodes the plutonium present in the aqueous phase is electrochemically reduced to Pu^{3+}. This electrochemical extraction offers the advantage that no redox-chemicals are required resulting in a remarkable decrease of radioactive waste volume. Another example is the mediated anodic oxidation of toluenes to benzaldehyde in which a dispersion of either pure toluene or toluene dissolved in an inert solvent like kerosene are processed in an aqueous electrolyte which contains the mediator. Frequently Ce^{4+} is used as mediator, which is recuperated from Ce^{3+} by anodic oxidation.

References

1. K. Mund, F. v. Sturm, Electrochim. Acta *20* (1975) 463

Further Reading

I. Rousar, K. Micka, A. Kimla, Electrochemical Engineering, Vol B Elsevier, 1986

CHAPTER 8

Electrochemical Cell and Plant Engineering

8.1
Materials Choice and Corrosion Problems

Metallic materials – cheap steels in particular – are preferred for the construction of process equipment for the chemical industry and also for electrochemical plants. But electrolytes are always corrosive towards metals. Because of their ionic conductivity and their protonic and/or basic properties they favour corrosion by formation of local elements for instance by cathodic O_2 reduction or H_2 evolution combined with anodic metal dissolution. Since the metallic elements which constitute the usual steels – in particular iron – are less noble than hydrogen, steels will always tend to corrode and the tendency to corrode will even be enhanced in oxygen-containing electrolytes unless the metals are not passivated by alloy-components like Cr, Mo or Ni, to speak only of the most common passivating additives, which tend to form dense and passivating oxide layers.

Even non-metallic materials of so different a nature as are ceramics and organic polymers very often do not withstand degradative action of (hot) electrolytes. Ceramics are attacked because some of their ionic constituents are leached out of the solid – especially by strongly caustic or strongly acidic aqueous electrolytes. Organic polymers may be attacked by dissolved reducing or oxidising agents like H_2 or O_2 or CO_2 or by other dissolved oxidising agents (Fe^{3+}, chromate, permanganate etc.); polyethers or polyesters are hydrolysed by water particularly in acidic electrolytes at elevated temperatures and at very low or high pH-values and organic solvents or organic additives to the electrolyte may give rise to deterioration of polymeric materials due to extractive uptake accompanied by swelling and softening.

Corrosion stability of any material used for electrolyzer construction usually is of higher importance than its mechanical properties like elasticity modulus, ultimate tensile strength or hardness. It is not uncommon in electrolyzer construction to make use of composite materials and to combine for instance the superior corrosion-stability of rubber or polymer coatings with the high mechanical strength of normal steel. For instance cell vessels for chlorine-alkali electrolyzers according to the diaphragm and amalgam technology are very often made of polymer or rubber-coated steels.

8.1.1
Metals

Corrosion of metals staying in contact with aqueous electrolytes is an electrochemical process comprising the deterioration of the metal by anodic dissolution of the metal or the formation of non-passivating layers of debris of metal oxides and metal oxide hydrates together with cathodic hydrogen evolution, in non-aerated and completely oxygen-free electrolytes. Oxygen reduction is the cathodic reaction, whenever there is access of gaseous or dissolved oxygen to the

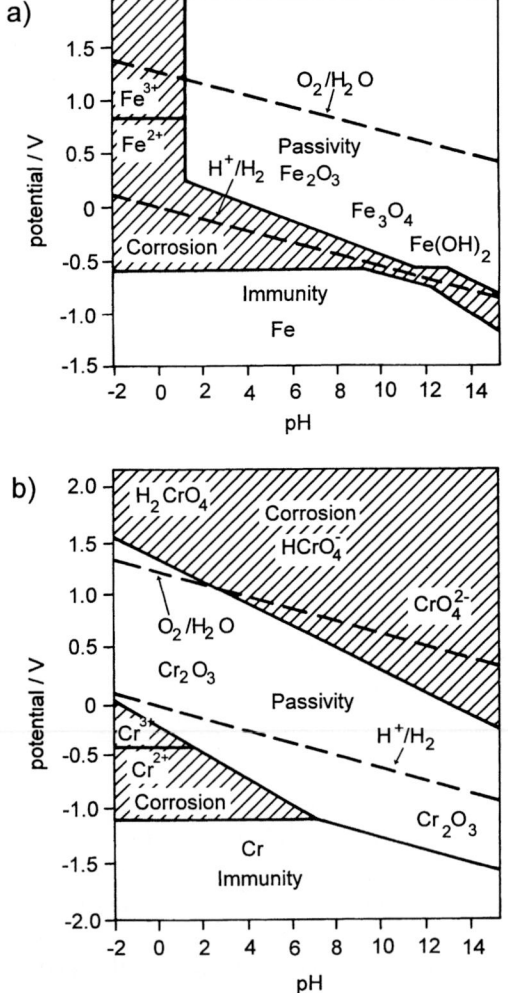

Fig. 8.1. Pourbaix diagrams of: **a** iron; **b** chromium

metal/electrolyte interface. For these two different types of cathodic reactions which supplement the anodic metal dissolution, passivation conditions are quite different because the H_2/H_2O or H_2/H_3O^+ couples and the O_2/H_2O or O_2/OH^- couples respectively establish equilibrium potentials which are separated by approximately 1.2 V. Correspondingly the thermodynamic conditions for formation (or breakdown) of passivating oxide layers may only be guaranteed for aerated solutions or solutions which contain another redox couple of more positive redox-potential than the H_2/H_2O couple.

Pourbaix diagrams for iron (a), chromium (b), nickel (c) and titanium (d), (in Fig. 8.1) are a means for estimating whether the respective pure metal may become passivated under given conditions of pH and redox-potential. If, for in-

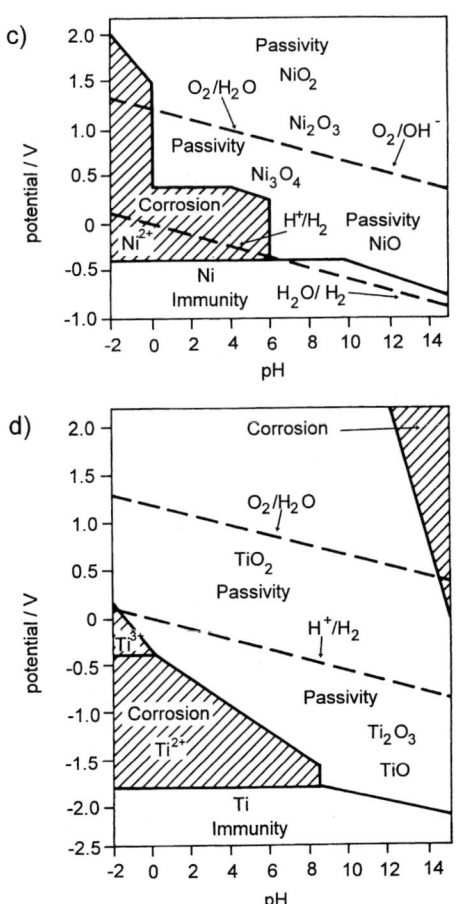

Fig. 8.1. Pourbaix diagrams of: **c** nickel; **d** titanium

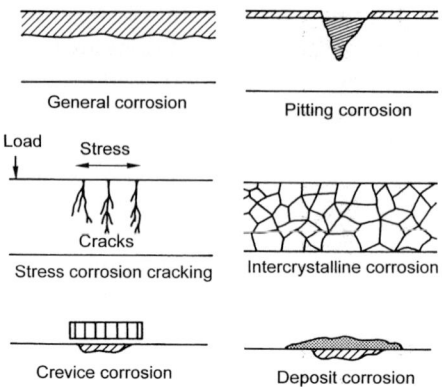

Fig. 8.2. Schematic presentation of different types of corrosion

stance, oxygen is present it may passivate the surface of the steel, because at potentials of the oxygen electrode the passivating oxide layer (Fe_3O_4) becomes stable. Data from Pourbaix diagrams are, however, not at all conclusive. A prerequisite of passivation is the formation of a dense and completely closed, insulating or semiconducting oxide or oxide–hydrate layer possessing a very low solubility and very low diffusivity for oxygen and metal ions. This layer due to its inherent hardness should withstand mechanical abrasion to a certain extent. Whenever these conditions are not fulfilled the metal or alloy is not expected to become passive. "Physical passivation" due to formation of a closed suface layer of low-solubility salts or precipitates ($PbSO_4$ on Pb) which do not exhibit semiconductivity and negligible solubility may – sometimes – be as effective as electrochemical passivation, but very often does not protect the metallic matrix perfectly.

Relevant for electrochemical corrosion of metals staying in contact with electrolytes are:
(i) surface corrosion;
(ii) pitting corrosion (in particular in Cl^--containing electrolytes);
(iii) crevice and deposit corrosion; and
(iv) stress corrosion cracking which becomes especially enhanced for pressurised electrolyzer vessels and gas and electrolyte pipes.

Figure 8.2 describes schematically these different types of electrochemical corrosion. Passivation against surface corrosion is most easily established but may be jeopardised, in particular in electrolyzers and in metallic electrolyte-filled pipes, by stray currents and uneven potential distribution. This leads locally to potential-induced passivity breakdown by establishing subpassive or transpassive conditions.

Table 8.1 collects physical properties and costs for a number of metallic materials and Fig. 8.3 presents schematically the applicability of these materials contacting different electrolytes, that means under oxidising or reducing condi-

8.1 Materials Choice and Corrosion Problems

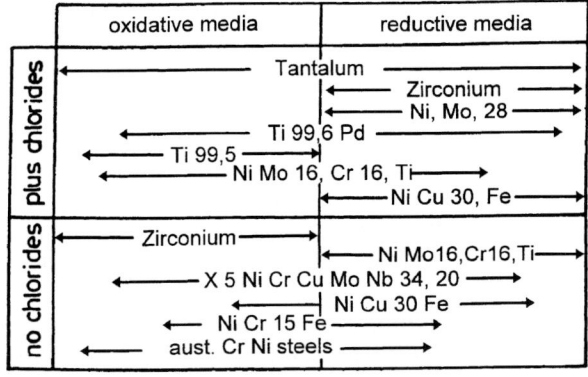

Fig. 8.3. Applicability of metallic materials in different electrolytic environment

Table 8.1. Characteristic data of some important metallic materials

Material	UTS[a] N/mm^2	Elast. Modulus N/mm^2	Density g/cm^3	Price[b] US$/kg
unalloyed steels	200–300	2·10^5	~ 7.8	0.5
stainless steels	200–300	2·10^5	~ 8.2	1.5–3
nickel	100		9.	3.8–4.7
titanium	420–650		4.5	6
zirconium	500–700		6.4	10
hafnium	500–1200	2·10^5	~ 13	200
tantalum[c]			16.6	200–350

[a] Ultimate tensile strength
[b] Price in US $/kg; calculated from prices valid for the Ger.Fed.Rep. 1997 with rate of exchange 1 US$= 1.7 DM
[c] Very soft and ductile material which may be used only for corrosion-protection coatings

tions in absence and presence of chloride ions – the latter causing pitting. As seen from Table 8.1 unalloyed steels are the cheapest metallic construction materials available, but their tendency to corrode limits their applicability. They may only be used for processing weakly to intermediately concentrated caustic electrolytes not containing halide, nitrate, sulfate and carboxylate ions at not too high process temperatures (50–70 °C). Therefore today alloyed steels and high-nickel alloys are much more important for the electrochemical engineering practice than unalloyed steels. Only pure nickel, titanium, zirconium or hafnium can meet the high demands for processes handling highly corrosive electrolytes and aggressive moist electrochemically evolved gases at high temperatures.

Chromium-alloyed steels possess better corrosion resistance in oxidative environments whereas nickel-base alloys are passivated more reliably in reductive environment. Titanium and zirconium as more passive, relative expensive valve metals are easily passivated by TiO_2 and ZrO_2 formation and are very stable in oxidising medium. They become unstable under reductive conditions. H_2 embrittlement, generated at potentials more cathodic than the H_2/H_2O equilibrium potential is particularly detrimental for these both metals. Oxygen or any oxidising agent dissolved in the electrolyte is already sufficient to cause the metal potential to shift towards the O_2/H_2O potential which stabilises the material effectively by passivation.

The corrosion resistance of titanium is improved a lot by alloying this metal with Pd. Palladium lowers the H_2-evolution overpotential and prevents H_2 embrittlement. Surface implanting of Pd results in comparable improvement. The presence or absence of chloride ions, which typically enhance pitting-corrosion for nearly every passive metal, define another limitation of the applicability of the different metallic materials exposed to the corrosive action of electrolytes.

Tantalum is singular in exhibiting an exceptionally high corrosion resistance under oxidative as well as under reductive conditions irrespective of the presence of chloride ions. But it is enormously expensive and additionally a relative soft material. Therefore it may only be used for cladding harder and less stable materials.

Lead is uniquely suitable whenever moderately concentrated sulfuric acid is to be processed. It is the physical passivation of the metal – which becomes covered by a dense, closed layer of $PbSO_4$ which makes it so suitable for cladding the walls of electrolyzers in which H_2SO_4-solutions are handled. $PbSO_4$ solubility in aqueous electrolytes, however, is not zero as shown in Fig. 8.4. Since dissolved Pb^{2+}-ions are reduced at the cathode and oxidised to PbO_2 at high anode potential at the anode there is a slow and steady transport of Pb to the electrodes and correspondingly slow dissolution of Pb claddings is experienced and cathodic as well as anodic electrolysis products are expected to become contaminated by this metal whenever Pb-clad electrolyzer vessels are used.

8.1.2
Carbon

In the chloralkali and chlorate industries carbon anodes are now nearly globally superseded by activated titanium anodes. It is in particular the expressed instability of carbon materials (of any kind) towards anodic oxidation, which renders this material dimensionally unstable as anode material, and the many difficulties and high costs involved in processing this clumsy, hard and brittle carbonaceous material which doomed its further use for processes working with aqueous electrolytes. The situation is somewhat different for organo-electrochemical processes using organic solvents (like alcohols) where the passivity of the usual metals and alloys – even of titanium – is either doubtful or their stability not guaranteed at all. There carbon still is being used – nowadays often in a solid, heterogeneous electrically conducting mixture of carbon and perfluoroethylene.

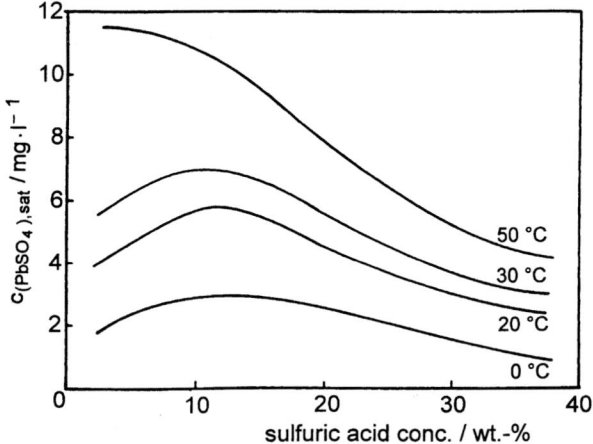

Fig. 8.4. Solubility of PbSO$_4$ in H$_2$SO$_4$-solution of varying concentration and temperature

For molten salt electrolysis (Al, Mg and Ti production) graphitised carbon is still the material of choice for electrodes and cell linings because it withstands the severely corrosive action of hot salt melts, aggressive gases like chlorine (but not CO$_2$ and H$_2$O vapours) and molten metals (but not K, which forms intercalation compounds). Carbon materials – with the exception of the very expensive glassy and pyrolytic carbon – are essentially polycrystalline graphites, which are produced from "green" materials, that means a mixture of coke and special pitches by the Acheson-process. Pyrolysis and subsequent heating to around 2000 °C yields in transformation of amorphous coke and carbonaceous structures into graphite.

Carbon as it is used for electrodes and electrolyzer linings etc. is not really a cheap material. Due to the demands of chemical purity of the raw materials and due to relative high costs of the Acheson-process, which is used for graphitisation the material costs range from 1–2 \$ kg^{-1} of carbon and go up to 10 \$ kg^{-1} for especially treated densified types of carbon.

8.2
Electrode Materials

The choice of materials for electrodes and electrode-supports is dictated by corrosion resistance, high intrinsic conductivity and material strength together with considerations concerning the price of the material and methods for shaping and processing the metals and their respective costs. The overpotential for the electrode process under consideration very often is lowered by choosing an appropriate catalytic coating on a corrosion resistant support like alloyed steel or titanium. In aqueous solutions or in electrolyte solutions of other protic solvents like methanol or acetic acid an additional argument for the choice of the cathode material is an enhanced overpotential for hydrogen evolution. High H_2 overpotential becomes very important whenever hydrogen evolution has to be avoided as a cathodic side reaction accompanying an intended electroreduction – for instance the hydrodimerization of acrylonitrile to adipodinitrile in the Baizer–Monsanto process, where cadmium-clad stainless-steel cathodes are state of the art.

8.2.1
Stainless Steel

Stainless-steel electrodes are the cheapest choice for cathodes and anodes for processes which use weakly alkaline or nearly neutral aqueous electrolytes. Used as cathodes for hydrogen evolution they are cathodically protected (immunity) at H_2 overpotentials exceeding –200 mV vs RHE and become passive under depolarised condition. However, as consequence of the Pourbaix diagram, Fig. 8.1 a, overpotential for H_2 evolution has still to be sizeable at practical current densities in order to guarantee cathodic immunity of iron. If catalytic coatings allow the hydrogen overpotential to decrease to less than –0.2 V, then iron and steel are no longer immune, become subpassive and corrode. Stainless steel used for oxygen-evolving anodes demands weakly alkaline, phosphate – containing solutions in order to become sufficiently passive.

On steel or stainless-steel, lead or cadmium coatings provide high hydrogen-evolution overpotentials which is often demanded for cathodic electroorganic syntheses.

8.2.2
Nickel

Nickel electrodes are much more expensive than steel electrodes. Therefore nickel is used as material for electrodes or electrode supports only in cases of enhanced passivity demands. Nickel coated steel electrodes (coating applied by electroless deposition) are sometimes a cost efficient compromise but this option does not meet very high quality demands because such coatings are always somewhat porous so that the steel as base-metal is still exposed to some extent to the electrolyte – for instance in advanced alkaline water electrolyzers. In strongly acidic electrolytes and in alkaline solutions containing complex-forming agents, nickel electrodes are no longer passivated. Nickel anodes prior to O_2 evolution form NiOOH in alkaline solution on their surface which mediates anodic oxidation of organic substrates very effectively, so that such anodes are today more and more used for anodic electrosynthesis processes. At nickel cathodes catalytic coatings of highly porous Raney nickel which may reduce hydrogen-evolution overpotential drastically can be easily applied by cathodically codepositing nickel and zinc on nickel supports followed by alkaline leaching of zinc – compare Chap. 7.

8.2.3
Lead

Lead is particularly suited for oxygen-evolving anodes in strongly acidic aqueous solutions in particular in aqueous sulfuric acid solutions because under operating conditions at O_2-evolution potential it is passivated by formation of lead dioxide. Still silver-alloyed Pb/PbO_2 anodes are preferred for metal electrowinning from aqueous H_2SO_4-solutions. O_2-evolution overpotential is, however, particularly high on PbO_2 and can be sizeably reduced by alloying silver to the lead. Unalloyed Pb/PbO_2 anodes are preferred for some anodic oxidations of organic substrates or inorganic redox couples serving as anodic mediators (Mn III/II, Ce IV/III, Cr VI/III) where O_2 evolution is an undesired side reaction.

8.2.4
Titanium

Titanium electrodes are state of the art as anodes for chloralkali electrolysis. The reason is threefold: First the excellent stability of titanium against surface and pitting corrosion in acidic and slightly basic aqueous solution cannot be met by any steel or nickel alloy. Second the application of RuO_2 coatings or other coatings containing metal oxides or metals of the platinum group allows to reduce remarkably the overpotential for anodic chlorine and anodic oxygen evolution. These coatings are particularly stable on titanium-base metal and adhere firmly to its surface. Third the price of titanium decreased strongly during the last decade so that costs are no longer prohibitive for a more extended use of Ti anodes. Surface-doped Ti anodes with RuO_2/IrO_2 coatings for O_2 evolution from acidic electrolytes in metal-electrowinning processes is becoming industrial practice. Ti metal cannot be used as cathode material for H_2 evolution because of hydrogen embrittlement of this metal under cathodic H_2 evolution.

8.2.5
Noble Metals

Noble metals, in particular platinum metals, are too expensive to be used as bulk materials for electrodes, except for very few special applications (Pt-wire anodes, are used for anodic peroxidisulfate formation). It is more appropriate to apply the respective noble metal as a coating to a suitable support according to techniques similar to preparation of RuO_2-activated Ti anodes or by cathodic deposition of the platinum metal on titanium from cyanide melts. Noble metals and Pt in particular may be used also in the form of thin sheets for instance on carbon electrodes fixed to the carbon surface by special conducting glues. Such electrodes are applied, for instance, for anodic Kolbe synthesis of dicarbonic acids.

8.2.6
Massive Carbon

Massive carbon electrodes are no longer of any practical importance for conventional aqueous electrolysis process because they are little stable and have been substituted nearly everywhere by coated titanium anodes. There is however one process in which a special type of carbon electrodes – amorphous carbon – is used namely in anodic fluorine generation and fluorination of organic compounds in electrolytes which are essentially based on hydrofluoric acid as solvent.

8.3
Electrode Design

Technical electrodes for processes in which neither gas evolution nor electrochemical consumption of gases is involved are generally planar electrodes made of smooth metal. Flat electrodes are typical for instance for hypochlorite or chlorate production or in the form of NiOOH-covered smooth or roughened nickel anodes for the anodic oxidation of organic compounds in caustic aqueous electrolyte. Very often also for gas evolution smooth electrodes are used as for instance for anodic O_2 evolution in zinc electrowinning plants, where PbO_2-covered Pb anodes are loaded with anodic current densities as high as 500 mA cm^{-2}.

8.3.1
Gas Evolving Electrodes

For more than fifty years electrodes intended to be used for gas evolution were modified by slits and holes which had the purpose to facilitate the escape of gas bubbles towards the rear of the electrodes where the electrolyte/gas emulsion ascends at the backside of the electrode. Such perforated electrodes have to be applied as pre-electrodes. As shown by the schematic flow pattern of (Fig. 8.5 a) the gas/electrolyte emulsion leaves the interelectrodic gap through the openings clearing the interelectrodic gap of the resistance enhancing gas bubbles. The holes must not be too small in order to prevent gas trapping. Expanded metal gauze (Fig. 8.5 b), louvered (Fig. 8.5 c) finned or slotted electrodes serve the same purpose and are particularly effective for bubble removal from the interelectrodic gap. The inclination of the flow-diverting fins has an optimum value of some 20–30° with lowest gas holdup and lowest electrolyte resistance.

Figure 8.6 shows how pre-electrodes/fore-electrodes are arranged in monopolar cells and bipolar cell stacks. Such pre-electrodes have the disadvantage to increase the effective cell width and conflict with the aim to keep cell volume and cell costs as low as possible. But they are often necessary for supporting gas-evolving electrodes. For more detailed information compare Chap. 10.

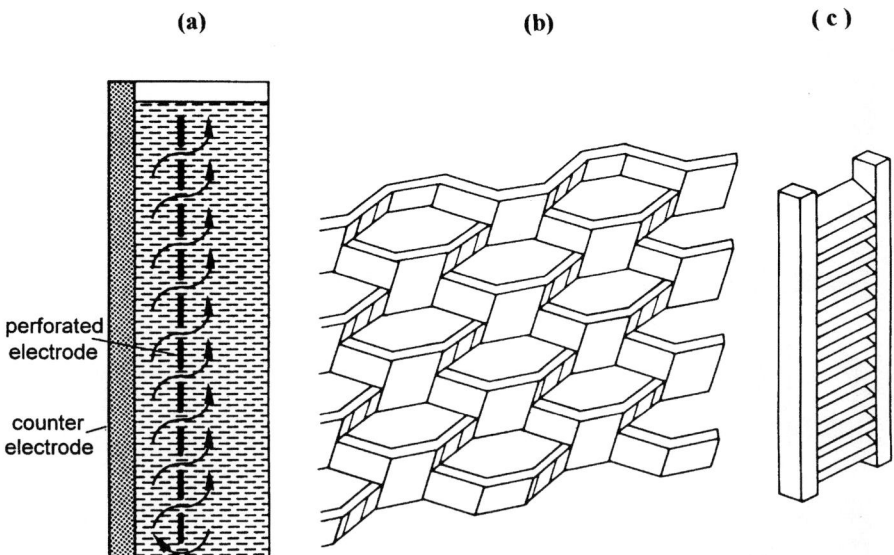

Fig. 8.5 a–c. Surface structured electrodes for gas evolution: **a** flow pattern of gas evolving electrodes supplied with slits or holes; **b** expanded metal gauze; **c** louvered electrodes

8.3.2
Gas Consuming Electrodes, Gas Diffusion Electrodes

Gas consuming electrodes are needed and are particularly developed for fuel cells but are beginning to find their practical application in electrolysis processes as well. Substitution of cathodic hydrogen evolution by cathodic oxygen reduction in chloralkali electrolysis and anodic H_2 oxidation instead of anodic O_2 evolution in metal electrowinning are still futuristic options but not too far away from technical realisation.

The main problem of gas consuming electrodes are the ample supply of the gas to the outer and inner surface of the nanoporous electrode particles which in the form of highly porous (inner pores 1–10 nm diameter) grains or particles (outer particle diameter 1 μm<d_p<10 μm) are forming a microporous layer with a thickness of no more than fractions of a millimetre (100–200 μm) in the old fashioned Raney nickel anodes of alkaline H_2/O_2 fuel cells (see Fig. 7.2 a). Such gas diffusion electrodes in which the hydrophilic electrode particles are bonded to each other by copious fibers of hydrophobic organic polymer (most often polytetrafluorethylene, PTFE or Teflon) constitute an interwoven dual system of hydrophobic micropores and hydrophilic nanopores. The hydrophilic nanopores are filled by the electrolyte which also forms a film on the outside of the electrode particles and establishes the ionically conducting connection between the electrode particles and the interelectrodic gap and the counter electrodes respective-

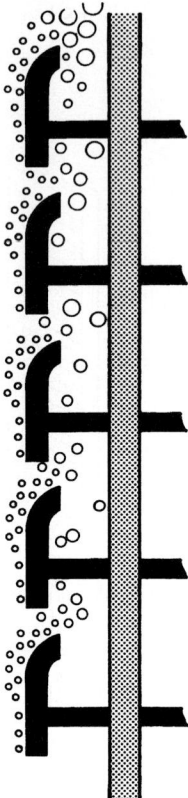

Fig. 8.6. So called preelectrodes in monopolar and bipolar cells

ly. The hydrophobic pore system supplies the gases to the surface of electrolyte film which is wetting each electrode particle in the electrode matrix. Mass transfer of gases dissolved in the electrolyte proceeds by diffusion across very thin electrolyte films and bulk diffusion into the highly porous electrode particles.

Figure 8.7 depicts schematically the detailed morphology of the up to date PTFE bonded gas diffusion electrodes of phosphoric acid fuel cells as a typical example for low temperature fuel cell electrodes, whose electrode and catalyst material is Pt-doped active carbon. The electronic current is transmitted through the electrode grains, which are in electronic contact with each other, towards a metallic or carbonaceous collector grid, fabric or mesh which backs and strengthens this delicate structure physically, whereas the ionic current travels through electrolyte pores and films covering the electrode particles. For proton-exchange-membrane fuel cells (PEMFCs) the electrode structure is very similar but the electrode is supported by the membrane.

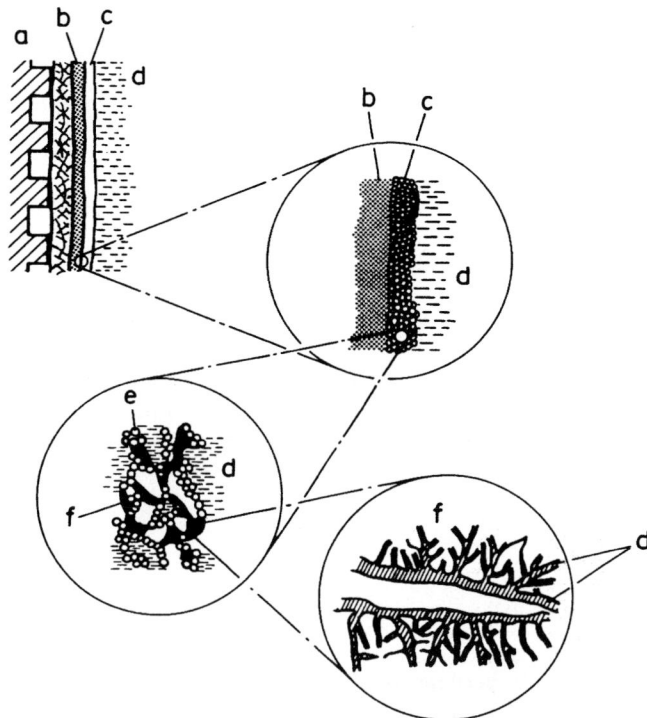

Fig. 8.7. Schematic of a gas consuming electrode, so called gas diffusion electrode, for low-temperature fuel cells composed of nanoporous, electrolyte flooded electrode particles with wide gas-filled pores between them. – **(a)** support (grid or tissue), **(b)** hydrophobic gas diffusion layer, **(c)** hydrophilic electrode layer, **(d)** electrolyte and (at highest magnification) electrolyte film on gas conducting pore walls and also in nanopores of **(f)** catalyst particles, **(e)** PTFE particles.
This electrode structure is typically used in phosphoric acid and membrane fuel cells.

8.4
Separators: Membranes and Diaphragms

For divided cells the choice of the separating diaphragm, Table 8.2 a, or membrane, Table 8.2 b, and its physical and mechanical properties together with its price are as important as the appropriate choice of electrode materials for proper functioning, life time and cost contribution for the respective electrolytic process. Whenever the separator has to separate anolyte and catholyte completely from each other, then only dense ion-exchange membranes can be used. If mixing of anolyte and catholyte is to be only partially prevented and the main reason for using a separator is to exclude intrusion of evolved gases into the counter electrolyte cycle then microporous diaphragms which in general are much cheaper than ion-exchange membranes will do the job.

Table 8.2a. Diaphragm materials

Design	Excess pressure for gas permeation MPa	Surface specific resistance (30 wt%, KOH, 90 °C) $\Omega\ cm^2$	Remarks
plastic reinforced asbestos (Polyvinyl pyridine)	0.03	0.3–0.4	not stable above 90 °C in caustic electrolyte
ZrO_2[a] filled polysulfone/Zirfon	>0.5	0.2–0.3	stable in concentrated KOH
Ni-net supported NiO	>0.02	0.1–0.2	price of NiO appr. 20 US$ m^{-2}
PPS-needle felt	<0.005	0.3–0.4	price approx. 20 US$ m^{-2}
PPS cloth			40–50 US$ m^{-2}

[a]Developer: VITO, Inco, Boeretang 200, B-2400 Mole, Belgium

Table 8.2b. Ion exchange membranes

Commercial Name	Manufactor	Type	Costs US$ m^{-2}
NeoSepta CM 1,2,X[a]	Tokuyama soda	perfluorinated cation exchange	–
NeoSepta AM 1,3,X[a]	Tokuyama soda	perfluorinated anion exchange	–
Nafion	Dupont	perfluorinated cation exchange	700–900
Nafion NE-455	Dupont	perfluorinated cation exchange 97% current efficiency at 33% KOH	–
Flemion[a]	Asahi Glass	perfluorinated strongly acidic cation exchange and strongly basic anion exchange	–
Selemion[a]	Asahi Glass	chemically particularly stabilized, highest permselectivity	–
Gore Select[a]	W.L. Gore Ass.	perfluorinated cation exchange reinforced by PTFE fabric	–
FuMA-Tech membranes[a]	FuMA-Tech	anion and cation exchange, particularly tailored to customers demand	–

[a]Costs depend on customers demands, technological purpose and the amount ordered

8.4 Separators: Membranes and Diaphragms

(a)

$[-(-CF_2-CF_2-)_n -CF-CF_2-]_x$
$\quad\quad\quad\quad\quad |$
$\quad\quad\quad\quad (O-CF_2-CF-)_m-O-CF_2-CF_2-SO_3H$
$\quad\quad\quad\quad\quad\quad\quad |$
$\quad\quad\quad\quad\quad\quad\quad CF_3$

$n = 5$ to 13
$x = $ ca. 1000
$m = 1$ to 3

(b)

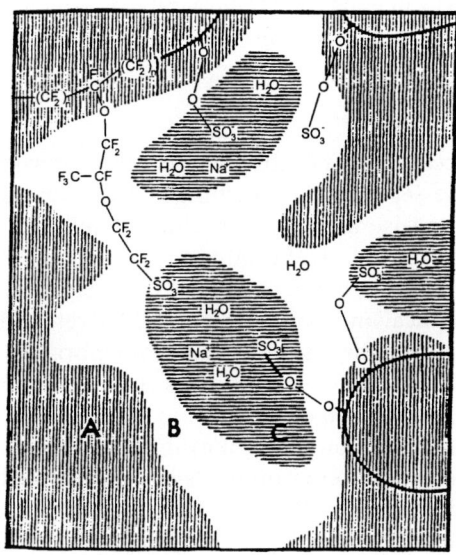

Fig. 8.8. a Schematic presentation of the chemical structure on Nafion. **b** Phase separation in water swollen Nafion – A hydrophobic area, B solvated sulfonic acid groups, C aqueous phase

8.4.1
Membranes

Ion exchange membranes are composed of organic polymers which possess ionogenic groups like sulfonic acids ($-SO_3H$) carboxylic acids ($-COOH$) or quaternary ammonium hydroxides ($-NR_3OH$) and which are called ionomers. The ionogenic groups are attached to an aliphatic backbone, which is perfluorinated in the case of Nafion, Neosepta and Flemion ionomers. Figure 8.8 a depicts schematically the chemical structure of Nafion, the membrane material used in chlo-

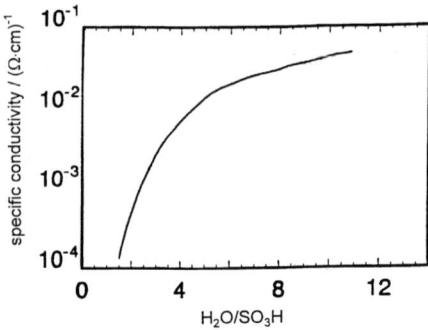

Fig. 8.9. Dependence of ionic conductivity of Nafion on the molar ratio of water to sulfuric acid groups

ralkali electrolyses and PEM fuel cells. Figure 8.8 b shows that the water-swollen ionomer undergoes a type of phase-separation by forming hydrophobic and hydrophilic clusters. The former comprise mainly the perfluorinated alky backbone the latter, mainly the solvated sulfonic acid groups.

For electrolysis processes as well the organic polymer as the ionogenic substituents have to be of highest chemical and thermal stability. Therefore mainly perfluorinated polymers are today used for cation-exchange membranes. Three different types of fluorinated cation-exchange membranes (namely containing carboxylate and sulfonate groups) are available mainly from only three different firms (Dupont: Nafion, Asahi glass: Flemion and Tokuyama Soda: Neosepta), which can withstand chemically aggressive environments as concentrated caustic potash, chlorine, oxygen and hydrogen. Recently cation exchange membranes are offered, too, by Dow chemicals, which possess lower resistance but at the expense of also lower perm selectivities. Anion exchange membranes, based on polymerised fluorinated alkyls with quaternary ammonium ions, are commercially available, but do not meet the high technical standards of perfluorinated cation exchange ionomers, in particular they still seem to lack chemical stability under severe chemical, particularly oxidative conditions. They are mainly used in electroosmosis processes, where temperatures are low and aggressive chemicals are not generated. The cation-exchange membranes, which are essentially PTFE-reinforced layered composites, are exhibiting astonishingly high permselectivities even at caustic alkali concentrations as high as 30 wt% (This is a prerequisite for obtaining 95% current yield in NaOH production). They are composed of very thin layers of highly permselective carboxylate ionomer and a thicker, less permselective sulfonate ionomer which has a lower permselectivity but a much higher ionic conductivity. But these membranes are still very expensive. The price for high performance membranes is approximately 700–900 US\$ m^{-2}. Their ionic conductivity is dependent on swelling that is water uptake. Table 8.2 b collects a selection of commercially available membranes.

Figure 8.9 shows for Nafion the dependence of the ionic conductivity on the molar ratio of water to sulfonic acid groups for protonated Nafion.

8.4.2
Diaphragms

Diaphragms are less demanding devices than ion-conducting, permselective membranes and accordingly much cheaper than ion exchange membranes. They can be obtained for 20–60 US\$ m^{-2}. Most common had been asbestos diaphragms produced from chrysotile as fabrics which might have been even reinforced by metal fibres. Very often asbestos diaphragms are formed in the electrolyzer by slurry coating a metal sieve, fabric or other porous metallic support. Asbestos is chemically stable in acidic and neutral aqueous solutions. Its application in alkaline solution, however, is limited to temperatures below 90 °C. Because of the health hazards involved in processing asbestos it is expected that mining of asbestos will be stopped before the end of this century. Therefore in the future asbestos will no longer be a reasonable option. Although not yet available commercially there exist now several options of chemically comparably stable and mechanically strong diaphragms which exhibit greatly enhanced chemical stability – compared to asbestos – against hydrolytic caustic and acidic attack at temperatures above 100 °C. Table 8.2 a summarises these new types of diaphragms. All of them, with the exception of PPS needle felt and PPS cloth have been developed recently in the framework of R and D programmes for improved alkaline water electrolysis. All these new diaphragms – again with the exception of PPS needle felt – are made of composite materials because the high demands with respect to low surface specific electrolytic resistance (highly porous and thin), high mechanical strength and high chemical stability can usually be met only by the combination of different materials.

8.5
Polymeric Materials for Cell Bodies and Electrolyte Loops

For the construction of light-weight fuel cells polymers have been used for more than 30 years. For the construction of electrolysis cells polymers had been introduced only recently. The reason is the limited chemical and thermal stability of most cheap mass produced polymers like polyethylene (PE), polypropylene (PP) or polyvinylchloride (PVC). For most of these materials the softening temperature is too low (being lower than 100 °C) and/or their chemical stability against caustic/acidic and oxidative/reductive attack is too weak.

Only more expensive materials like polyepoxide resins, e.g. polypropylene oxide (PO) or even more polyphenylensulfon (Polysulfon), polyphenylensulfide (PPS, Polysulfide, Ryton) polymides (PI) and the fluorinated polymers PFA and PTFE are combining high chemical stability with (relatively) high softening temperatures. Most recently developed polyether ketones are very expensive but combine very high glass transitions (>200 °C) with high chemical stability.

Table 8.3. Properties and approximate prices of some polymeric materials

Polymer	Abbreviation	Max. temperature, °C without creeping	Highest temp., °C for utilizing	Density g cm^{-3}	Price[a] US\$ kg^{-1}
polyethylene high density	PEHD	45	40	0.95	0.9
polyethylene low density	PELD	–	40	0.88	0.8
polypropylene	PP	60	55	0.91	0.9
polystyrene	PST	75	60	1.04	0.9
High density polystyrene	HDPST				1.0
polyvinylchloride	PVC	75	60	1.40	0.64
poly-fluoroethylene- propylene	FEP	105	120	2.1	~3
poly-perfluoroalkyl-vinylether	PFA	160	200	2.1	~4
polytetrafluoroethylene	PTFE	160	220	2.2	~4.5
polyarylethersulfone[b] (polysulfone, UDEL)	PS	180	120	1.25	~3.5
Polyphenylensulfide[c]	PPS	260	230	1.6	5–15
Polyetherether ketone (LUVOCOM)[d]	PEEK				77

[a] Price in Germany mid 1997. Rate of exchange: 1 US \$ equal 1.7 DM, Source: Kunststoff Information (KI), D – 61,350 Bad Homburg
[b] Source: AMOCO
[c] Source: Philips petroleum International, price swing according to different qualities
[d] Source: Lahmann and Voss, Hamburg

Whereas polysulfones and polysulfides possess a high degree of crystallinity and are therefore hard – even somewhat brittle – PFE and PTFE show typical creeping behaviour which is undesired for construction of rigid electrolyzer components or electrolyzer cell structures. The crystallinity of polysulfones and polysulfides cause some difficulty for processing these materials (for instance by extrusion) and allows only the production of thinner sheets and almost prohibits the production of thicker and more massive parts.

Chlorinated polymers like PVC are not used for the construction of electrolysis cells because of their low softening temperatures (lower than 80 °C). They are, however, preferred for pipes and components through which highly aggressive electrolytes or electrochemically evolved gases are transported at temperatures below 60 °C. They exhibit a relatively good chemical stability and are relatively cheap materials which can be easily and cheaply processed and shaped. Very cheap and nonetheless rigid glass-fibre reinforced polyester material may only be used for low temperature storage tanks for electrolytes because the polyester is susceptible to hydrolysis. Applying additional internal PVC-linings to glass fibre reinforced vessels allows to overcome this difficulty. Table 8.3 gives an overview of some selected polymer materials together with their costs.

8.6 Gaskets

For the tightening of electrolyzers, bipolar stacks and tubes which conduct liquids and gases, gaskets made of electrically insulating materials are needed. Typical gasket materials possess either rubber elasticity or only limited elasticity together with low compressibility. Therefore either special crosslinked polymers with rubber-like compressibility and elasticity or normal but filled polymers and especially polymers with little or no tendency to creep under pressure are used for gaskets, sealings and O-rings. Physical and chemical stability against the electrolyte and aggressive gases which are generated in the respective electrolyzer is of ultimate importance. In particular changes of material properties with increased working temperatures limit their applicability and have to be taken into account. For instance Klingerite (rubber-bound asbestos) is sufficiently stable against caustic solutions at room temperature, so that it can well serve for static gaskets at ambient temperatures but is attacked and destroyed very soon at 90 °C by 30 wt% KOH.

For ambient temperatures there exists a host of cheap unfilled and filled organic polymers (polyethylene, polypropylene, polypropylene oxide, polyvinylchloride) which may serve as flat sealings and likewise a number of rubber-elastic materials like acrylonitrile-butadiene, chlorbutadiene-, butyl/chlorobutyl- or silicone-rubbers are available.

For temperatures not exceeding 60 °C and for aggressive electrolytes and gases the choice is restricted to chloropolymers and rubbers and to fluorinated polymers like polyvinylidenefluoride, polyvinylidenefluoride ethers, polytetrafluoroethylene and fluorinated and perfluorinated rubbers (Viton and Kalrez produced by Dupont). Fluorinated polymers are the only choice for temperatures exceeding 160 °C with a ceiling temperature of 200 °C. Fluorinated polymers – with the exception of the fluorinated rubbers – creep under pressure and have to be applied either in "filled" form (the filler may be carbon, brass, talcum and other inorganic materials) or creeping has to be prevented by special constructive means like "encasing" and/or applying very thin foils or gaskets. Sometimes a foil of the chemically very stable fluorinated material may be used as a thin protective layer or blanket on a gasket made of chemically less stable but mechanically and thermally more satisfactory material. There are indeed no general rules available, which would allow a straightforward choice of an appropriate gasket for a respective purpose (combination of electrolyte/gas/temperature). The upper temperature limit for the most stable gasket materials (fluoro- and perfluoro polymers) is about 200 °C. For extraordinarily corrosive conditions and particularly for working temperatures above 200 °C laminated carbon may be used which being electrically conducting must be combined with laminated non-conducting inorganic materials if the gasket separates different parts of the electrolyzers which are at different electrical potential. Above 300 °C –temperatures which might become important for novel molten salt electrolysis processes or advanced low-temperature fuel cells – there does not yet exist any reasonable option for gaskets or sealing materials. For such cases, the sealings made of low temperature materials have to be kept cool or have to be applied only at the cooler, outermost parts of the electrolyzer. A second not really satisfying option is the "wet seal", which makes use of the capillary forces in a finely porous matrix which is able to keep the molten electrolyte in the matrix even against a moderate pressure difference.

8.7
Electrodes

8.7.1
Horizontal Electrodes

Horizontal electrodes are used in the amalgam technology for chloralkali electrolysis, in the Hall–Heroult process for aluminium electrowinning and had been used in the so called AlCOA-process for electrowinning of aluminium from alkali chloride melts containing $AlCl_3$, which, however, today is no longer practised as it turned out to be not a viable electrolysis.

In all these cases designing an electrode structure which ensures easy and effective gas release and prevents the build-up of electrolytically insulating gas pillows is indispensable. For aluminium electrolysis one is not able to produce by an appropriate shaping of the carbon anodes which possess a considerable surface of approximately 1 m^2, a flat surface because the electrodes are corroding steadily and are steadily changing their shape, so that in a sense shaping of the electrode surface and gas-pillow build-up are self-adjusting. For the amalgam

8.7 Electrodes

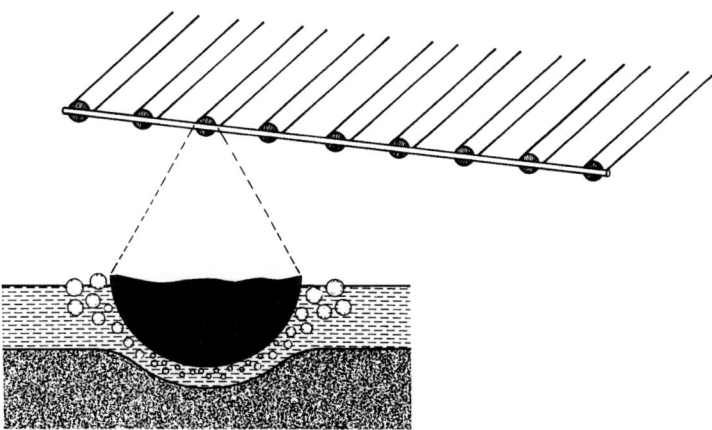

Fig. 8.10. Schematic of a horizontal, gas-evolving grid electrode slightly dipping into the mercury allowing easy escape of evolved chlorine in mercury cells and adjustment of small anode/cathode distance of less than 1 mm, so that essentially a zero-gap cell is established

process, the world-wide introduction of RuO_2-activated Ti anodes, which are flat grid- or wire-electrodes or expanded metal electrodes, to keep a very narrow distance between the wires of the grid anode and the amalgam surface (Fig. 8.10). The evolved chlorine ascends rapidly around the grid or wires which are dipping into the mercury bed and cause a vigorous movement and copious exchange of the electrolyte which on one hand facilitates gas removal and on the other the exchange of the depleted by fresh electrolyte.

8.7.2
Membrane Electrolyzer

Figure 8.11 depicts schematically the cell geometry of a membrane water electrolyzer. The liquid electrolyte is substituted by a fluorinated cation exchange membrane. There are no externally applied electrodes, only porous current collectors, but the electrode metal (platinum or other platinum metals) is chemically deposited in the form of dispersed nanoparticles in the membrane surface by counterdiffusion of a reducing agent like Sn II, KBH_4 or hydrazine and an aqueous or organic solution of the noble metal, for instance H_2PtCl_6. These finely dispersed electrodes are then additionally reinforced by cathodic deposition of Pt-metal on the membrane surface. In this way a highly dispersed electrode is formed in the membrane surface from which the evolved gases may escape easily through the microporous current collector towards the gas collecting channels grooved into the bipolar plate which contacts the metallized membrane surface.

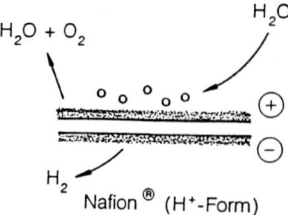

Fig. 8.11. Schematic of a membrane electrolyzer

8.8
Cell and Electrode Design

8.8.1
Zero Gap Electrolysis Cells

In order to reduce in diaphragm-divided electrolysis cells the width of the electrode gap to an absolute minimum one contacts the diaphragm from both sides with the respective electrode in the so called zero-gap geometry. Zero-gap cells can of course only be constructed by corrugated electrode structures using the preelectrode or current collector technique to allow for easy release of gas bubbles towards the backside of the electrodes and access of fresh electrolyte.

Figure 8.12 a explains schematically by a cut through a zero-gap cell composed of perforated toil electrodes pressed out either side of a diaphragm that due to unequal current distributions electrolysis typically proceeds to a certain

Fig. 8.12. a Schematic of a zero-gap cell composed of a diaphragm squeezed between two perforated plate electrodes demonstrating unequal current density distribution with finite current density also at the backside of a perforated electrode. **b** Cut through an electrode/diaphragm/electrode (EDE) unit with porous, sintered electrodes

extent also at the rear of the electrode structure. Zero-gap cells have been successfully introduced for alkaline water electrolysis already since more than 30 years. Recently the technology found application, too, for the chloralkali electrolysis membrane process. Figure 8.12 b shows a cut through the zero-gap cell of an advanced alkaline water electrolyzer in which porous sintered-nickel electrodes are integrated into a EDE (electrode/diaphragm/electrode) unit together with a sintered-ceramic diaphragm.

8.8.2
Vertical/Horizontal Electrodes

One of the most important issues in cell design is keeping the internal cell resistance low. The electrolyte flux must be managed in a way, which provides sufficient supply of reactants evenly across the electrode surface and to maintain high mass transfer rates. Additionally cell construction should account for easy access to and exchange of cell components. Very often the electrochemical engineer has to find a compromise. For instance in bipolar stacked cells easy access to single cells is sacrificed to the other issues mentioned above. Vertical electrodes (Fig. 8.13 a) are generally preferred to horizontal electrodes, because of controlled electrolyte flux, easy gas release, easy access to singular electrodes and their ease of exchange and – for bipolar cell – simple way of stacking. Horizontal electrodes are only used in fuel cells and in processes in which one of the electrodes is a liquid metal as in the amalgam chloralkali and in aluminium electrowinning electrolysis (Fig. 8.13 b).

8.8.3
Divided/Undivided Monopolar/Bipolar Cells and Modes of Electrolyte Flow

Separation of anolyte and catholyte by a porous diaphragm or ion conducting membrane should be avoided whenever possible, because separators are costly and tightening of a divided cell is difficult and encounters a host of mechanical and corrosion problems. Monopolar cells (Fig. 8.13 a, b or c) are either operated as separate units or as troughs in which anodes and cathodes alternately inserted are switched in parallel (Fig. 8.13 c). Most chloralkali electrolysis technologies and aluminium electrowinning work with monopolar single cells and almost all metal winning and electrorefining electrolyses are operated in monopolar trough cells. In bipolar cells (Fig. 8.13 d) each electrode works on one side as a cathode and on the opposite side as the anode. Bipolar cells are stacked to large units, whose total voltage adds up to hundred or several hundreds of volts.

The electrolyte flux in undivided trough cells is usually meandering slowly from the entrance to the exit. In divided cells – for instance in water or in membrane chloralkali electrolysers – anolyte and catholyte are kept separate and are separately collected (Fig. 8.14 b) in contrast to the flux pattern of undivided cells shown in Fig. (8.14 a). A particular case are diaphragm cells, which keep gases evolved at the anode and cathode separate from each other and which prohibit

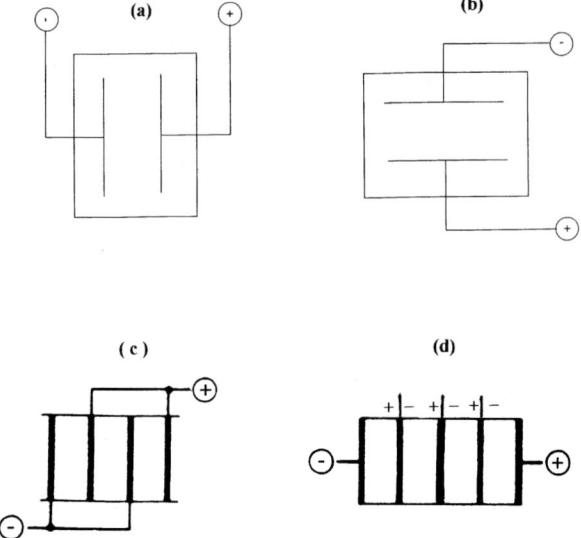

Fig. 8.13 a–d. Electrode geometries: **a** vertical electrodes; **b** horizontal electrodes – usual in all processes using a liquid metal cathode; **c** electricity supply to all cathodes and anodes in monopolar trough cells; **d** schematic of bipolar cell stack

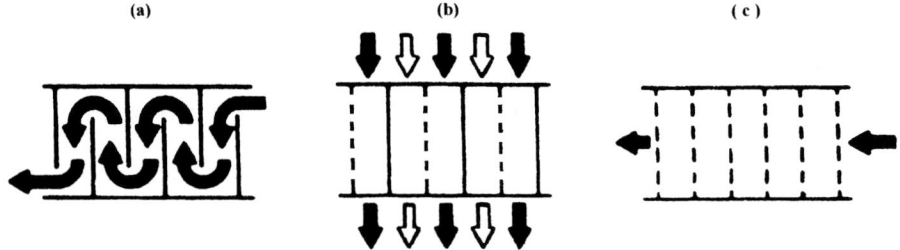

Fig. 8.14 a–c. Electrolyte flow pattern: **a** through undivided multielectrode trough cell; **b** through divided cell; **c** through diaphragm-divided cell with electrolyte flow-through

mixing of anolyte and catholyte, but permit due to a small hydrostatic head a slow and controlled flux of the electrolyte through the cell from the anode to the cathode as shown in Fig. (8.14 c).

8.8.4
Special Cell Designs

During the last twenty years a considerable number of special cell designs have been developed with the intention to reduce ohmic voltage drops across the cell

8.8 Cell and Electrode Design

on one hand and to increase space time yields on the other hand. A particular problem involved in electrolyzers for organo-electrosyntheses with non-aqueous electrolytes is the low electrolytic conductivity of salts, acids and bases in non-aqueous solvents of low dielectric constants. In order to avoid excessive energy consumption and Joule's heating of the electrolyte it is necessary to reduce drastically the width of the interelectrodic gap. Developing capillary-gap cells of hitherto unknown small electrode distance serves mainly this purpose but enhances also the space time yield (ρ) of the electrolyser

$$\rho = a_e \frac{iM\phi^e}{v_e F} \quad (8.1)$$

The higher the space time yield, the lower are, in general, the specific investment costs. The space time yield is the product of a design factor – the volume specific electrode area a_e and two factors determined by the electrode reaction under consideration namely current density and current efficiency. In many organoelectrosynthesis reactions the electrochemical conversion rate and hence the practical current density is charge transfer controlled and often has to be kept very low in the mA cm^{-2} range. The electrodeposition of metals in contrast is very often close to mass transfer controlled, which holds in particular for electrolytic depletion of toxic metal ions like Ni^{2+}, Cu^{2+}, Pb^{2+} from waste water streams. In the latter case mass-transfer-limited current densities become also very low because of low concentration levels, which have to be achieved in effluent purification.

For such cases a cell design which allows for a high-volume specific electrode surface is required in order to achieve acceptable space time yields. In Fig. 8.15 a the volume specific electrode areas of a number of different cell designs is plotted versus the "characteristic" length of the respective cell design which generally is the width of the interelectrodic gap.

Lowest specific electrode areas exhibit classical plate cells with electrode distances of typically 0.5–5 cm.

For an assumed current efficiency of unity the space time yield can be expressed by the so called current concentration i*, the volume related curren $i^* = a_e \cdot i = I/V; \quad (8.2)$

$$\rho = i^* M / v_e F \quad (8.3)$$

The current concentration i* is plotted in Fig. 8.14 b vs the characteristic length of the same electrolyser types as in Fig. 8.14 a. To estimate these a_e i values, typical current densities for well known applications of these cells are chosen. Very high current densities, e.g. 10 kA m^{-2} in the chloralkali electrolysis, can be applied only in a limited number of cases. The use of cells with high specific electrode area like trickle-tower and packed-bed cells usually is restricted to cases of low current densities. Some special cell concepts are listed in Table 8.4, where each cell design is

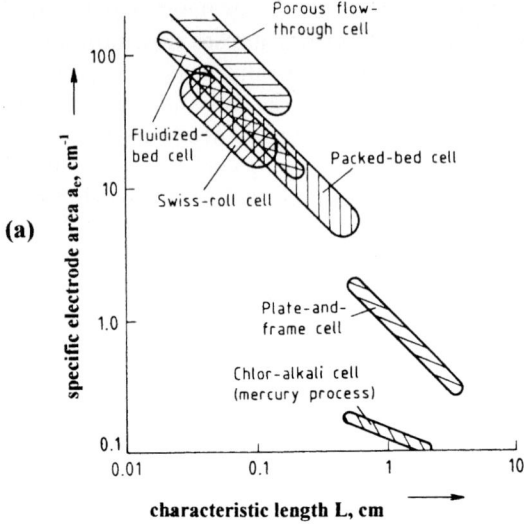

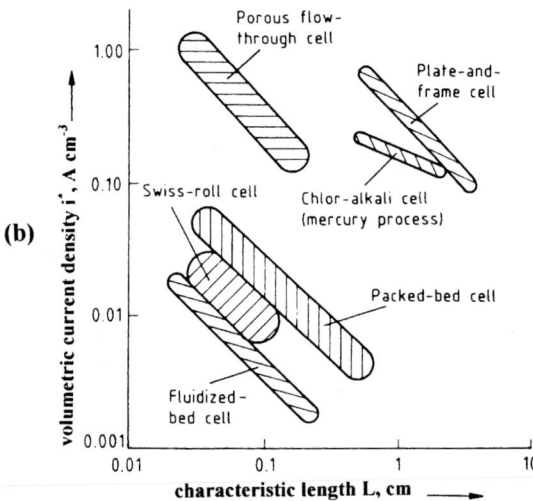

Fig. 8.15. a Volume-specific electrode area a_e vs characteristic length, which in most cases is the interelectrodic gap width for different types of electrolysis cells. **b** Current concentration, i^*, measured in amperes per volume against characteristic length of the electrolyzers, which are dealt with in Fig. (8.15 a). i^* is directly related to space–time yield: $\rho = i^* M / v_e F$

qualitatively characterised with respect to ohmic voltage drop, mass transfer rate, and specific electrode area. If mass transfer is the main problem, then relatively simple means for the enhancement of mass transfer will allow to maintain traditional cell geometries.

8.8 Cell and Electrode Design

Table 8.4. Special concepts of cell design

Cell	Ohmic voltage drop	Mass transfer rate	Specific electrode area
Cells with turbulence promoters	high	high	small
Rotating-cylinder electrode cell	high	high	small
Trickle-tower cell	high	high	large
Capillary-gap cell	low	low	small
Pump cell	low	high	small
Swiss-roll cell	low	high	large
Zero-gap cell	low	low	small
Solid-polymer electrolyte cell	low	low	small

In the classical parallel-plate cells, which are usually applied as filter-press-type arrangements, turbulence promoters can be introduced easily into the interelectrode gap or into the diaphragm-electrode gap to enhance mass transfer. Mass transfer promoters in the form of rods, studs or fins perturb the evolution of laminar velocity profiles with steadily increasing extension of concentration boundary layers. Figure 8.16 depicts schematically a plate electrode equipped with rods as turbulence promoters.

For channel flow – as for pipe flow – the critical Reynolds number at which laminar flow changes to turbulent flow is approximately 2300 which is drastically reduced by such turbulence promoters. The effect is a sizeably enhanced mass-transfer rate and steady and spatially equal mass transfer coefficients. This is very important since under laminar flow conditions k_m changes with $L^{-1/3}$ (L is the length of the electrode – compare Chap. 5).

Instead of rods, one may also use coarse polymer tissues filling the interelectrodic gap for promoting turbulence. Nonconducting particles in the form of packed or fluidized beds enhance mass transfer considerably but do not offer additionally a high volume specific electrode area. Therefore packed-bed electrodes which provide both are even more efficient in enhancing space time yields. Another possibility to en-

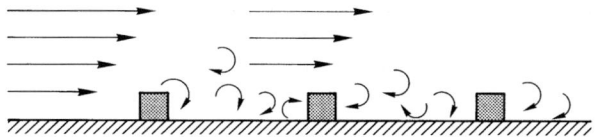

Fig. 8.16. Laminar flow disturbance by studs embedded into a planar electrode are working as turbulence promoters

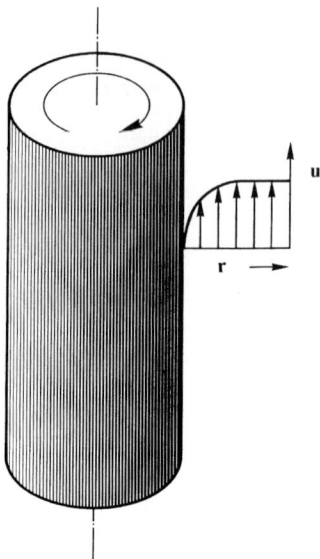

Fig. 8.17. Schematic presentation of a rotating cylinder electrode with steep velocity gradient generated at the cylinder surface

hance mass transfer is provided by rotating electrodes. Figure 8.17 depicts schematically the rotating-cylinder electrode which due to a steep velocity gradient at the electrode surface induces turbulence and high mass transfer rate. Here the fast relative velocity of the electrolyte is generated by the fast rotation of the electrode whereas the electrolyte is essentially at rest – thus sparing a lot of energy dissipation induced by friction and turbulence generation of the circulated electrolyte (A drawback is the electrical contact across a rotating cylindrical collector).

The trickle-tower cell depicted in Fig. 8.18 is easy to construct and scale up which also offers favourable mass transfer rates together with relatively high volume specific electrode area a_e. It is composed of several layers of conducting particles in form of Raschig rings or discs separated from each other by insulating polymer mesh. The electrode particle of each layers are acting as bipolar electrodes because the trickle tower is operated with little electrolyte hold up, so that along the electrolyte film running down along the surface of the conducting discs or rods, the electrolyte current generates an ohmic potential drop which is higher than the minimum cell voltage of the performed electrolysis process.

If the void between the particles is large enough, that means if the mesh is coarse enough, the trickle tower can be used to produce gas.

The trickle-tower is restricted to cell reactions which can be performed in undivided cells and it is particularly useful for performing multiphase reactions for instance the anodic epoxidation of propylene via anodic bromohydrination of this olefin which is fed into the reactor in gaseous form. It is electrochemically absorbed by the electrolytic formation of hypobromite:

8.8 Cell and Electrode Design

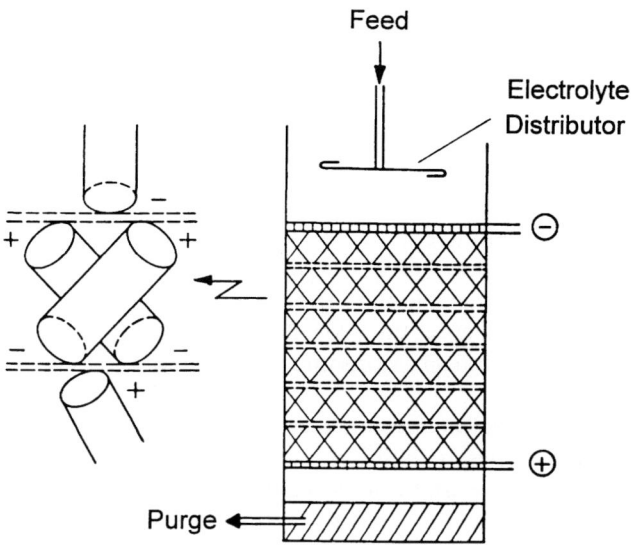

Fig. 8.18. Schematic of trickle-tower cell

$$2Br^- - 2e^- \rightarrow Br_2 \tag{8.4}$$

$$2H_2O + 2e^- \rightarrow H_2 + 2OH^- \tag{8.5}$$

$$Br_2 + 2OH^- \rightarrow Br^- + BrO^- + H_2O \tag{8.6}$$

and reactive absorption of the gaseous propylene

$$\left(CH_2 = CH - CH_3\right)_g \rightarrow \left(CH_2 = CH - CH_3\right)_{solv} \tag{8.7}$$

$$\left(CH_2 = CH - CH_3\right)_g + BrO^- + H_2O \rightarrow CH_2Br - CHOH - CH_3 + OH^- \tag{8.8}$$

followed by hydrolytic epoxidation

$$CH_2Br - CHOH - CH_3 + OH^- \rightarrow \underset{\underset{O}{\backslash\ /}}{CH_2 - CH - CH_3} + Br^- + H_2O \tag{8.9}$$

Also electrochemical cleaning of effluent gases is performed with trickle-tower electrolyzers

8.8.5
Capillary Gap Cells

A cell design combining ease of construction and scale up with a small interelectrode gap is the capillary gap cell, which was dealt with in Chap. 5. It consists of a stack of bipolar rectangular or circular-electrode disks, see Fig. 5.9 a,b, separated by nonconducting spacers. The electrolyte enters the circular stack via a central channel and is radially distributed between the electrodes. It enters the rectangular stack from one side and flows along the electrodes. A certain advantage of the concentric vs the rectangular capillary gap cell is its easier construction and that there is no need for gaskets at the two sides of the stack. The modern form of the undivided cell of the Baizer–Monsanto process today is a rectangular gap cell (see Chap. 12).

Although bipolar capillary gap cells are designed as undivided cells, they can be operated in a way which avoids electrochemical reconversion of a product at the respective counter electrode. This can be achieved by carefully matching of disk geometry and flow velocity to ensure that the diffusion layer thickness, which increases along the electrode radius or length respectively, does not extent to the counter electrodes. By this means the products formed at the electrodes are prevented from contacting the counter electrode. However, the products must be removed, e.g., by extraction or by other means before the electrolyte can be recycled again into the electrode gap. Whereas the scaling up of the stack height and the number of electrode disks is simple, the outer disk radius of a circular capillary gap cell cannot be enlarged without limit. The slower linear flow velocity along the radius lowers the local mass transfer coefficient, which lowers the mean current density and therefore the space–time yield does not increase appropriately as the outer radius is scaled up. The decreasing mass transfer coefficient along the radial flow direction is avoided by the pump cell: A rotating disc electrode around a central shaft is placed between two stationary electrode disks. By this means a tangential velocity component, which is caused by disk rotation and which increases along the radius, is superimposed on the radial flow velocity component, which decreases along the radius. This modified flow pattern keeps variation of the local mass-transfer coefficient and current density with the radius small. The number of electrodes can be easily increased by always placing one stationary electrode between two rotating electrodes. Operation is bipolar, alleviating the need to contact the rotating electrodes. The rotation serves also to pump the liquid through the gaps of the stack (see Fig. 8.19).

8.8.6
Swiss-Roll Cell

An effective, simple cell design for undivided narrow gap cells is offered by the Swiss-roll cell. Two metal foils acting as electrodes are separated from each other

8.8 Cell and Electrode Design

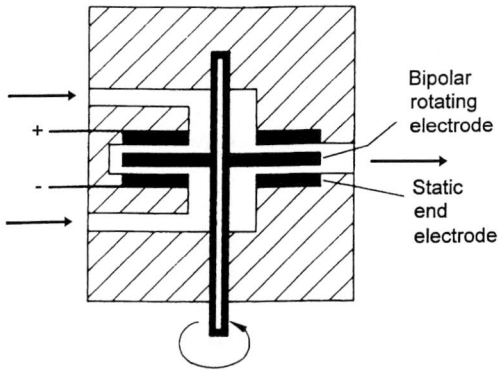

Fig. 8.19. Schematic of a rotating circular capillary gap cell also named pump cell

by two insulating plastic meshes or tissues that act as turbulence promoters and are rolled up (Fig. 5.9 c). This cell simultaneously provides large specific electrode area and low ohmic voltage drop. Because of the thinness of the electrodes and their relatively high electrical resistance a nonuniform current distribution results if the electrode contacts are positioned at the same end. However, this can be avoided by contacting the electrodes at opposite ends: one electrode contact is in the centre of the roll, and the other is at its periphery. Scaling up of the electrode lengths then does not cause any trouble, but this is at the expense of a higher total cell voltage.

The main disadvantages of the Swiss-roll cell are: (a) relative low mass-transfer rates which, however, can be enhanced by the incorporation of polymer mesh and (b) the expressed sensitivity of this device towards clogging by debris and unavoidably formed polymeric sludge. Short circuiting of the cell by conducting particles, stemming from the electrolyte loop, captured in the very thin interelectrodic gap is another hazard. This type of cell has nonetheless been developed to technical scale using nickel-sheet electrodes for producing a precursor of vitamin C (Chap. 12).

8.8.7
Cells with Three-Dimensional Electrodes

The packed-bed and fluidized-bed cell which became so important for electrochemical removal of noxious effluents and for the recuperation of traces of metals from effluent streams are cells which establish an incomparably high space time yield for the electrochemical conversion of highly diluted solutions. They have been dealt with in Chap. 7, under dispersed systems.

8.9
Power Supply for Electrochemical Plants

Usually the grid supplies alternating current of 50 or 60 Hz with a voltage of from 120 V upwards. For electrolysis the voltage must be transformed and the current rectified.

8.9.1
Rectifiers

Today only silicon diodes are used for power rectifiers because of their high efficiency, reliability and relatively low costs. They are quite flexible with respect to the actual currents drained from them. They are available in different size the respective maximal currents ranging from 50 to 500 or even 1000 A. Across the barrier layers of such silicon diodes, the voltage drop, which is one of the mayor reasons for energy losses in rectifiers, amounts to only approximately 1 V. To remove the dissipated heat rectifiers are either cooled by circulated air or for high currents they are cooled by a special water cooling.

8.9.2
Transformer Wiring

Larger industrial electrolysis plants are generally supplied by high tension three phasic alternating current. The rectifier equipment usually consists of:
– transformer capable of variable output voltage with adequate compensation for changing input voltage;
– silicon rectifiers;
– constant-current control gear;
– transducers for metering and control;
– control panels;
– insulators and safety devices;
– cooling equipment; and
– ancillary safety and monitoring equipment.

Each set of rectifiers is connected through high-voltage switchgear to the three-phase supply. Smaller units use a 10–30 kV supply, but large units can be connected into the high-voltage power system (>100 kV).

The transformer usually is either operated in interphase transformer connection, (Fig. 8.20 a), or in three-phase bridge-connection mode, (Fig. 8.20 b). Both figures demonstrate the pulsed currents in two of the primary coils, and in one of the secondary coils, the voltages in the secondary coils and the resulting dc-voltage with its hum. The power of such transformers is generally controlled by changing the inductance of the primary coils synchronously. The power specific costs for rectifier units decreases sizeably with increasing power approaching a level of approximately only 40 US\$ kW^{-1} for plants of 100 MW size and larger.

8.9 Power Supply for Electrochemical Plants

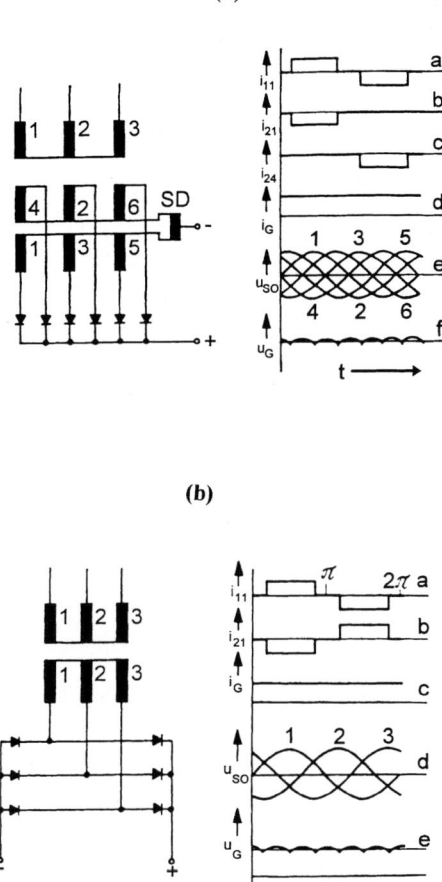

Fig. 8.20. Wiring of transformers supplying semiconducting rectifier devices: **a** interphase transformer connection; **b** three phase bridge connection

8.9.3
Further Equipment

In most electrochemical plants the additional equipment which is necessary for product separation, purification and conditioning includes heat exchangers, distillation and absorption columns, blowers and compressors. This equipment for normal "chemical" unit operations is not at all a negligible quantity. Whereas for instance in chloralkali electrolysis plants the costs for cells, power supply and electrolyte loop amount to approximately 75% of total investment costs and the rest is expended for non-electrochemical parts, the situation for typical electroorganosyntheses processes is quite different. There are known a number of cases where the electrochemical part of the total process demands a share of less than 20% of total investment costs. Heat exchange, separation steps and equipment for supplementing unit operations are however not dealt with in this book, since they are thoroughly treated in other textbooks on chemical engineering. But it must be stressed that for process development these traditional chemical process engineering tools are indispensable for the electrochemical engineer. The costs of the conventional equipment can be greatly influenced by the chosen mode of operation of the electrolyzer as for instance a lower selectivity of the reaction would demand a greatly enlarged demand for product recovery and purification. According to the general economic rules by skilfully engineering and designing the whole process and choosing properly the electrochemical operating conditions one has to aim at the cost optimum of the product. Taking into account the constraints by environmental protection considerations is, of course, indispensable in this context.

Further Reading

K. Kaesche, Die Korrosion der Metalle, Springer-Verlag, Berlin, Heidelberg, New York 1979

A. Rahmel, W. Schwenk, Korrosion und Korrosionsschutz von Stählen, Verlag Chemie, Weinheim, New York, 1977

E. Heitz, R. Henkhaus, A. Rahmel, Korrosionskunde im Experiment, Verlag Chemie, Weinheim, Deertield Beach, Basel 1983

K. Scott, Electrochemical Process Engineering, Plenum Press, New York 1995

CHAPTER 9

Process Development

9.1
Scope and Purpose of Laboratory and Pilot Plant Measurements

Designing an electrochemical process, constructing the respective technical electrolyzer and defining its optimal operating conditions demands the knowledge of the underlying fundamental electrode kinetic data and detailed, reliable information on fluid flow and mass transfer including residence-time distribution for different optional reactor designs. Therefore initially the electrode kinetics of a given electrochemical reaction is usually investigated in typical laboratory devices – in particular at rotating electrodes – which have nothing to do with electrode geometries and cell configurations chosen at a final stage of the technical realisation. Typical devices for electrode kinetic measurements are supposed:

(a) to define exactly the prevailing mass transfer conditions and to allow for variations of mass transfer rates across at least one but possibly two orders of magnitudes namely from 10^{-4} cm s^{-1} up to an upper limit of 10^{-2} cm s^{-1};
(b) to assure an even current density distribution across the entire electrode surface;
(c) to allow for corrections of the ohmic potential drop in the electrolyte to obtain IR-corrected current/voltage curves.

Apart from electrode kinetics a second important question to be investigated prior to designing a process and the respective electrolyzer is the estimation of the mass transfer rate for any choice of electrolyzer configuration and residence time in the electrolyzer. For this purpose downscaled versions of the respective cell geometries may be investigated with the aim to obtain generalised adimensional correlations relating mass transfer to fluid flow and current density distributions to relevant geometrical and electrochemical quantities of the cell and the electrode/electrolyte system. For many electrolyzer geometries such correlations are found in the literature. But one should take into account that by downscaling, fluid-dynamic inlet and outlet disturbances become more prominent and yield in general in effectively too high mass-transfer rates.

The purpose of all these investigations concerning electrode kinetics and mass transfer is to obtain a reliable basis for modelling the electrolyzer as an electrochemical reactor. Eventually based on a sound electrolyzer model one would aim at modeling the whole process. Further one tries in a downscaled pilot plant to learn by experimentation something about long-term effects like electrocatalyst deterioration, corrosion of electrolyzer and electrolyte-loop components and accumulation of impurities and side products in the loop.

A completely different, purely empirical approach – which is not founded on physicochemical information like electrode kinetics and mass transfer – but which might be justified equally well for complex systems – consists in statistically investigating the effect of parameter variations on the outcome of the process – this again being performed in a downscaled version of the cell. This approach dispenses with any physical and physicochemical background, knowledge and interpretation of the process and confines to the aim of developing a purely mathematical model correlating process output parameters like current efficiency, reaction selectivities, specific energy consumption with process input parameters like current density and mass transfer rates, process temperature reactant concentrations and raw material utilisation. For this type of investigation it is important to choose the proper factorial design of the experiments. But it is often decisive to use not a downscaled but a real version of the electrolysis cell because only such a cell exhibits the real and effective properties like mass transfer rate and current density and their spatial distribution across the electrode surface, or for instance residence time distribution of the electrolyte in the cell. These properties often cannot be extrapolated satisfactorily from downscaled cells to cells of real size since edge and inlet and outlet disturbances are much more pronounced for smaller than for larger cells.

For simple electrochemical reactions as for instance for anodic chlorine evolution from highly concentrated brine, which are only little influenced by mass transfer this circumstances can be neglected and for them modelling may become a relatively easy task. For electroorganic synthesis reactions however, in the course of which parallel and consecutive chemical reactions can and almost always do compete with the main reaction one has to be much more careful in modelling and in collecting a set of sufficiently complete and reliable fundamental data for physical or mathematical modelling.

9.2
Laboratory Methods

9.2.1
Steady-State Measurements of Current Density Potential Correlations

9.2.1.1
General Remarks

Current density potential correlations at single electrodes are the basis of the prediction of current density / cell voltage correlations. It is important to note, that in many cases these single electrode current / potential curves are decisively determined in particular by the electrode material but also the respective choice of the electrolyte and the temperature. The current voltage curves of anodic chlorine evolution at carbon and RuO_2-activated titanium anodes is just one example. It is therefore essential in the measurement of current voltage curves to establish as closely as possible the conditions, which are likely to prevail in the technical process.

9.2.1.2
Measuring Devices

Steady state measurements of current voltage curves demand the existence of well defined fluid dynamic and mass transfer conditions. Since mass transfer conditions are well defined and equal for the whole surface of rotating disc electrodes at known rotation speed this electrode is often preferred to any other measuring device although many measurements have been reported, too, in the literature on current voltage curve measurements with rotating cylinder electrodes and stationary electrodes in stirred electrolytes. Any other experimental assembly as electrodes in flow channels, stationary electrodes in stirred tanks etc. are less practical and exhibit less well defined mass-transfer conditions and should therefore not be taken into consideration. Mass transfer to rotating disc electrodes has been dealt with in Chapt. 5 and for laminar flow is described by the Levich equation which being transformed into adimensional correlation of Sh to Re reads

$$Sh = 0.60 Sc^{0.33} Re^{0.5} \qquad (9.1)$$

and is valid for $10^2 < Re < 10^4 - 10^5$ with the radius r of the disc as the relevant length for Sh and Re and with the definition $Re = r^2\omega/\nu$

Substituting $i_l = v_e F k_m c$ one obtains for mass-transfer limited current densities,

$$i_l = 0.6 v_e F D^{2/3} r \nu^{-1/6} \omega^{1/2} c \qquad (9.2)$$

9.2.1.3
Evaluation of Rotating Disc Measurements

Plotting i for smaller depolariser concentrations ($c < 2.10^{-2}$ mol dm^{-3}) vs the electrode potential allows to measure i_l. This helps to determine preliminarily whether the process is mass transfer or reaction controlled and in case it is mass-transfer controlled, how many electrons (v_e) are consumed by the electrochemical conversion. Plotting $i_l \omega^{-1/2}$ according to Eq. (9.2) vs $\omega^{-1/2}$ gives a line parallel to the abscissa provided the electrode reaction is mass-transfer controlled.

As shown in Fig. 9.1, starting at low rotation speeds with a higher $i_l \omega^{-1/2}$ value which levels off to half of the initial value or less at high rotation speeds indicates either that in a sequence of electrochemical charge transfer, chemical reaction and further charge transfer the intermediate chemical reaction becomes rate de-

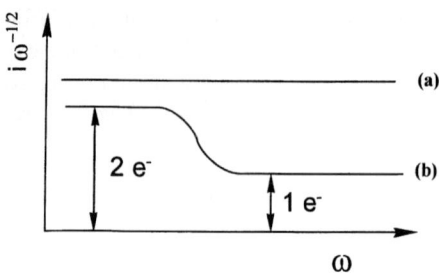

Fig. 9.1a,b. Rotating-disc measurements: **a** constant value of $i_1\omega^{-1/2}$ against $\omega^{1/2}$ is indicative of a simple charge transfer reaction; **b** if $i_1\omega^{-1/2}$ decreases from a high value at low rotation speed to a lower value at high ω-values, this is indicative of interfering chemical reactions with lower rate following a first charge transfer in a ECE (electrochemical-chemical-electrochemical) sequence

termining and relatively too slow at high rotation speed lowering the number v_e of electrons transferred from for instance two at lower to one at higher rotation and mass-transfer rates or

$$A \xrightarrow{k} B \tag{9.3}$$

$$B + v_e \longrightarrow \text{products} \tag{9.4}$$

that the electrochemical reaction is preceded by a chemical reaction, which becomes rate determining at higher rotation speeds according to Eqs. (9.3) and (9.4). Already by such relatively simple measurements the investigator obtains an insight into the kinetics of the electrochemical reaction. The methodology of a detailed electrode kinetic investigation has been outlined in Chap. 4.

It is, however, important for reliable process modelling, that the kinetic data are determined across that range of concentrations and mass transfer rates which is expected also to prevail in the cell under technical conditions so that it is senseless to restrict rotating disc electrode measurements only to low concentrations, which is very often done because low concentration simplify the measurement of mass transfer limited current densities. For instance reactions of higher order competing with first order reactions often hardly can be detected at lower levels of reactant concentrations. This circumstance is often neglected. As an example the reader may refer to controversial results published on the electrode kinetics of the cathodic hydrodimerisation of acrylonitrile to adipodinitrile (Monsanto process) which proceeds only with sufficiently high concentrations of the depolariser, according to Eq. (9.5):

$$2CH_2=CHCN + 2e^- + 2H^+ \rightarrow NC-CH_2-CH_2-CH_2-CH_2-CN \tag{9.5}$$

because only at higher acrylonitrile concentration the key reaction, the coupling of acrylonitrile radical anions to acrylonitrile,

$$CH_2=CHCH+e^- \longleftrightarrow (CH_2=CHCN)^{\cdot -} \tag{9.6}$$

$$CH_2=CHCN+(CH_2=CHCN)^{\cdot -} \rightarrow (NC-\dot{C}H-CH_2-CH_2-\overline{C}H-CN) \tag{9.7}$$

becomes sufficiently fast compared to fast first order protonation of the radical anion which is followed by further reduction to proprionitrile.

$$(CH_2=CHCN)^{\cdot -}+H^+ \rightarrow CH_3-\dot{C}HCN+e^-+H^+ \rightarrow CH_3CH_2CN \tag{9.8}$$

In this case confining the measurement of current voltage curves to acrylonitrile concentrations of lower than 10^{-2} mol dm^{-3} would never allow to reveal the electrode kinetics of the coupling reaction which is the intended electrosynthesis reaction.

Considering the range of mass-transfer rates which predominates in technical electrolyzer cells, one has to account for mass-transfer rates from several 10^{-4} cm s^{-1} to upper values of 10^{-2} cm s^{-1} at the most. Therefore rotating disc measurements should concentrate on this range of k_m since even under free convection conditions in larger electrolyzers k_m does not fall below 10^{-4} cm s^{-1} and on the other hand values above 10^{-2} cm s^{-1} are hardly achieved under strong, forced convection in such cells. Applying Eq. (9.2) to these limits with D=10^{-5} cm^2 s^{-1} and the kinematic viscosity v=10^{-2} cm^2 s^{-1}, which are typical for aqueous electrolytes, means that using the rotating disc electrode in aqueous electrolytes, one should vary the quantity ω from lower values of fractions of 1 s^{-1} to values not exceeding several thousands per s.

9.2.1.4
Current–Voltage Correlation for Competing Reactions by Non-Electrochemical Methods

It must be strongly stressed that current voltage curves most often do not allow to obtain a clear insight into the details of the electrochemical and in particular the chemical kinetics of involved reactions as far as competing reactions or for instance the occurrence of side and consecutive reactions are concerned. Therefore the measurement of current voltage curves have to be supplemented in a later stage of the investigation by "preparative" current–voltage curves based on the chemical analysis of product mixtures obtained by long-term but small-scale batch electrolysis under controlled mass transfer and current density conditions. As an example Fig. 9.2 shows such preparative current voltage curves in the form of a plot of current yields for different products against anode potential for the anodic oxidation of benzene at lead dioxide anodes. Figure 9.2 is a more detailed version of Fig. 4.18 b demonstrating additionally that not only oxygen

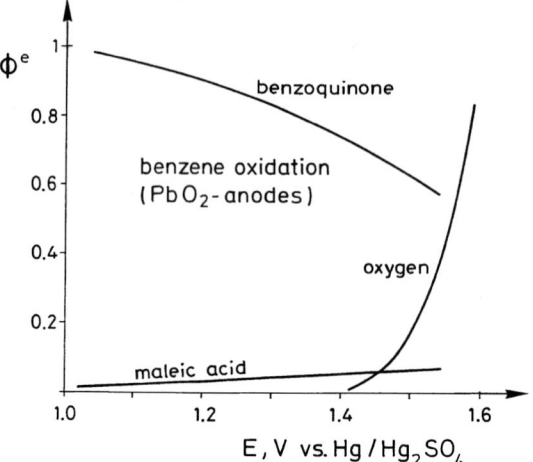

Fig. 9.2. Product current efficiencies for anodic oxidation of emulsions of benzene in 1 mol dm^{-3} H$_2$SO$_4$ at PbO$_2$-anodes (according to J. Electroanal. Chem., *70*, 333 (1976))

is evolved at higher potentials but that also a further oxidation of benzoquinone to maleic acid has to be taken into account.

9.2.1.5
The Ring Disc Electrode

Electrochemical reactions sometimes due to a sequence of charge transfer and homogeneous chemical reactions produce reactive chemical species like metal ions of reduced or enhanced charge, radical ions or radicals obtained by reduction or oxidation of organic molecules or organic ions, which possess lifetimes comparable to diffusion times typical for prevailing mass-transfer conditions. In such cases it is desirable to determine the fast kinetics of the chemical reaction in which these species are converted. Rotating ring disc electrodes are easy to use for this purpose. Figure 9.3 a depicts schematically the assembly and flow pattern of a rotating ring disc electrode. Any species being generated at the disc electrode is carried along with the outward flowing electrolyte and on its way outward it diffuses away from the electrode surface toward the electrolyte bulk – and if it is a reactive intermediate – simultaneously reacting according to its specific chemical reaction rate. Figure 9.3 b describes schematically the concentration profile of such intermediates. Since the mass transfer rate at the disc is independent on the distance from the centre, the concentrations close to the electrode are uniformly layered parallel to the disc. Only at the edge of the disc there develops a concentration discontinuity. Fig. 9.3 b depicts the depolarizer concentration profile at mass transfer limited current density.

9.2 Laboratory Methods

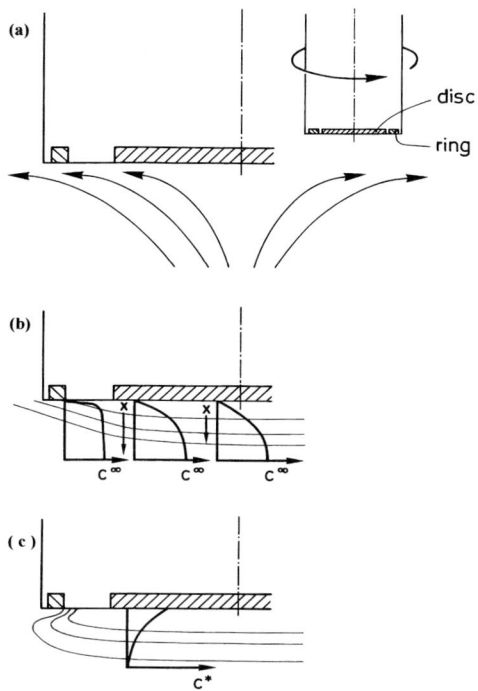

Fig. 9.3a–c. Rotating-ring disc electrode: **a** flow pattern at ring-disc electrode; **b** concentration distribution of converted depolarizer with concentration contour lines; **c** concentration distribution c^* of reactive intermediate which is reconverted with mass-transfer-limited current density at the ring (c^* at ring surface equals zero)

If the ring is polarised to potentials which assure complete reversal of the electrochemical reaction performed at the disc, mass-transfer-limited current densities for reconversion of the intermediate into the starting material prevail and the ring current will allow to determine the amount of intermediate which survived the time necessary for convective transport of the intermediate from the edge of the disc to the ring.

The schematic profiles of Fig. 9.3 c are taking account of this condition assuming the concentration of the intermediate to be zero at the ring surface since there the intermediate is reconverted with mass-transfer-limited current density. The faster the reaction of the intermediate in the electrolyte the steeper the concentration profiles perpendicular to the disc surface and in the gap between disc and ring and the smaller the limiting current density measured for reconversion of the intermediate. Therefore for reactive electrogenerated species the so called collection efficiency of the ring

$$N = \frac{i_l(\text{ring})}{i(\text{disc})} \qquad (9.9)$$

Table 9.1. Collection efficiencies N_0 for different ring/disc electrode geometries

	1	2	3
r_1/cm	0.387	0.348	0.364
r_2/cm	0.398	0.386	0.378
r_3/cm	0.405	0.438	0.484
N_0	0.097	0.262	0.404

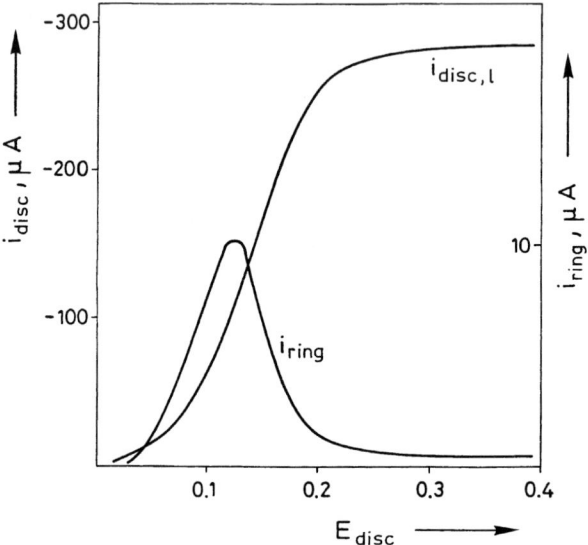

Fig. 9.4. Disc current i_{disc} and mass-transfer-limited oxidation current for Cu I oxidation at the ring of a rotating-ring disc electrode (according to Albery and Hitchman: Ring-Disc Electrodes, Oxford Univ. Press 1971)

is smaller than the collection efficiency N_0 for nonreactive products and becomes an immediate measure for the reactivity of the reactive electrode product. If the product of the disc reaction is stable (for instance for the reaction $Fe(CN)^{4-} \rightarrow Fe(CN)^{3+} + e^-$ the collection efficiency, which is determined by the gap width, ring width and disc diameter but does not depend on the rotation velocity, can easily be measured and this measurement is taken as reference for the measurement of life times and reaction rates respectively of unstable intermediates.

Table 9.1 gives three examples of measured collection efficiencies in order to show that N_0 increases as the width of the ring (r_2-r_3) increases and that the ring current is one tenth to one half of the disc current. Figure 9.4 depicts disc and ring current for the reduction of Cu^{2+}-ions in dependence on the disc potential. Since the ring potential was set to 0 V, Cu^+ ions generated at the disc by 1e-reduction of Cu^{2+} ions are reoxidised at the ring. The ring current indicates what

9.2 Laboratory Methods

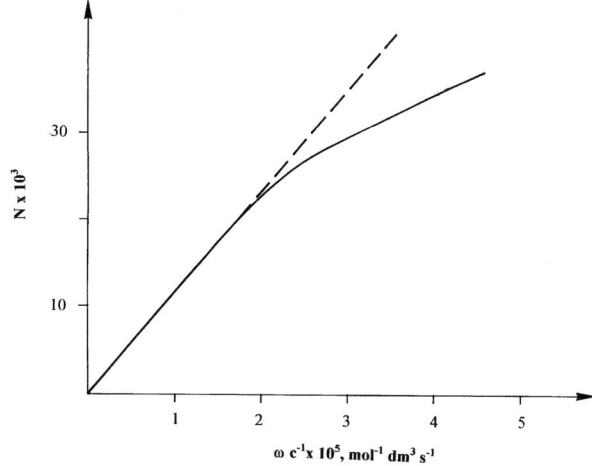

Fig. 9.5. Presentation of dependence of collection efficiency of disc-ring electrode for bromide oxidation/bromine reduction on rotation speed and concentration c_0 of scavenger (allyl alcohol). At high rotation speed the collection efficiency deviates from the straight line because it approaches the value measured without reaction (according to Albery and Hitchman ring: Ring-Disc Electrodes, Oxford University Press, 1977)

percentage of the initially drawn cathodic disc current goes into Cu^+ formation – the rest being consumed by cathodic deposition of metallic copper at the disc. This is an important information for Cu-electrowinning and refining.

In a similar manner decreasing collection efficiency with decreasing rotation speed and hence increasing flow time across the ring/disc gap is indicative of reaction times which compare to flow times. Figure 9.5 shows this by a plot of collection efficiency for anodically generated bromine

$$Br^- \rightarrow \frac{1}{2} Br_2 + e^- \tag{9.10}$$

which reacts according to a second order reaction with allyl alcohol.

$$CH_2 = CH - CH_2OH + Br_2 \rightarrow CH_2Br - CHBr - CH_2OH \tag{9.11}$$

The plot of N against ($c_{alcohol}^{-1}$) shows the reaction to be first order with respect to allyl alcohol according to Eq. (9.12). At high rotation speed the collection coefficient approaches that for stable electrode products. This is the reason for the deviation from the linear increase of N vs ωc^{-1} for higher values of the quantity ωc^{-1}. From the straight line in Fig. 9.5 the rate constant of reaction at Eq. (9.11) can be determined making use of Eq. (9.12):

$$\frac{N}{N_0} = \frac{0.339 r_2^2 D^{1/3}}{r_1^2 v^{1/3} k c_{alcohol}} \omega \tag{9.12}$$

With r_1 and r_2 the outer radius of the disc and the inner radius of the ring being known the rate constant k for reaction of bromine with the alcohol is determined, (k = 6.2×10^5 M^{-1} s^{-1}).

9.2.2
Non-Steady State Methods

9.2.2.1
General Remarks

In the steady state the respectively slowest step determines the total rate and therefore steady state measurements permit only conclusions concerning this step. In order to investigate the faster reaction steps, it is necessary to study the system under investigation as it develops with time upon sudden or periodic changes of process parameters. An understanding of the time-resolved behaviour of the system (depolariser/electrode/electrolyte solution) is often of some importance in industrial reaction control especially whenever there exists a complex sequence of different competing reactions or electrode deterioration due to corrosion or catalyst fowling due to formation of surface films, for instance by polymers.

For non-steady state measurements one uses either measurements with the electrode potential as the commanding quantity (stepped or continuously changed with time) and current transients are recorded or an imposed current density is the commanding quantity and then transients of the electrode potential are measured. Cyclic voltametry is one of the most popular non-steady state methods. Cyclic voltametry, however, is always bound to use low concentrations of the depolariser – typically 10^{-3} mol dm^{-3} or 10^{-2} mol dm^{-3} at the most – and therefore very often cannot provide the kinetic information necessary to understand a technical process, operating at depolariser concentrations which are typically by two orders of magnitude higher or even higher than that (c >1 mol dm^{-3}). Cyclic voltametry is, however, a potent tool in fundamental electrochemistry dealing with typical reactions of electrode surfaces as cathodic hydrogen adsorption or anodic oxide layer formation.

9.2.2.2
Potentiodynamic Polarisation Curves

Potentiodynamic measurements form a transition between steady-state and non-steady-state methods. The cell is provided potentiostatically with a potential which changes linearly with time, and the resulting current density is registered with a fast response x-y recorder.

$$i = f\left(E(0) \pm \dot{E}t\right) \tag{9.13}$$

Where the scanning rate $\dot{E}$ of the potential is small (quasi-stationary) the curve obtained resembles steady-state curves.

9.2.2.2.1
Cyclic Voltammetry and Linear Potential Sweep Method

If the concentration of the electrochemical reactant is so small that at sufficiently high polarisation the transport of the species to the electrode surface becomes rate-determining, the potentiodynamic method with defined convection gives a current density curve with mass transfer limited current density (curve a in Fig. 9.6) characterised by the half-wave potential $E_{1/2}$. When working on a plate electrode with a higher potential sweep rate in a stagnant solution, however, the decrease in limiting current density which is typical for a stagnant solution will

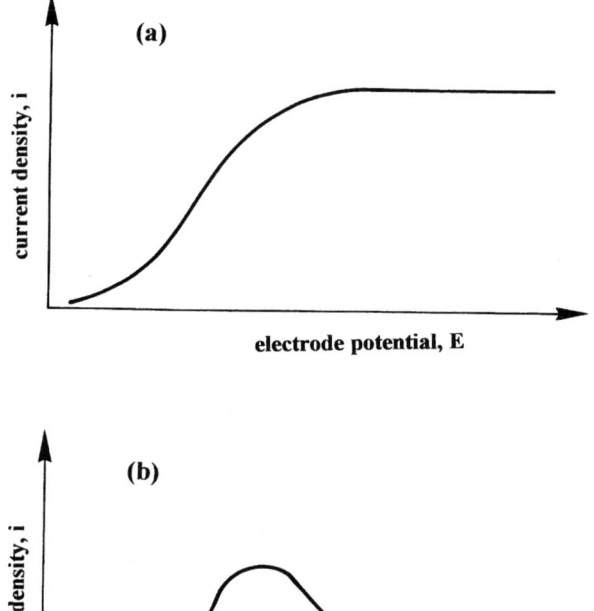

Fig. 9.6a,b. Schematic presentation current voltage correlation of electrode reactions measured as sweep voltammograms at electrodes: **a** with controlled convection; **b** in stagnant electrolyte

Table 9.2. Diagnostic criteria for the evaluation of cyclic voltammograms

Criterion		Interpretation
$dE_p / d\lg \dot{E}$	=0	reversible charge transfer
	=30 mV	irreversible activation controlled reaction
$\dfrac{i_{p,\text{forward}}}{i_{p,\text{backward}}}$	=1	reversible electrode reaction
	>1 to ∞	irreversible electrode reaction
$\dfrac{i_p}{\dot{E}}$	=const	elementary reversible or irreversible electrode reaction
	$= f(\dot{E})$	preceding or consecutive chemical or catalytic steps

override the increase in current density with rising potential, resulting in a non-steady-state curve (curve b in Fig. 9.6) with a peak in current density indicated by the peak potential E_p and the allocated peak current density, i_p. Although the mathematical theory of these curves can only be applied strictly for one experimental run (linear potential-sweep method) scanning voltammetry is often used repetitively in the form of cyclic voltammetry. In this technique the potential range between two reversal points is repeatedly traversed on a linear time scale, alternating in cathodic and anodic direction. Voltammograms which do not change further with higher numbers of repetition are generally obtained after just a few runs. They contain interesting information in their often complex appearance with a series of peaks about the mechanism of the electrode reactions taking place and adsorption and desorption processes at the electrode. Convolution potential sweep voltammetry corrects such curves from time dependent towards time – independent mass transfer has greatly extended the usefulness of cyclic voltammetry, and it is one of the most frequently applied electrochemical techniques.

In order to obtain cyclic voltammograms a potentiostat is used, the potential of which is controlled by a voltage scan generator. The current density vs potential graphs are recorded with the aid of a fast response x-y recorder or, at high potential sweep rates $\dot{E} > 1 V/s$, with an oscilloscope or a transient recorder. In order to evaluate the cyclic voltammograms the peak potentials E_p and peak current densities i_p are determined as well as their dependence on the potential sweep rate. The most important criteria for various types of reactions are listed in Table 9.2. For a detailed discussion of complex reaction mechanisms see the literature (Macdonald, Brown, Gileadi, Adams, Bard).

9.2.2.2.2
Initial Polarisation Curves

Where there prevails mixed activation and diffusion control or where chemical reactions precede or follow the charge transfer reaction, the kinetic parameters (charge transfer coefficient, exchange current density) cannot be determined from steady-state polarisation curves. If the potential and current density are determined immediately after a potential or a current density step respectively (e.g. switching on or off the circuit), an initial polarisation curve is obtained, the shape of which is determined solely by the kinetics of the charge transfer reaction.

Potentiostatic and galvanostatic initial measurements are equivalent, however, the methods of data acquisition are different. In the case of galvanostatic initial measurement, overpotential transients are obtained, and the linear part can be extrapolated to zero time. With potentiostatic initial measurement, the current density vs. time curves cannot be extrapolated directly; instead $1/i$ is plotted against $t^{1/2}$ and extrapolated to $t = 0$.

9.2.2.3
Square-Wave Pulses

With the aid of a square-wave generator, producing rectangular current pulses of any desired length it is possible to investigate electrode kinetics by means of potential transients as the response to the applied current pulse (see Fig. 9.7). Galvanostatic current pulses are widely used. A very useful method for the study of electrode kinetics are double current pulse methods (Gerischer) working with two different successive square wave current pulses. In this method a short pulse of a higher current density for rapidly charging the double layer precedes the main pulse used for measuring potential transient and transition time τ_1. Another double pulse method applies, after reaching the transition time, a second reversed pulse of the same current density with reversed polarity. The second transition time, τ_2, for the reversed pulse is measured. The relationship between the two transition times enables the calculation of the rate constant of

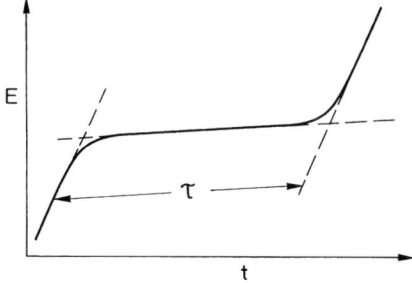

Fig. 9.7. Schematic presentation of a potential time transient after galvanostatic pulse

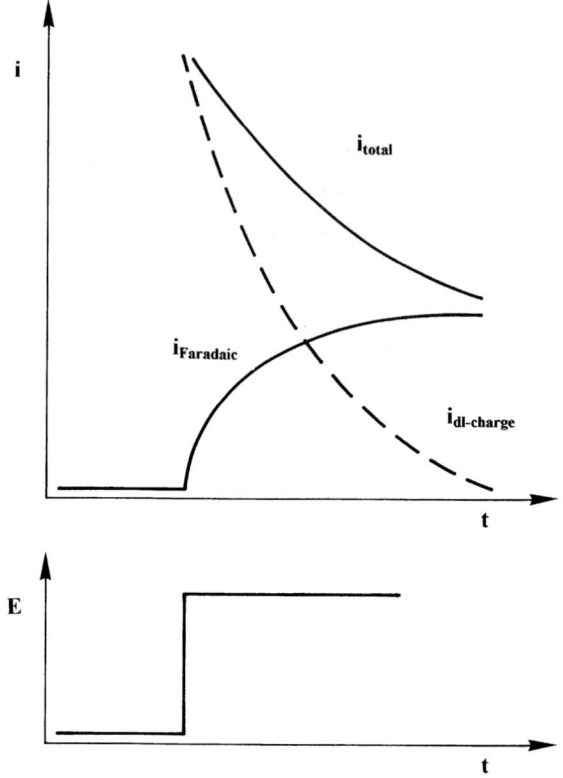

Fig. 9.8. Schematic presentation of a potentiostatic current transient. Charging the electrode potential is a relaxation process associated with the relaxation time $\tau = RC_{dl}$. Only after charging of the double layer, the Faradaic current is measured reliably

consecutive chemical reactions following the initial charge transfer. In the application of all pulse or step techniques one has to consider that each experimental arrangement has some lower time limit for working reliably, where at shorter times the observed variable is so greatly influenced by the characteristics of the electrical circuit that information about the electrode process cannot be extracted. Figure 9.8 explains the principles of the measurement of current transients following a potentiostatic step. One has to observe in this case that the sudden step of an electrode potential is not really possible as changing the electrode potential is a charging process always associated with the relaxation time RC_{dl} with R the electrolyte resistance between the Luggin-capillary tip and the electrode and C_{dl} the double layer capacitance of the electrode.

The simplest transient technique is to interrupt the current after a steady state has been established. This may be done also periodically. For an irreversible charge-transfer-controlled reaction the observed potential decay caused by

9.2 Laboratory Methods

Faradaic discharge of the double layer after current interruption is described by the Frumkin equation.

$$\eta_t = \eta_0 - \frac{RT}{\alpha F} \ln\left\{1 + \frac{it}{bC}\right\} \tag{9.14}$$

b=Tafel slope, C=electrode capacitance, i=current density before current interruption.

Current stepping is advantageous in eliminating the IR drop as shown in the following chapter. Furthermore, information about Tafel slope and electrode capacitance can be extracted from such potential decay curves according to Eq. (9.14).

9.2.2.4
Eliminating the IR Drop

In an electrolysis cell with finite current density an ohmic potential drop ΔU_Ω between the tip of the Luggin probe and the working electrode cannot be avoided (see Fig. 9.9). This IR drop increases as the current density increases and add to the electrode potential.

$$E = E_{electrode} + IR_{electrolyte} \tag{9.15}$$

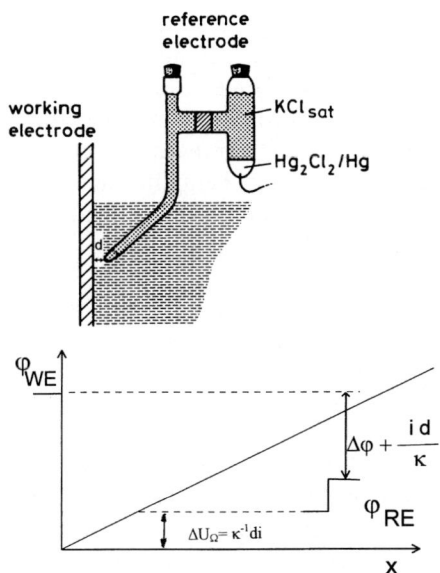

Fig. 9.9. Connection of the reference electrode by means of a Luggin capillary diminishes the ohmic potential drop ΔU_Ω to $\Delta U_\Omega = \kappa^{-1} d\,i$ with κ electrolyte conductivity, i applied current density and d distance of capillary tip to electrode surface

In order to correct or eliminate this inherent systematic error, the following procedures can be adopted.

9.2.2.4.1
Galvanostatic Methods

(1) The potential is measured for various distances between the Luggin probe and the working electrode, and extrapolated to zero distance.
(2) The cell current is periodically interrupted for a short time, and the potential is measured during this short period of current interruption. This demonstrates another specialised use of square wave pulses.

Using interruption techniques implies that it is impossible to measure the potential immediately after the current interruption due to the time characteristics of the applied electronic circuit. This difficulty can be overcome by a computer evaluation of the potential time transient which has to be sampled by a fast transient recorder (sampling rate about 2 MHz).

In general, one has to be cautious with the application of interrupter techniques for the elimination of the IR drop since only electrode systems with a uniform current distribution yield reliable results.

9.2.2.4.2
Potentiostatic Procedures

In what is known as IR compensation, a resistor is placed in series with the cell, with a resistance as close as possible to the effective electrolyte resistance between the Luggin probe and the electrode surface. The voltage drop through this resistor opposes the total voltage between the working and reference electrodes, so that the only potential that is controlled by the potentiostat is the difference between these two, the actual electrode potential, which is kept constant. However, this method poses several problems due to the instability of the operational amplifier at certain values of the compensating resistance, and very high resistances in particular.

9.3
Pilot Plant Methods

9.3.1
General Considerations

In order to model an electrochemical process it is not sufficient to measure the relevant electrode kinetic data and to make use of the generalised adimensional correlations concerning mass transfer in order to define the mass transfer conditions in the chosen electrolyzer. There are many additional circumstances which the electrochemical engineer has to account for in order to design electrolyzers, process flow sheets and to select auxiliary equipment. He has to make use of available data und gather himself experimentally special information on material behaviour, in particular concerning electrode fowling or corrosion, the generation and accumulation of side products, or the necessity and technical means for recycling of non-converted raw materials to name only some of the most important problems to be investigated. Usually a down scaled pilot plant or miniplant is therefore constructed serving as a token to investigate these problems of long term behaviour with reasonably low financial efforts.

Experimental model- or pilot plants reproduce industrial processes on a smaller scale trying to match the behaviour of a big plant in its operation conditions as closely as possible. Mass transfer under in-cell conditions in particular for more complicated cell designs as packed or fluidized bed cells or for more complicated types of electrolytes like two-phase emulsion electrolytes very often need detailed experimental investigation in pilot plants as does the particular residence time distribution. The latter is important to account for side reactions like hydrolysis or solvolysis of the starting material or – even worse – the product which might interfere severely with the electrochemical reaction impairing yield and selectivity of the process.

9.3.2
Mass Transfer Measurements

Electrolyzer cells with multiple flow diverters like cells exhibiting unusual flow patterns cannot be simply modelled in their mass-transfer behaviour with one of the well known adimensional correlations. Another case which is not easy to handle theoretically would be cells in which gas evolution e.g. cathodic hydrogen evolution or anodic oxygen evolution occur as a side reaction improving mass transfer conditions a lot.

For both cases one is interested to obtain a knowledge of the mean mass transfer coefficient – if possible – or of uneven local distribution of the mass transfer conditions and moreover of the residence time distributions. Like for fundamental measurements on mass transfer a simple redox system is used to measure mass transfer conditions by measuring overall and local limiting current densities for a practical cell. For local mass transfer measurements one inserts into the electrode surface microelectrodes being electrically isolated from the surrounding main electrode but their potentials being operated in parallel to the main electrode, thus enabling the determination of local mass-transfer coefficients. Care must be taken to avoid a sizeable depletion of the concentration of the probing system whenever the electrolyte is flowing slowly through the cell. Therefore a pulsed operation of the big and the inserted electrodes is advised. The pulse length has to be long enough to allow for establishing everywhere steady-state conditions with correspondingly developed diffusion layers ($\tau > D k_m^{-2}$) but short enough to avoid sizeable concentration depletion. Off times have to be chosen long enough to avoid depletion $\tau_{off} > 0.1 \tau_{res}$ with τ_{res} being the mean residence time of the electrolyte in the cell $\tau_{res} = V_{cell} V_{el}^{-1}$. The measurement of the mass transfer at simultaneous gas evolution is more involved because one has to take into account the superposition of at least two dif-

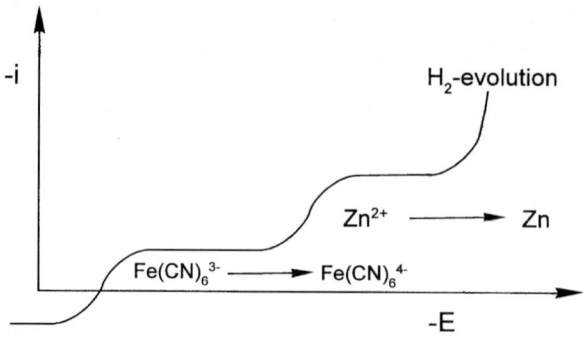

Fig. 9.10. Superposition of current voltage curves of probing system, A (e.g. $Fe(CN)_6^{4-}/Fe(CN)_6^{3-}$, main reaction (e.g. zinc deposition), B, and electrochemical gas evolution (e.g. hydrogen evolution) as a side reaction, C

ferent electrode processes with their superimposed respective current/voltage curves, because the gas evolving process exhibits its own curent potential curve.

Figure 9.10 explains the situation for cathodic zinc deposition from acidic electrolytes accompanied by cathodic hydrogen evolution. In order to measure the mass-transfer-limited current density of the probing system $\left(\dfrac{Fe(CN)_6^{3-}}{Fe(CN)_6^{4-}}\right)$, it is necessary to switch the cathode potential for a very short time from the working potential to a more positive potential at which the mass transfer limited current density for $Fe(CN)_6^-$ reduction can be measured separately. Since this switching causes an interruption of the hydrogen evolution with a subsequent drop of the effective mass transfer, the current density of the indicator system relaxes and should be reextrapolated to the time of switching because usually immediately after switching the signal is disturbed and allows no unambiguous reading.

9.3.3
Determination of Residence Time Distributions

Residence-time distributions had been pointless in most classical industrial electrolysis processes like chloralkali electrolysis, water electrolysis or metal winning/refining. But in chlorate production, see Eqs. (9.19) and (9.20), which is essentially based on a homogeneous reaction

$$2NaOH + Cl_2 \rightarrow NaClO + NaCl + H_2O \tag{9.19}$$

$$3NaClO \rightarrow 2NaCl + NaClO_3 \tag{9.20}$$

residence time distributions might be important. For organic electrosynthesis reactions the situation may be even more critical. For instance, acrylonitrile is hydrolysed by OH⁻ to cyanoethanol, Eq. (9.21):

$$CH_2=CH-CN+H_2O \xrightarrow{OH^-} HOCH_2-CH_2-CN \qquad (9.21)$$

so the residence time distribution in the cathode chamber (the catholyte becomes basic) is decisive for the material loss due to hydrolysis and hence the obtained yield.

Residence time distribution measurements are performed either by the pulse method yielding in differential residence time distribution or by the displacement method which yields in integral residence-time distribution – see Chap. 5.

9.4 Mathematical Modelling and Optimisation by Factorial Design of Experiments

9.4.1 Introduction

Optimisation of electrochemical processes as any chemical process has the purpose to reduce production costs to a minimum. Optimisation of electrochemical as well as chemical processes is usually performed on two different levels.

(i) On the pilot plant and prepilot plant level all possible means of reaction engineering and electrochemical engineering are exploited in order to design intrinsic process parameters which assure optimal performance of the electrolysis process providing:
   – lowest consumption of electrical energy for instance by proper cell design and choice of stable electrocatalysts;
   – high selectivity of the electrochemical reaction together with high current efficiencies;
   – optimal adaptation of effluent composition to the demands for electrolyte work up procedures to name only one of the most important issues;
   – sufficiently high longevity of electrolyzer and electrolyte loop components.
(ii) On the overall process engineering level based on a given cell design and performance and a given generalised process flow sheet, optimisation has the aim to reduce unit production costs by finding the most favourable operation conditions of the relevant process variables as current, residence time, temperature and so on. It is the second type of optimisation "cost optimisation" which is the subject of this chapter.

9.4.2 General Procedure for Optimum Finding by Experiment

The fundamental assumptions underlying any experimental optimisation procedure is that:

1. there must exist various solutions to the problem under discussion, any of which may be chosen;
2. it must be possible to assign a quantitative assessment to each of the solutions, indicating to what extent it contributes to achieving the target.

There are several possible targets in optimisation. In chemical engineering the aim is usually to find the operational procedure which achieves the maximum profit or incurs the minimum cost (economic optimum). Other possible targets are highest possible yield, highest purity, greatest operational safety or minimum environmental pollution (technical optimum). The object of optimisation is solely to extract from many possible solutions the one that most nearly realises the desired optimum. The individual steps towards solving the optimisation are listed below.

1. Definition of a target quantity and the aim of optimisation.
2. Establishing a *mathematical* model of the system concerned (physicochemical, statistical or stochastic model):

$$A = f(p_1, p_2, \ldots, p_n) \tag{9.22}$$

3. Formulation of the target quantity in relation to the variables, with regard to the mathematical model target function:

$$z = f_1(p) + f_2(p) + \ldots + f_n(p) \tag{9.23}$$

4. Determination of an extreme value for the target function, using a suitable mathematical procedure (optimisation method):

$$\frac{dz}{dp} = \frac{df_1(p_{opt})}{dp} + \frac{df_2(p_{opt})}{dp} + \frac{f_n(p_{opt})}{dp} = 0 \tag{9.24}$$

5. Formulation of the optimum value of p as a function of the remaining parameters:

$$p_{opt} = f(p_1, \ldots, p_n) \tag{9.25}$$

9.4.3
Factorial Design of Experiments

According to the outlined, generalised approach the development of electrochemical processes, the transfer of laboratory results onto an industrial scale (scale-up) and the optimisation of production plant demands a quantitative assessment of the relationships between the different characteristic reaction engineering quantities and the adjustable process variables, p_i.

Reaction engineering quantities may be for example product yield, current efficiency, specific energy consumption and fractional conversion, and the variables are for instance current density, cell voltage, temperature, concentrations and flow rate of electrolyte. The relationships between these parameters are on the whole

9.4 Mathematical Modelling and Optimisation by Factorial Design of Experiments

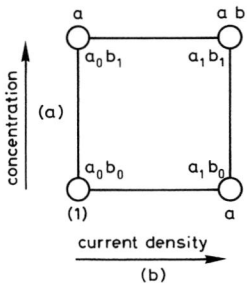

Fig. 9.11. Factorial 2^2 design

very complex and influenced by several factors which cannot easily be quantified, such as electrode material, current density distribution non-ideal residence distribution or impurities in the electrolyte, and it is therefore almost impossible to formulate the relationships on the basis of chemical or physical laws. As had been already pointed out this situation leads to the need for experimental investigation of the whole process, especially on a pilot plant scale simulating the industrial plant. Since these experiments are very costly and time consuming, methods are required providing a maximum of information with a minimum number of experiments. A particularly effective but simple method is the factorial design of experiments. The following description of this method is deliberately confined to a very simple case, but one, which can still be used to deal with practical problems.

In factorial design of experiments, the variables whose influence on a particular quantity is being investigated are referred to as factors. The values of these factors which are set for each experiment are called levels. It is important to note that these levels are not completely arbitrary. On the contrary they are usually chosen on basis of already collected knowledge. For instance a certain electrosynthesis might have zero current efficiency at 1 mA cm^{-2} but already at 15 mA cm^{-2} it is greater than 50%. Then 15 mA cm^{-2} might become the low level of i and 50 and 100 mA cm^{-2} medium and high respectively. For a study of the influence of n factors at m levels, a m^n factorial design is employed. m^n indicates the number of experiments required. An example for the use of a simple 2^2 factorial design is the study of the influence of current density i (factor A) and concentration of reactant c (factor B) on the current efficiency ϕ^e. Each of the two factors (concentration and current density) is assigned a low (a_0, b_0) and a high level (a_1, b_1). These four values give $2^2=4$ combinations of 2 factors, as shown in Fig. 9.11. These combinations of parameters are taken as settings for experiments to determine the current efficiency experimentally.

The results of the experiments (current efficiency) for each combination of levels are expressed by the symbols given in Table 9.3..

The effect E_a of the factor A on the result is derived from the difference between the mean values of the results for high and low levels of factor A:

$$E_a = 1/2(a+h) - 1/2((1)+b) \tag{9.26}$$

Table 9.3.

Combination of factor levels	Symbol for the result
a_0, b_0	(1)
a_1, b_0	a
a_0, b_1	b
a_1, b_1	h

E_a represents the mean difference in the result in the progression from a low to a high level of factor A. The effect of factor B, E_b, is derived in a similar way:

$$E_b = 1/2(b+h) - 1/2((1)+a) \tag{9.27}$$

Besides these two main effects of the factors, there is often an interaction E_{AB} between the factors. This is calculated according to the equation

$$E_{AB} = 1/2(ab+(1)) - 1/2(b+a) \tag{9.28}$$

In the presence of a mutual interaction of both factors the effect of a factor in the region between the lowest and highest levels on the result is interrelated with the level of the other factor. Taking the example of current efficiency, this interaction can also be explained physicochemically. If the polarisation curve shows an obvious limiting current region for the reaction under investigation, and if the current density at concentration c_1, for example, is always lower than the limiting current density arising from c_1, current density has no effect on current efficiency. If, on the other hand, the concentration is reduced to c_0, at which value the limiting current density is smaller than the current densities applied, current efficiency will decrease with increasing current density.

In order to determine this relationship by theoretical means, the kinetic parameters of all the reactions (equilibrium potential, exchange-current density, charge-transfer factor and limiting-current density) for the electrode material and electrolyte in question must be known. This is hardly ever the case and therefore an empirical equation to calculate the current efficiency must be established from experimental results by an appropriate factorial design. Factorial design of experiments is so advantageous because four experiments of the 2^2 design, above, are sufficient for the formulation of an equation which can be used to calculate results in the whole parameter range.

$$a_0 < A < a_1 \tag{9.29 a}$$

$$b_0 < B < b_1 \tag{9.29 b}$$

The coordinates z_A and z_B are defined such that they run from -1 to $+1$ in the parameter range under observation (see table 9.3). If x_0 represents the mean of the experimental results, every value of x within the parameter range under observation is expressed by the equation:

$$x = x_0 + 1/2 E_A z_A + 1/2 E_B z_B + 1/2 E_{AB} z_A z_B \tag{9.30}$$

E_A, E_B and $E_{A,B}$ are the effects calculated as explained above. Since this type of equation is based on the use of statistical methods, they are also referred to as statistical models. For example, taking the influence of current density i (factor A) and concentration c (factor B) on current efficiency, the coordinates z_A and z_B are defined as follows:

$$z_A = \{i - 1/2\,(a_0 + a_1)\}/\{1/2(a_1 - a_0)\}, \tag{9.31 a}$$

$$z_B = \{c - 1/2\,(b_0 + b_1)\}/\{1/2(b_1 - b_0)\}. \tag{9.31 b}$$

The reliability of the prediction is somewhat limited by the fact that it presupposes a linear relationship between the result of the experiment and the variation of the factors A and B. The validity of this assumption can be tested by introducing an additional experiment into the factorial design shown in Fig. 9.11 by choosing the centre point. It is permissible to assume linear relationships if the mean of the 4 results of the factorial design is the same as the result of the central experiment. Whenever this is not the case, linearity can often be achieved by modifying the results of the experiment (e.g. lg x, $1/x$ or x^2 instead of x). If this still does not work, more complex factorial designs must be used, allowing for generally a non-linear regression of the results.

9.5
Cost Analysis

9.5.1
Composition of Productions Costs

Total costs, TC, per year of electrochemical processes are composed of:
(i) fixed costs, FC which are mainly determined by the invested capital, since the expended total investment defines a fixed amount of money per year for interests and capital amortisation, the so called annuity;
(ii) variable costs, VC, consisting mainly of the yearly expenditure for electrical energy, raw materials and other utilities; and
(iii) total maintenance costs, MC, which often may be included into the fixed costs since maintenance costs (including repairs) are almost independent of the operation variables which influence TC, the most important of which will be shown to be the current density.

Total yearly costs are summed up correspondingly:

$$TC = FC + VC + MC \tag{9.32}$$

The relevant issue, however, is not the total expenditure TC but the expenditure per unit mass of production T_m, which has to be minimised.

$$T_m = TC/P \tag{9.33}$$

P means the yearly productivity of the plant which is the given target for a given production unit at the time of its construction although this quantity may change during the lifetime of a plant due to later plant improvements, changed market situations or an unforeseen change of important costs as for instance energy costs.

9.5.2
Total and Specific Investment Costs

Total investment costs include not only the investment for the electrochemical parts of the plant like cells, rectifiers and busbars but have to account also for usual "chemical" process equipment. For instance in chloralkali electrolysis plants (see for instance the largely simplified flow sheet, Fig. 9.12) total invest-

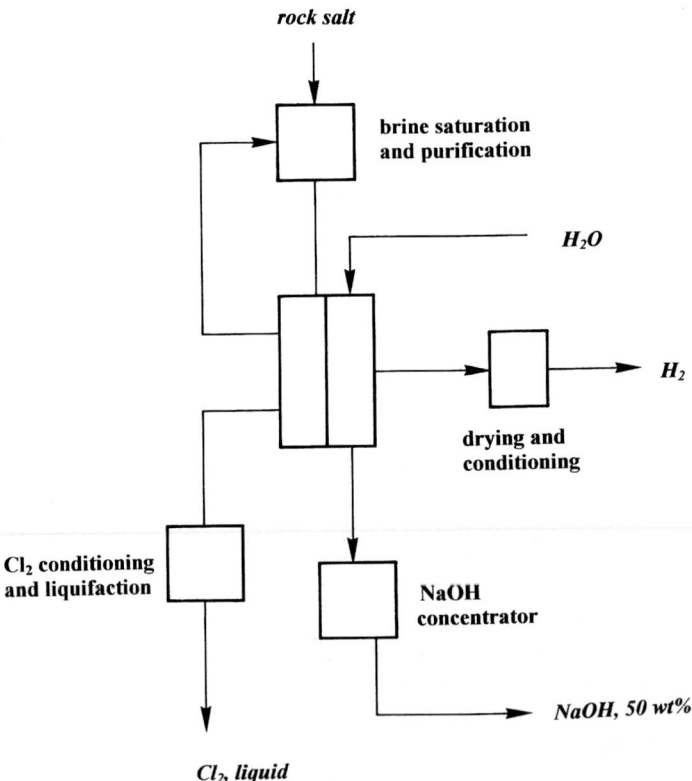

Fig. 9.12. Simplified flow sheet of a chloralkali-electrolysis plant of the membrane technology in simplified (block) notation

ment costs include all the equipment for brine make-up and purification, for chlorine drying, and (most often) liquefaction and storage of chlorine as well as for hydrogen purification, drying and storage. The money invested in the non-electrochemical parts of a chloralkali electrolysis plant is not negligible but amounts in modern membrane electrolysis plants to more than 30%.

For electroorganosynthesis processes the expenditure for the performance of chemical, thermal and separation unit operations becomes even more important so that for instance for the improved Monsanto process (adipodinitrile process) the investment for cells, rectifiers and bus bars contribute only approximately 10% to the total investment costs.

Therefore the total fixed costs reasonably can be subdivided into an electrochemical part FC_{el} and a chemical part FC_{chem}

$$FC = FC_{el} + FC_{chem} \tag{9.34}$$

as can be the variable costs and the total costs

$$VC = VC_{el} + VC_{chem} \tag{9.35}$$

$$TC = (FC + VC + MC)_{el} + (FC + VC + MC)_{chem} \tag{9.36}$$

as well as the specific costs

$$T_m = \frac{TC}{P} = \frac{\{(FC + VC + MC)_{el} + (FC + VC + MC)_{chem}\}}{P} \tag{9.37}$$

For the large, well known electrolysis processes like chloralkali electrolysis both parts of the costs can be assumed to be independent of each other to a good approximation since performance characteristics of the cells – in particular the cell current density – do influence the details and layout of the chemical part of the plant only to a minor extent. For organoelectrosynthesis processes this is not always the case. In this type of process applied current density very often influences process selectivity, raw material utilisation and other product parameters which by themselves are calling for changes in the chemical part of the plant constructed for product recovery and purification, solvent recycling and so on.

9.5.3
Cost Optimisation with Respect to Current Density

The electrochemical part of the investment costs (fixed costs) and the variable costs are both changing with the chosen current density, although in opposite sense. Whereas investment costs are decreasing with increasing current density, as the productivity per cell increases and the number of cells which are necessary to maintain a given yearly production decreases, the cell voltage increases with current density and hence the energy costs which constitute the main part of the variable costs increase.

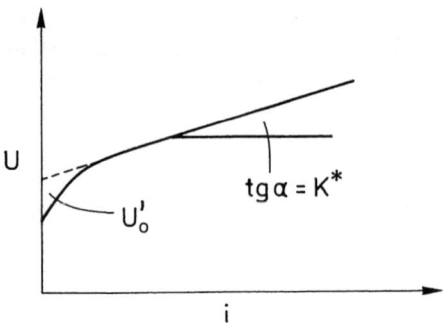

Fig. 9.13. Schematic presentation of cell current-voltage curves defining the extrapolated quantities U_0' and K^*

A simplified treatment will clarify this situation. The simplified assumptions are:
a) the costs of electrolyzer cells are simply given by the necessary electrode surface, A, which means that there is no sizeable economy of scale with respect to the number or size of cells installed;
b) the current voltage curve of the respective electrochemical process can be linearized within the range of current density which is technically relevant – this approximation is explained by Fig. 9.13 and reads

$$U = U_0' + R^* i \tag{9.38}$$

R^* is the effective surface specific cell resistance including ohmic and charge transfer resistances often called the k-value. With these assumption FC can be approximated by

$$FC = \frac{AaI^*}{P} \tag{9.39}$$

with A the total electrode area and a the surface specific costs of the electrochemical equipment – the cells in particular – per unit electrode area and I^* is the yearly percentage of interests and capital return. Since

$$Ai\left(\frac{M}{v_e F}\right) = P \tag{9.40}$$

one obtains

$$FC_{el} = \left(\frac{v_e F}{M}\right) aI^* i^{-1} = Ci^{-1} \tag{9.41}$$

For the electricity costs VC_{el} one obtains with e the specific energy costs

9.5 Cost Analysis

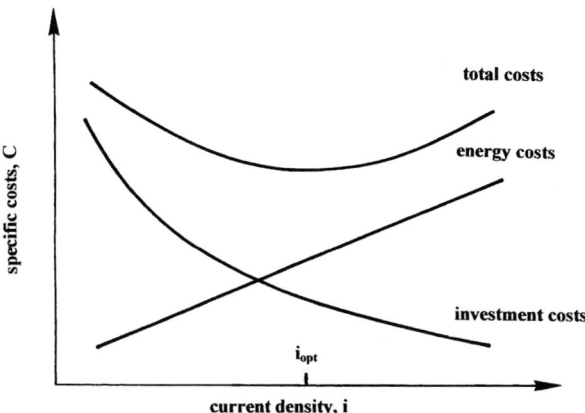

Fig. 9.14. Schematic presentation of optimisation of the current density accounting for a linear increase of energy and an inverse decrease of investment costs with increasing current density

$$VC_{el} = M^{-1}(v_e FU)e = M^{-1}v_e F(U_0' + R^*i)e = D + Ei \quad (9.42)$$

and adding to the total cost per unit mass of product

$$TC_{el} = Ci^{-1} + D + Ei + MC \quad (9.43)$$

one obtains optimal current density by differentiation of TC with respect to i:

$$i_{opt} = \left(\frac{C}{E}\right)^{1/2} \quad (9.44)$$

Figure 9.14 shows the situation by plotting the fixed costs and the variable costs per unit mass of product against the current density i. As the electricity costs e increase and the cell cost per unit electrode area a and interests plus capital return I* decreases, the optimum current density decreases.

A more detailed discussion would have to take into account the economy of scale due to increase of the plant size that means scale-dependent specific costs for the electrolyzer cells

$$a = const \cdot A^\alpha \text{ with } \alpha < 1 \quad (9.45)$$

However, this is a second order effect as for electrochemical plants contrary to chemical plants α is smaller than unity but still comes close to 1 because an extension of the electrode area A is most often brought about by adding cells of identical design and size to each other. This does, however, not hold if electrolyzer cells of significantly different size and power consumption are compared.

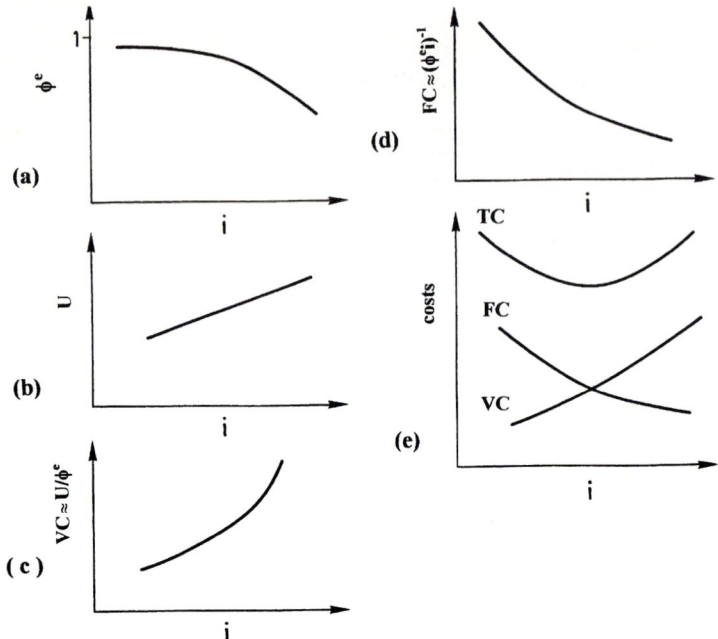

Fig. 9.15 a–e. Explanation of current density optimisation, when the current efficiency varies with current density. Current density dependence of: **a** current efficiency; **b** cell voltage; **c** variable costs defined by U/Φ^e; **d** fixed costs according to $(\Phi^e i)^{-1}$; **e** i-optimum by superposition of energy and investment costs

For water electrolyzers we note that for 100 kW units the power specific cost are at least five times as high as for 1 MW electrolyzers.

The same holds with respect to the size dependence of power specific costs of rectifiers, transformers and electrical auxiliary equipment. Again the scaling-exponent comes close to one for this type of equipment.

The very broad range of electricity costs, ranging from fractions of 1 USc kWh^{-1} under very favourable conditions in the US or Canada to costs of close to or even more than 10 USc kWh^{-1} in Europe or Japan preclude a detailed generalised quantitative treatment of the optimal choice of the current density. Nonetheless it must be mentioned that provided cheap off-peak electricity is offered for instance during night time or holidays the mean costs of production can be reduced to a remarkable extent by varying the current load of the plant and hence current density in the cells according to this opportunity.

9.5.4
Optimisation of Non-Selective Electrolysis Processes

Often current efficiencies are lower than unity because – for instance – a cathodic metal deposition is accompanied by hydrogen evolution, or membrane chloral-

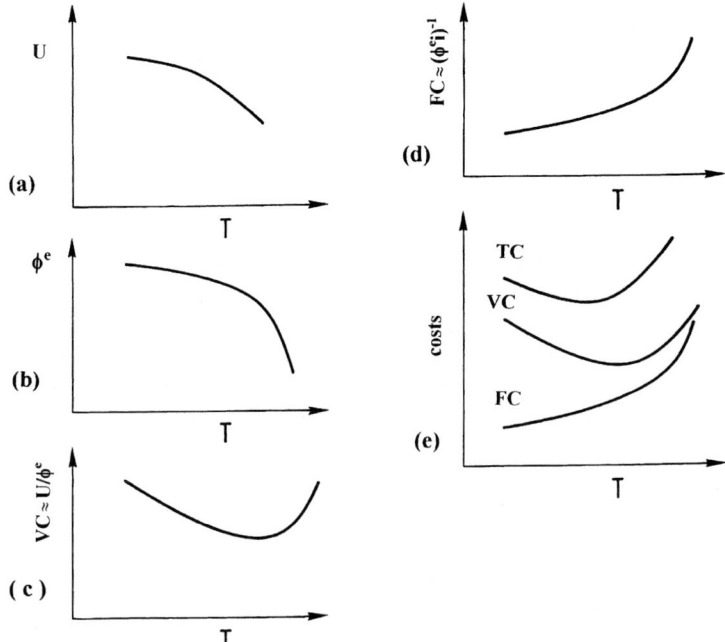

Fig. 9.16 a–e. Explanation of temperature optimisation at fixed current density if the current efficiency varies with temperature. Temperature dependence of: **a** cell voltage; **b** current efficiency; **c** ratio U/Φ^e; **d** $(\Phi^e i)^{-1}$; **e** T-optimum by superposition of energy and investment costs

kali electrolysis is accompanied by back diffusion of hydroxyl ions from the cathode into the anode chamber thus diminishing the current efficiency of caustic production. In organoelectrosynthesis processes current efficiencies lower than 1 are even more frequent.

We will treat subsequently in schematic and mathematically non-explicit form the possibilities to optimise either the process costs with respect to current density or with respect to temperature, provided the current yields are dependent on one of these two process variables.

9.5.4.1
Current Density Against Current Efficiency

Commonly current efficiencies decrease with increasing current density as depicted schematically in Fig. 9.15 a. Since to a good approximation cell voltages rise linearly with current density (Fig. 9.15 b) the ratio U/Φ^e which is the relevant quantity determining variable costs per unit mass of product VC increase stronger than proportional to i (Fig. 9.15 c). The cell area and hence fixed costs per unit mass of product FC decreases according to $(i\Phi^e)^{-1}$ (Fig. 9.15 d) and the sum of both terms define the optimal current density (Fig. 9.15 e).

9.5.4.2
Temperature vs Current Efficiency

Current efficiencies usually are progressively impaired at increased temperatures. Simultaneously cell voltage and correspondingly the consumption of electric energy per unit mass of product is decreasing because of progressively improving electrolyte conductivity and thermal activation of the electrode processes. Figure 9.16 a depicts schematically this decrease of the cell voltage whereas Fig. 9.16 b describes schematically the decrease in current efficiency with rising temperature which is brought about by a higher effective activation energy for the parasitic electrode reactions compared to the main reaction. Figure 9.16 c describes the variation of the quantity U/ϕ^e which mainly determines the variables costs VC per unit mass of product whereas Fig. 9.16 d depicts the variation of the quantity $(\phi^e i)^{-1}$ which mainly determine the fixed cost per unit mass of product FC. The sum of both cost contributions then define the optimal process temperature (Fig. 9.16 e).

9.5.5
Examples Including Influences of Process Parameters on the Equipment for Non-Electrochemical Unit Operations and Corresponding Costs

Hitherto an influence of the chosen current density or the process temperature on the investment to be expended for the "chemical" parts of the process was not taken into account. For more complicated process flow sheets as are very often encountered for electroorganic synthesis processes this is no longer justified. Figure 9.17 explains schematically that very often recycling of the electrolyte and (sometimes very precious) unconverted raw material is necessary as is the separation of side products and the purification and conditioning of the main product. Such loops are often indispensable in order to render the economic feasibility of the process. In such cases the influence of the process parameters on current efficiency and selectivity and the resulting necessary changes of the

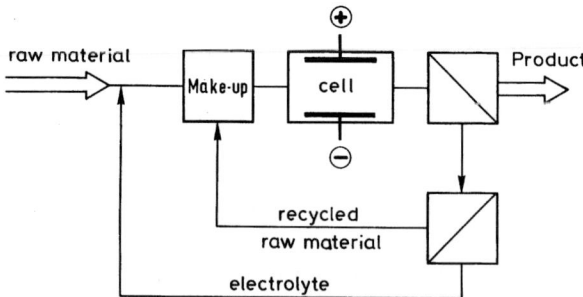

Fig. 9.17. Schematic presentation of electrolysis process including make-up and separation units for recycling of unspent raw material and electrolyte

work-up equipment have to be taken additionally into account. This gains utmost importance in cases where the costs of the electrochemical equipment (cells, rectifiers and transformer, busbars, etc.) become only a smaller fraction of the total equipment. In such cases not only current density and process temperature but additionally electrolyte composition and – most important – degree of conversion (utilisation) of the raw material have to be considered since selectivities and current yields very often deteriorate with increasing utilisation. In these cases the cost of material losses and the efforts to be expended for product purification caused by poorer selectivities have to be traded off against lower energy consumption and lower equipment cost for electrolyte / product separation at higher utilization of the raw material or for solvent and electrolyte recovery which are almost always brought about by distillation or rectification of the cell electrolyte in energy-consuming and investment-intensive columns.

Further Reading

D.D. Macdonald, Transient Techniques in Electrochemistry, Plenum Press, New York, 1977
E.R. Brown, R.F. Large, Chap. 6 in Physical Methods of Chemistry, Part II A, A. Weissberger and B.W. Rossiter, Wiley Interscience, New York, 1971
E. Gileadi, E. Kirowa-Eisner, J. Penciner, Interfacial Electrochemistry, an Experimental Approach. Addison-Wesley Publishing Company, Reading 1972
R.N. Adams, Electrochemistry at Solid Electrodes, Marcel Dekker, 1969
A.J. Bard, L.R. Faulkner, Electrochemical Methods, Academic Press, New York, 1972
L. Sachs, Statistische Auswertungsmethoden, Springer Verlag, Berlin, Heidelberg, New York, 1969
D.J. Pickett, Electrochemical Reactor Design, Elsevier Scientific Publishing Company 1977
M.S. Peters, K.D. Timmerhaus, Plant Design and Economics for Chemical Engineers, Mc Graw Hill, 1981
G.D. Ulrich, A Guide To Chemical Engineering Process Design and Economics, John Wiley, New York 1984
T.E. Fahiday, Principles of Electrochemical Reactor Analysis, Elsevier, Amsterdam 1988
R.E.W. Jansson, Economic Driving Force in Electroorganic Synthesis in Electrochemical Cell Design, R.E. White Ed., Plenum Press, New York 1984
W.L. Mc Cabe, J.C. Smith, Unit Operations of Chemical Engineering, Mcgraw-Hill Kogakushu Ltd. Tokio, Düsseldorf, London, 1956

CHAPTER 10

Industrial Electrodes

10.1 Catalytically Activated Electrodes

Until the middle of the 20th century it was usual to operate technical electrolyses processes without caring much about electrocatalysis. But today catalytically activating electrodes by doping the electrode surface with an appropriate electrocatalyst became general state of the art, because energy savings by reducing electrode overpotentials are now imperative for improved process economy. Electrocatalysts are usually used in industrial electrolyzers as coatings on a supporting metal. The best known example is RuO_2 on titanium electrodes (Table 10.1). In selecting electrocatalysts for technically applied electrodes the electrocatalytic activity is only one important criterion among several others. Criteria to be fulfilled are:
(1) electrocatalytic activity;
(2) high intrinsic electronic conductivity of the catalyst;
(3) long term stability due to chemical stability or chemical inertness and physical stability that means hardness and good adherence to the electrode support;

Table 10.1. Development of catalytic coatings at ICI Mond Division

Formulation	Main constituent	Comments
Mond Pt, Pt/Ir metal	Electroplated Pt metal, Pt or Pt/Ir from paint platinizing formulation	problems with passivation and adherence
Improved Pt metal coating	Pt electroplated in the presence of organic compounds	non-passive under Hg- cell conditions
Mond 1	RuO_2/TiO_2	Coating composition depends on cell
Mond 2	RuO_2 overglaze	–
Mond 3	RuO_2/SnO_2	–
Mond 5a/5b	RuO_2/TiO_2 with added refractories	Preferred coating in mercury cell

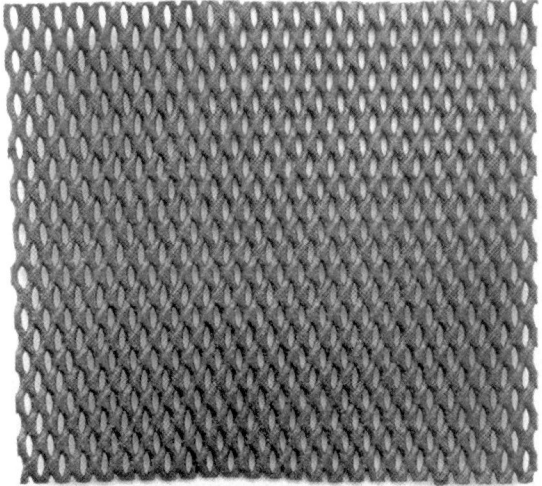

Fig. 10.1. Typical RuO_2-activated extended metal titanium electrode used in diaphragm and membrane electrolyzers

(4) existence of simple and cheap methods for applying the catalytic coating;
(5) chemical compatibility of coating and supporting metal;
(6) high specific surface of the coating; and
(7) low cost.

The number of relevant electrode reactions in aqueous electrolytes is very limited. Almost any process has to do either with hydrogen or oxygen evolution and perhaps chlorine generation. Therefore only a small number of catalysts and catalyst systems play a significant role in industrial electrolyses. It is obvious that electrocatalytic activation of industrial electrodes is only one of many issues in designing electrodes for industrial processes. But most other issues as shaping, integrating the electrode into the cell or its fabrication, are very specific. As an example Fig. 10.1 shows a typical RuO_2-activated titanium extended metal electrode for chloralkali electrolysis according to the diaphragm process.

10.2
Functioning, Longevity and Application of Electrocatalyst Coatings

Electrocatalysts, as far as they are composed of highly active precious metals or materials of relatively low electric conductivity (e.g. semiconducting oxides), are usually applied to metal supports, constituting the electrode proper, in the form of thin coatings amounting to catalyst loadings of only a few mg cm^{-2} or less. Long term stability and long term adhesion to the support of such coatings is as important as their catalytic activity. Longevity of electrocatalyst coatings might be threatened by a number of different effects.

The four most important are (i) poisoning, (ii) corrosive dissolution, (iii) erosion and (iv) loss of internal surface of porous coatings by sintering and Ostwald ripening.

Poisoning can only be reliably mitigated by using carefully purified electrolytes, raw materials and (in fuel cells) anode and cathode gases. Corrosion is most often a serious problem, whenever by non-steady conditions, for instance in off-times, redox reactions of the catalyst surface which are due to floating and uncontrolled potential render the catalyst material thermodynamically unstable or more soluble in the electrolyte than under operating potential. For instance cobalt in oxides of the trivalent state (e.g. in spinels like $NiCo_2O_4$ or Co_3O_4) is insoluble in aqueous, caustic solution, but this typical oxygen-evolution catalyst becomes unstable at more cathodic potential reacting with water by release of O_2 and forming the divalent state, which is soluble due to formation of the complex $Co(OH)_4^{2-}$. Erosion by itself and coupled to corrosion may become a serious problem whenever electrochemical gas evolution is the main reaction. Erosion is also a problem if the electrode is exposed to rapidly flowing electrolyte – in particular if the electrolyte contains dispersed solid matter. It can only be counteracted by modifying the structure and composition of the coating increasing its strength and adherence and by influencing gas bubble formation in a way which removes the physical strain due to bubble overpressure and violent two-phase flow. Most reliably can this be achieved by displacing by rapid electrolyte flow bubble formation off the catalyst surface into the electrolyte solution or by avoiding bubble formation in pores by reducing the mean pore diameter of porous coatings below a critical value.

Loss of internal surface of highly porous metal coatings by slow sintering can sometimes be prevented to a certain extent by stabilising the porous structure by so called internal or external dispersion hardening that means by interdispersion of a second electrocatalytically inactive, oxidic highly dispersed powder which hinders creep or surface diffusion of the metallic catalyst material which would lead to agglomeration and surface loss. Whenever the enhanced catalytic activity of the catalyst is due to so called active sites, that means to exposed crystal defects or dislocations these sites will only be active on long term, if processes which would lead to healing or recrystallisation and accordingly to deactivation have an activation energy in excess of 100 kJ mol^{-1}. Such high activation energies would render at 100 °C a retardation of any surface relaxation processes to effective relaxation times in excess of several years. This cannot be achieved with relatively soft metals like gold, silver or copper but is possible with harder transition and platinum metals. Catalysis and electrocatalysis by active sites, is the rule with catalytically active metals. The surfaces of oxidic redox catalysts, however, which are typically accelerating anodic oxygen or chlorine evolution or are mediating the anodic oxidation of organic substances, act catalytically as a whole not depending on active sites. They catalyse the respective reactions with every surface metal atom which may undergo respective redox reactions and is exposed to the electrolyte. For this type of electrocatalysts physically reinforcing the coatings and strengthening their adherence to the support is imperative. Application of appropriate interlayers (e.g. SnO_x between RuO_2 and Ti electrodes) or mixing the catalyst with inert valve metal oxides became the method of choice for catalyst strengthening.

Very often one has to compromise between these targets, for instance sacrificing higher catalytic activity to better long term stability. The deterioration of catalytically activated electrodes is a slow and steady process giving rise to a steadily increasing overvoltage. Introducing a time dependent term for the cell voltage of an electrolysis cell, $\Delta U(t)$,

$$U_{cell} = U_0 = \eta_\alpha - \eta_c + IR + \Delta U(t) \tag{10.1}$$

takes account of this effect and allows to treat the change of electrocatalyst performance with time in a quantitative way. For fuel cells $\Delta U(t)$ is negative and diminishes the cell voltage whereas for electrolysis cells the cell voltage increases with time. Ageing of electrocatalysts is not only caused by fouling or poisoning but also by the limited chemical stability causing slow dissolution of the catalytic coating or its deterioration due to chemical reactions, or also by physical detachment due to destruction of an intermediate layer between the support and the catalyst.

10.3
Design of Industrial Electrodes

Apart from microscopic structural features of catalytic coatings the macroscopic structure and electrode shape can vary considerably according to the respective purpose and constructive details of the electrolyzer. The most important distinction with respect to electrode form, shape and function is whether the electrode is monopolar or bipolar on one hand and whether its surface is planar or corrugated on the other hand.

10.3.1
Monopolar Electrodes and Current Density Distribution on Their Surface

Monopolar electrodes are usually connected to the bus bar from one side only (usually the upper part which is not immersed into the electrolyte). This one sided current connection (Fig. 10.2), due to ohmic voltage drops in the electrode gives rise to uneven potential and current density distribution on the electrode surface which limits the length of monopolar electrodes.

The ohmic potential drop along the height of a top-connected electrode with breadth y and thickness x sums up from the lower to the upper end. If at height h there exists the current density $i(h)$, the differential increment dI of current collected over dh is

$$dI = y\, i(h)\, dh \tag{10.2 a}$$

the total current at height h is:

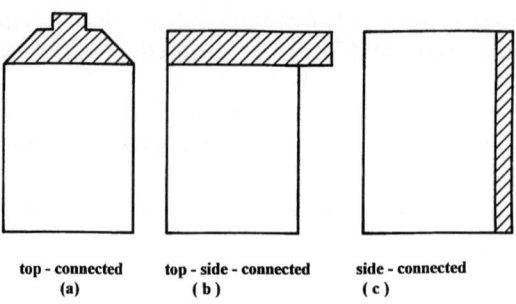

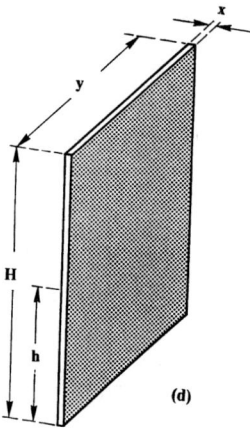

Fig. 10.2a–d. Schematic presentation of a monopolar electrode immersed in the electrolyte and different options for connecting such electrodes: **a** top-connection; **b** top–side connection, **c** side connection in cell stacks; **d** schematic explanation of calculating the current density distribution in a top-connected horizontal electrode

$$I_h = y \int_0^h i(h)\, dh \tag{10.2 b}$$

Along dh the ohmic potential drop dU_Ω in the electrode amounts to

$$dU_\Omega = \frac{I \kappa_{elode}^{-1} dh}{xy} \tag{10.2 c}$$

κ_{elode} is the electric conductivity of the electrode metal.
Integration from the lower end to h results in

$$\Delta U_{\Omega,h} = \frac{\kappa_{elode}^{-1}}{x} \int_0^h \int_0^h i(h)\, dh\; dh \tag{10.3 a}$$

10.3 Design of Industrial Electrodes

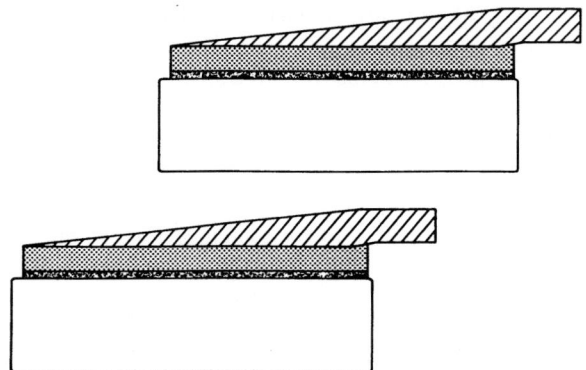

Fig. 10.3. Schematic of a side connected electrodes with narrowing connectors – ICI cell FMC 21

One obtains for the local electrode potential, E_h, at height, h, the difference between the electrode potential at the busbar $E_{h=0}$ and the ohmic potential loss $\Delta U_{\Omega,h}$:

$$E_h = E_{h=0} - \frac{\kappa_{elode}^{-1}}{x} \int_0^h \int_0^h i(h) \, dh \, dh \quad (10.3\,b)$$

The current density at height, i(h), is an explicit function of the electrode potential, the current voltage curves of both electrodes, electrode distance and electrolyte conductivity, so that a straight-forward solution of Eqs. (10.2 b) and (10.3 b) cannot be given. The most important result of the quantitative treatment of this problem is a sensible decrease of current density towards the lower end of a monopolar electrode. If for constructive reasons the busbar connection is situated not in the middle or along the breadth of an electrode but at one of its corners, then lateral ohmic potential drops have additionally and explicitly to be accounted for. Therefore the monopolar ICI chlorine electrolyser, for instance, has a particularly shaped current connection which is broad at the bus bar and which narrows towards the opposite edge (Fig. 10.3) with the aim to render current density distributions more evenly.

10.3.2
Electrodes for Bipolar Electrode Stacks

Bipolar electrodes, which are operated as anodes and cathodes on either side, most often do not comprise the anode and cathode together in one sheet of metal, but are usually, according to different electrode potentials and corrosion conditions, composed of two different metal sheets for anode and cathode, or are formed from a metal toil coated from one side by a particular other metal or metal oxide. Bipolar electrodes transmit the current from one cell to the next directly and often the respective electrodes are attached as "fore-electrodes" to the bipolar plate. Figure 10.4 b shows schematically a set of bipolar plate and foreelectrodes and demonstrates the use of intermediate metallic springs which allow to adjust the proper distance between foreelectrodes and the bipolar plate. The current is transmitted through the connecting springs and they provide an equal current distribution preventing excessive lateral ohmic voltage losses within the electrode. Figure 10.4 c gives as an example the elegant solution for establishing the equally distributed contact spots across the electrode surface. Lurgi Company used waffled steel sheets in its pressurised water electrolyser, the protrusions of the sheets carried the anode and cathode.

10.3.3
Gas Evolving Electrodes

Electrochemical gas evolution creates particular engineering problems because the electrode shape and cell designs have to account for the minimisation of bubble-induced additional voltage losses (compare Sects. 5.4.6 and 5.4.7). Using louvered electrodes helps a lot in conveying the gas bubbles towards the backside of the electrodes where they cannot interfere with the ionic current. Another measure is to use corrugated electrode structures like extended metals, grids, mesh electrodes or perforated plate electrodes with the aim to provide an outlet for convective bubble removal towards the electrode's back side (compare Chap. 8, Fig. 8.5 a–c). Particularly effective for bubble removal are these electrode structures, if the electrodes are touching the diaphragm. Figure 10.4 a demonstrates this so called "distance-free" electrode/diaphragm/ electrode configuration for perforated plate electrodes in a cut through a zero distance cell equipped with perforated plate electrodes as shown in the view of Fig. 10.4 d. In order to prevent gas clogging of the holes of the perforated electrodes, the hole diameters of the perforated plate electrodes have to surmount the mean bubble diameter which may be considerably different for the anode and cathode gas, e.g. 0.1 mm for hydrogen and 0.7 mm for oxygen in alkaline water electrolysis at 90 °C.

Gas-evolving electrodes which are positioned horizontally, compare Fig. 8.10, permit easy escape of the gases if they are composed of thin wires or grids. Figure 10.5 a shows a typical titanium electrode and Fig. 10.5 b shows a number of different designs of corrugated electrodes which, if activated by RuO_2, are used for anodic chlorine evolution. The buoyancy of the gas is driving the vertical flow of the bubble/electrolyte mixture in a horizontal mercury cell very efficiently. Therefore for grid electrodes used in amalgam–chloroalkali electrolyzers the bubble effect is negligible, almost non-detectable.

10.3 Design of Industrial Electrodes

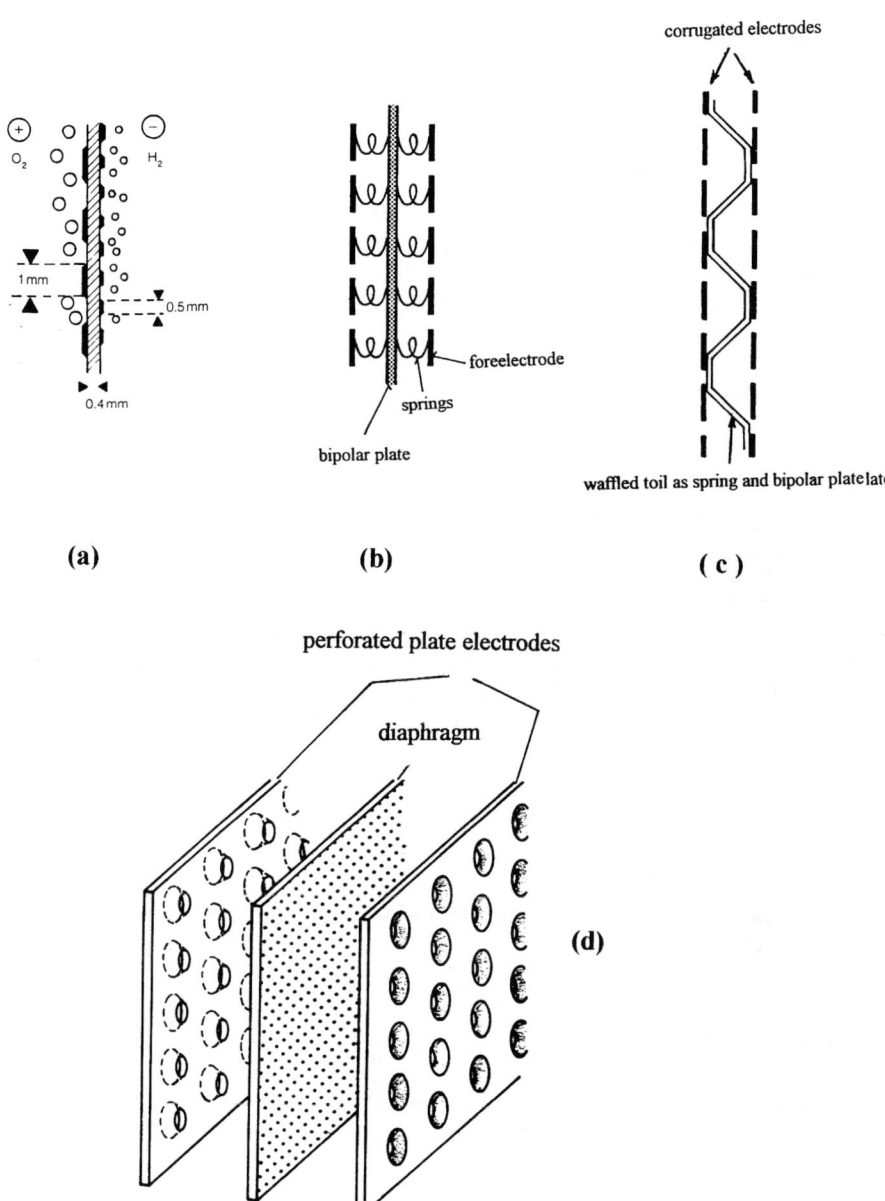

Fig. 10.4. a Zero-distance arrangement of vertical electrodes at either side of a diaphragm. The diameters of the holes in the perforated electrodes have to be larger than the mean bubble diameter in order to prevent gas clogging. **b** Schematic of bipolar plate with springs carrying fore-electrodes. **c** Waffled toil as spring and bipolar plate with attached corrugated electrodes. **d** Detached EDE (electrode–diaphragm–electrode) unit consisting of the diaphragm and two perforated-plate electrodes which corresponds to schematic **a**

10.4
Structural Features of Electrocatalysts for Gas Evolving and Gas Consuming Electrodes

A high specific surface of the catalyst increases its effective catalytic activity as increasing the true surface means reducing the true current density. The necessity of a high specific surface becomes less important, if the Tafel slope b of the current voltage curve is small because b determines the decrease in overvoltage due to a tenfold increase of the catalyst surface. If b is as small as, for instance 0.03 V/decade, an increase of the surface roughness by a factor of 1000, according to Eq. (10.4), would bring about an improvement in overvoltage by only 90 mV.

$$i_{eff} = i_{geom}(a^*)^{-1} \tag{10.4}$$

with a^* equal to the surface factor

$$a^* = \frac{\text{true surface}}{\text{geometric surface}} \tag{10 a}$$

Electrocatalytic RuO_2 coatings on titanium anodes for chlorine evolution possess only microroughness whereas macroscopically viewed, they have a relatively smooth appearance and the Tafel slope for anodic Cl_2 evolution amounts to 30 mV. But the situation is different for other gas evolving or gas consuming reactions – in particular for cathodic hydrogen evolution in chloralkali or water electrolysis. Such reactions have to cope with a relatively high Tafel slope of 60–120 mV and need a much higher specific surface for effective overvoltage reduction – compare, for instance, Chap. 7.

10.5
Electrocatalytically Activated Dimensionally Stable Chlorine-Evolving Electrodes

10.5.1
Technological History

Chloralkali electrolysis has been commercially performed for more than 100 years. The anodic reaction

$$2Cl^- \rightarrow Cl_2 + 2e^- \tag{10.5}$$

can be and had been performed at oxidic anodes (magnetite anodes), but since the beginning of this century carbon anodes, particularly Acheson graphite anodes had been used. Neither magnetite nor graphite catalyse reaction, Eq. (10.5), particularly well. Additionally graphite is not stable but is oxidised to CO, CO_2 and is, even worse, also chlorinated to perchlorinated compounds (chlorine butter).

In 1968 Henry Beer invented the dimensionally stable anode, DSA, which consisted of a supporting titanium anode covered with a catalytic layer consisting mainly of a mixture of TiO_2 and RuO_2. Due to the joint efforts of Beer's and of the De Nora Company the dimensionally stable chlorine anode revolutionised the technology of chloralkali electrolysis. In particular the exchange of the amalgam technology by the membrane technology which initially was enforced in Japan by legislation (1976–1986) profited from the already existing technology of anodic Cl_2 evolution at DSAs.

10.5.2
Electrocatalysis and Selectivity of Anodic Chlorine Evolution at RuO_2-Anodes

Figure 4.15 compares the current voltage curves of anodic chlorine evolution measured at graphite and RuO_2 anodes in brine. The RuO_2-anode is at a clear advantage. At pH 2 (the lowest pH value maintained in chloralkali electrolysis anolytes) the equilibrium potential of the oxygen anode is more cathodic than that of the chlorine electrode (+0,96 V vs NHE compared to +1.26 V). Therefore the RuO_2 coatings must catalyse chlorine evolution more than oxygen evolution. For this reason particular care must be taken in modifying the catalyst formulations with the aim to increase the catalytic activity for the Cl_2 reaction over that of the O_2 reaction. For chlorine evolution of dilute brine (30 g dm^{-3} for NaOCl-production) or sea water (Cl_2 production on board ships for disinfecting) an increase of the Cl_2 selectivity is brought about by adding small amounts of Pt and IrO_2 to RuO_2.

10.5.3
Preparation and Formulation of the Coatings

The method for the preparation of RuO_2 coatings today is still the same as patented by Beer. The metals whose oxides are to form the catalyst coating are dissolved in appropriate form (chlorides, organometallics) in high boiling organic solvents of low volatility forming a so called "paint". Typical examples of metal compounds are $RuCl_3$, H_2PtCl_6, $CoCl_2$ – the latter two may be added to modify the selectivity – and titanium tetrabutylate or tantalum trichloride (these might be added to improve adherence and strength of the coating by forming dispersed TiO_2 or Ta_2O_5 particles or mixed oxides containing TiO_2 and RuO_2). There follows a firing process in air at ~400 °C which converts the precursor paint into the oxide coating – see below.

10.5.4
Improvement of Adhesion and Strength of the Coatings

Good adhesion and resistance against erosion is of utmost importance for the long term stability of RuO_2 coatings. Therefore as a first step of coating preparation the TiO_2 layer of μm-thickness which usually covers titanium metal and which would prohibit the formation of a low-resistive and tight contact between support metal and coating has to be etched away by oxalic or hydrofluoric acid so that only a very thin oxide layer remains onto which the coating is applied. An oxidic additive to RuO_2 has to be supplied which assures a good adhesion, as well as an improved conductivity of the interlayer between the titanium metal and the coating, together with a greater physical strength, hardness and erosion stability and simultaneously prevents the anodic oxidation of the titanium support which would generate insulating, non-conducting TiO_2 layers. TiO_2 which crystallises in the rutile lattice added to RuO_2 by proper formulation of the catalyst paint forms semiconducting mixed RuO_2/TiO_2 phases and stabilises mechanically and chemically the electrocatalyst without significantly affecting its catalytic activity. The TiO_2 content of the coating may vary from 30 to 70 mol%. Also Ta_2O_5 is added to RuO_2 as stabilising component. IrO_2, a catalytically active additive, seems also to improve the wear rate, and is added in particular for DSAs which are used for O_2 evolution. Typical examples of high boiling organic solvents as solvents to form the paint are n-butyl or t-butyl alcohol, turpentine oil or similar low-volatility liquids.

The paint is applied to the sandblasted, etched titanium electrode by brushing, dipping or spraying. The solvent is evaporated at 100–200 °C and the metal compounds now covering the Ti electrode are converted to the respective oxides by firing the electrodes in air at temperatures ranging from 400 to 600 °C – preferentially at 450 °C. The firing temperature must be kept below 600 °C as above this temperature an undesired growth of non-conducting TiO_2 interlayers on titanium is unavoidable. Optimal temperature is 450 °C, as below this temperature oxide formation is incomplete rendering Cl-containing, less active and more soluble coatings and above 450 °C the mixed oxides lose more and more residual chemically bound water and become more ordered and catalytically less active.

10.5.5
Design of Cells Using DSAs

The change from dimensionally unstable carbon anodes to dimensionally stable titanium anodes permitted dramatic innovations in cell design, operation conditions and reduction of the energy consumption of chloralkali electrolysis. With metal electrodes almost any anode design can be realised. Figure 10.5 a depicts a typical anode for mercury cells which is immersed horizontally into the brine and is composed of the anode proper – a RuO_2 coated grid or extended metal sheet – supported by a so called x-bar which serves as current distributor being fixed to another support structure which carries the electrode and as current collector is a thick copper rod (protected by so called titanium riser tubes). Figure 10.5 b shows a collection of different anode designs which are fabricated from RuO_2-coated extended titanium bars or louvered titanium foils. Double sided anodes are typically used in monopolar diaphragm and membrane cells in which they are fixed at the cell bottom standing upright, an expanding spring allowing to adjust the anode–cathode distance within the cell.

So called zero gap membrane cells in which cathode and chlorine evolving anodes are touching the cation-exchange membrane, which separates the anode from the cathode compartment are also state of the art.

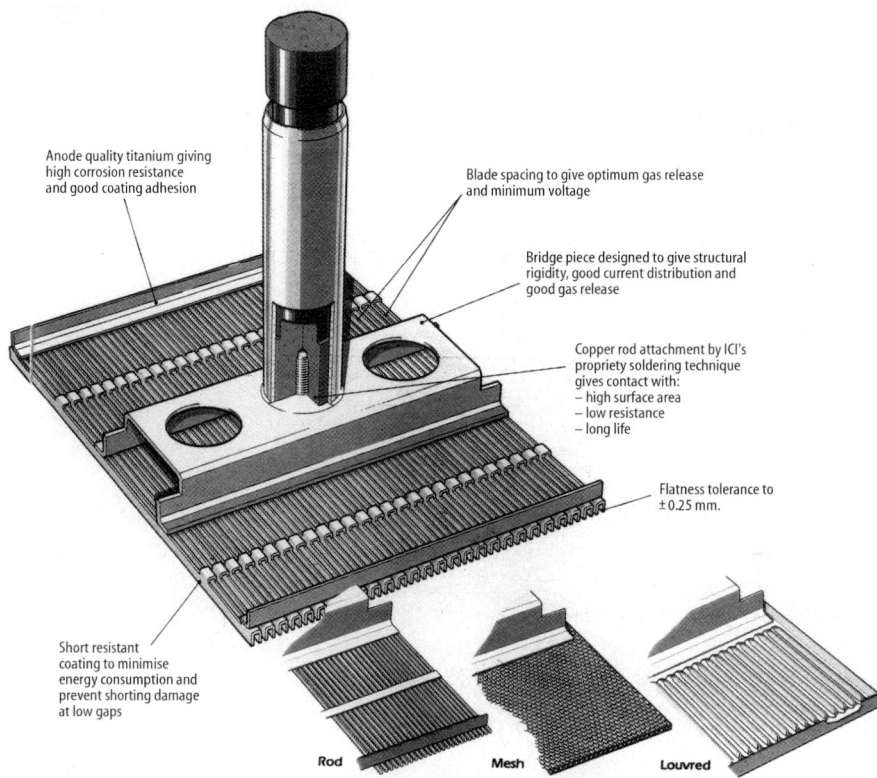

Fig. 10.5. a Typical electrode for chloralkali process according to the amalgam process. Courtesy of ICI

10.5.6
Lifetime of Dimensionally Stable Chlorine Evolving Anodes

The lifetime of RuO_2-coated anodes have been substantially improved by intelligent coatings formulations. At lower current densities (2–3 kA m^{-2}) which are usual in diaphragm and membrane cells the coatings are lasting at least five years, and even lifetimes of up to ten years are reported amounting to a production of 300,000 kg Cl_2 m^{-2} during the whole lifetime. In mercury cells with current densities of 10 kA m^2 the lifetimes amounts to approximately 3–4 years as the wear rate of the coating increases approximately linearly with current density. As the wear rate is supposed to be connected with slow oxidative dissolution of RuO_2 (RuO_4 formation), today it is believed that a substantial improvement in lifetime beyond these values may never be achieved.

Fig. 10.5.b Titanium electrode patterns used for chloralkali and other gasevolving electrolyses Courtesy of ICI

10.5.7
DSAs for Chlorate and Hypochlorite Production

Hypochlorite and chlorate production both rest on reacting the anodically evolved chlorine in dissolved form with cathodically generated hydroxyl ions, Eqs. (10.6)–(10.9):

$$2Cl^- \rightleftarrows Cl_2 + 2e^- \text{ (dissolved)} \quad \text{(anode)} \quad (10.6)$$

$$2H_2O + 2e^- \rightleftarrows 2OH^- + H_2 \quad \text{(cathode)} \quad (10.7)$$

$$Cl_2 + 2OH^- \rightleftarrows Cl^- + OCl^- + H_2O \quad \text{(hypochlorite)} \quad (10.8)$$

$$3ClO^- \rightleftarrows 2Cl^- + ClO_3^- \quad \text{(chlorate)} \quad (10.9)$$

For this reaction the pH value must be close to but not exceed a value of 5, that means that the pH value is remarkably higher than that for evolution of gaseous chlorine (pH 2) in usual chloralkali electrolysis, and that the competition of anodic oxygen evolution due to the pH dependence of the oxygen equilibrium potential becomes a serious issue as the chlorine potential is still more anodic than the oxygen potential at pH 5 than at pH 2. Therefore the catalytic activity for anodic O_2 evolution has to be diminished vs. Cl_2 evolution by addition af Pt, PdO and IrO_2 to RuO_2.

10.6
Oxygen Evolving Anodes

10.6.1
Technical Processes

Oxygen evolution is the counter electrode reaction in electrolytic hydrogen production from alkaline and in cathodic electrowinning of metals from acid aqueous solution. Examples for the latter are zinc, copper, tin or lead electrowinning. Water electrolysis usually is performed in alkaline solutions (30 wt% KOH), whereas metal electrowinning electrolytes are strongly acidic (approx. 1 mol dm^{-3} H_2SO_4). Therefore the electrode materials and electrocatalysts are quite different for oxygen evolution in alkaline water electrolysis and metal electrowinning. The electrode support for alkaline water electrolysis which used to be mild steel is today stainless steel or nickel for advanced water electrolyzers in which one needs materials of improved corrosion resistance. For oxygen evolution from acidic electrolytes the traditional anode material used to be lead – passivated by in-situ formed PbO_2 layers. Today these anodes are becoming superseded by catalyst-coated titanium anodes.

10.6.2
Electrocatalysis of Oxygen Evolution in Advanced Alkaline Water Electrolysis

10.6.2.1
Coatings Containing Cobalt and Iron Oxides

Cobalt oxides are unique insofar as in different oxides (Co_3O_4, Ni_2CoO_4 etc.) which possess a relatively low solubility in alkaline solution as a precondition to form stable catalyst coatings, the redox potentials of Co II/Co III and Co III/Co IV oxides respectively are relatively close to the oxygen equilibrium potential and therefore are able to catalyse anodic oxygen evolution according to Krasil'shikov's mechanism (compare Chap. 4).

Figure 10.6 compares the current voltage curve of nickel anodes with Co_3O_4 and Fe_2O_3-coatings. Clearly iron-III-oxide is a better catalyst than cobalt spinel. It has, however, a relatively high solubility in aqueous concentrated KOH solution so that due to cathodic iron deposition, which is expected to occur around

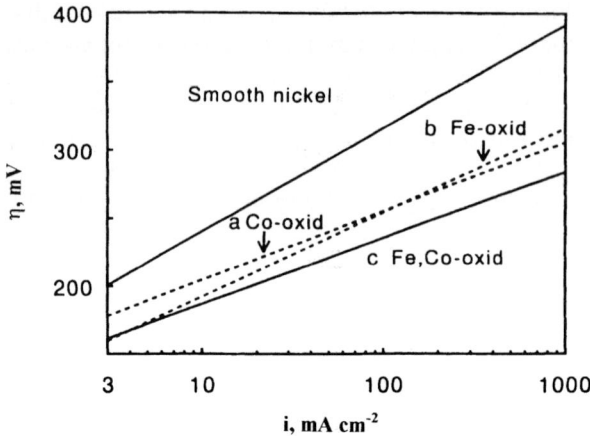

Fig. 10.6. Comparing Co_3O_4 and Fe_2O_3 and mixed Co/Fe-oxide coatings for anodic oxygen evolution from alkaline solution

−200 mV vs RHE, the catalyst is readily transferred from the anode to the cathode under normal operation conditions. This renders Fe_2O_3 inherently unstable in alkaline water electrolyzers. It is, however experienced, that in coatings which contain iron and cobalt together, both oxides are stabilized by forming mixed oxide phases with a lowered solubility of iron and cobalt. Under these conditions the good catalytic activity of iron oxide is preserved and the longevity of the coating is improved to a tolerable level of at least several thousand hours. Figure 10.6 shows also oxygen evolution current voltage curves measured at mixed Fe_2O_3/Co_3O_4 coatings.

10.6.3
Electrocatalysis of the Anodic Oxygen Evolution by Raney-Nickel Coatings

Raney nickel is used as an effective and cheap electrocatalyst for cathodic hydrogen evolution (see Chaps. 4 and 7). It may also be used as an electrocatalyst for anodic oxygen evolution as on its oxidised surface Ni II/Ni III oxyhydrates are formed, which are electrocatalytically active. Due to the high porosity of Raney nickel, which creates by anodic oxidation a rough surface with roughness factors approaching 1000, such oxidised Raney-nickel coatings are catalytically more active than oxidised surfaces obtained from smooth nickel electrodes and are comparable in their catalytic activity to Co_3O_4/Fe_2O_3-coated anodes.

10.6.4
Catalyst-Coated Titanium Electrodes for Oxygen Evolution From Acid Solutions

Traditionally anodic oxygen evolution from acid solution in metal electrowinning electrolyses – in particular from aqueous electrolytes containing sulfuric acid – had been performed at lead anodes, which are passivated and stabilised against corrosion by a self-forming, passive, coating of PbO_2. Oxygen evolution at PbO_2 anodes demands overpotentials of more than 0.7 V even at low current densities not exceeding 100 mA cm^{-2}. Therefore it is highly desirable to find an electrocatalytically active coating on Ti-anodes, which allows to reduce the oxygen overpotential sizeably offering at the same time an anode substrate, which, being a valve metal, is passivating and not sensible to corrosive attack. It must be kept in mind, however, that passivation of titanium in sulfuric acid is only safe and stable at lower temperatures. Consequently temperatures of metal-winning and metal-plating electrolyses with anodic O_2 evolution have to be operated below 40 °C if titanium anodes are used. Usually these types of electrolyses are operated at relatively low current densities allowing for relatively low process temperatures. RuO_2 is an excellent electrocatalyst in acid solutions for anodic oxygen evolution. However, it is unstable, being oxidised and carried away in the form of relatively volatile RuO_4. Therefore RuO_2 needs to be stabilized. Pt/Ir-oxides with added TiO_2 and Ta_2O_5/Nb_2O_5 as mechanical stabilisers are today used for anodic oxygen evolution in acid electrolytes. Still using this type of anodes is restricted to relative pure electrolytes for instance for a tin, or zinc plating bath for anticorrosive steel-sheet plating in the automobile industry. Catalytic coatings composed of IrO_2 and Ta_2O_5 show a life time optimum at 70% IrO_2/30% Ta_2O_5 (wt). Exchanging tantalum for titanium as base metal improves the stability of the combination support/coating a lot but at considerably higher costs.

Platinum metal oxide coated electrodes reduce the oxygen-evolution overpotential by approximately 0.5 V allowing to spare more than 30% of the energy costs which would be consumed by using conventional PbO_2 (Ag) anodes. This is good reason to extend now the use of IrO_2-coated anodes for plating processes to large scale electrowinning processes like Cu electrowinning and Zn electrowinning.

The membrane technology of water electrolysis is essentially water electrolysis in highly acid medium. It cannot dispense with noble metal oxides at the anode. Highly dispersed IrO_2 and mixtures of RuO_2/IrO_2 have successfully been applied. Because of the high and expensive catalyst loading there had not yet opened up a market for this type of electrolyzer, but the technology has been demonstrated to be, technically spoken, very effective with a small number of 100 kW electrolyzer stacks.

Summarising the situation of anodic O_2 evolution from acidic solution it may be stated, that today RuO_2/IrO_2 coated titanium anodes mechanically stabilised by TiO_2/Ta_2O_5 play already a significant role in large electroplating processes operating on relatively clean acid solutions. Their use also is beginning to extent into the field of metal electrowinning processes wherever relative highly purified electrolytes are used for instance for cathodic zinc deposition.

10.7
Hydrogen Evolving Cathodes

10.7.1
Technoeconomical Significance of Cathodic Hydrogen Evolution

Cathodic hydrogen evolution is the counter reaction to anodic chlorine evolution. It is expected to become superseded by the energy saving cathodic oxygen reduction only after a decade of further development and only if energy costs will continue to increase. The chloralkali electrolysis process is by far the most important source of electrolytically generated hydrogen. Cathodic hydrogen evolution is also of some significance as counter electrode reaction in anodic organoelectro syntheses as for instance in the Kolbe electrosynthesis of sebacic acid or the anodic oxidation of toluenes to benzaldehydes.

In chloralkali electrolysis and conventional water electrolysis the catholyte is strongly alkaline (c_{NaOH}, c_{KOH}=30 wt%) whereas in organoelectrosynthesis it is neutral or acidic depending whether a divided or undivided cell is used. In alkaline water electrolysis and chloralkali electrolysis the electrolyte is less corrosive than in processes which use acidic electrolytes and therefore the traditional cathode material was mild steel or stainless steel and was recently exchanged in advanced electrolyzers by nickel. Immunity of iron is granted only at cathodic H_2 overpotentials exceeding –200 mV, whereas modern catalytic coatings maintain overpotentials less than –200 mV. Since iron III oxide has a finite solubility in alkaline solutions, neither the necessary long-term stability nor the demanded purity of the electrolyte – in particular with respect to poisoning of cation exchange membranes by ferrous or ferric ions can be expected with stainless steel cathodes. But nickel is immune and $Ni(OH)_2$ is much less soluble than Fe_2O_3. Therefore nickel became the material of choice for electrodes, current collectors and cell bodies in advanced chloralkali electrolysis technology at the cathode side.

In acidic aqueous electrolytes the stability of nickel is not maintained and the cheapest valve metal, titanium, is chosen as cathode material. The stability of this metal, however, is questionable in presence of fluoride and complexing organic anions, so that particular care should be taken if choosing titanium for organoelectrosynthesis processes performed in organic solvents. Since titanium as cathode material is subject to hydrogen embrittlement, it has to be protected by devoted usually proprietary coating technologies against the attack of atomic hydrogen.

10.7.2
Electrocatalyst Coatings for Hydrogen Evolution from Alkaline Solution

10.7.2.1
Technically Applied Coatings

Today the following catalyst coatings are technically relevant:

1) nickel sulfide (NiS_x);
2) Raney nickel;
3) nickel alloys containing molybdenum;
4) coatings containing RuO_2 or platinum metal coatings, in particular Ru; and
5) doped nickel oxide.

Due to their relatively low costs the most frequently applied electrocatalysts are nickel sulfide and Raney nickel for alkaline conditions.

10.7.2.2
Nickel Sulfide Coatings

Two different methods are used to prepare nickel sulfide coatings: cathodic deposition of NiS_x and chemical sulfidization of nickel electrodes. The galvanic deposition process is based on cathodic deposition of nickel from baths containing sulfur donors like thiocyanate ion, thiourea or thioglycol which according to Eq. (10.10) are cathodically decomposed delivering sulfur.

$$Ni^{2+} + xRS + 2e^- \rightarrow NiS_x + xR \qquad (10.10)$$

The deposit does not contain sulfur in a well defined stoichiometry but is a mixture of different nickel sulfides, in particular Ni_2S_3, and amorphous nickel and nickel sulfide which is cathodically decomposed according to Eq. (10.11):

$$NiS_x + 2xe^- \rightarrow Ni + xS^{2-} \qquad (10.11)$$

Microscopic and spectroscopic investigations (SEM and XPS) reveal the relatively fast change of the chemical composition of nickel-sulfide coatings upon the onset of cathodic hydrogen evolution. At 90 °C all nickel sulfide phases present in a nickel-sulfide coating of several micrometer thickness are reduced to nanoporous nickel within several days to a week's time, so that these coatings ressemble in their microscopic structure more or less Raney nickel coatings.

Cathodically released sulfide anions are anodically oxidised to sulfate anions which enhance surface corrosion and stress corrosion cracking of mild and stainless steels and enhance surface corrosion even of nickel. Therefore nickel-sulfide coatings can only be used in non-pressurised electrolyzers working at relatively moderate temperatures (80 °C).

10.7.2.3
Raney-Nickel Coatings

10.7.2.3.1
Precursor Alloys and Fabrication of Coated Cathodes

Raney nickel, which more than 30 years ago had been introduced as an anodic electrocatalyst for alkaline fuel cells by Winsel and Justi, is used since more than

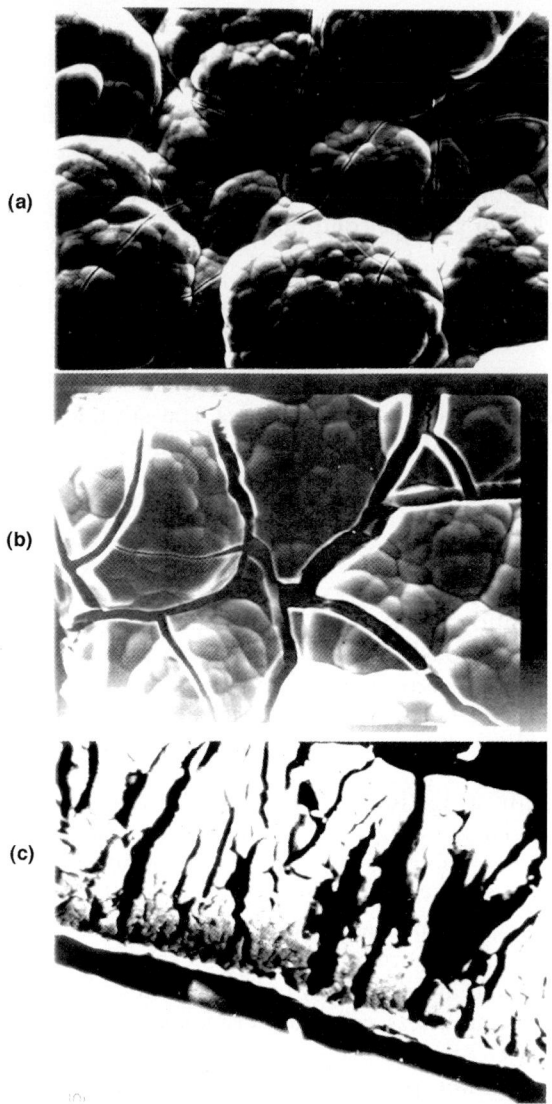

Fig. 10.7a–c. Surface of Raney nickel coating prepared by cathodic codeposition of nickel and zinc: **a** precursor; **b** surface of leached coating; **c** cut through the leached coating

20 years as catalytic coating for hydrogen evolving cathodes in chloralkali electrolysis and alkaline water electrolysis. Raney nickel is a nanoporous, pyrophorous nickel quality, initially invented and applied by Raney for catalytic hydrogenation of organic compounds. Raney nickel is obtained from precursor alloys, e.g. from nickel alloyed with a high percentage of non-noble metals (preferentially Al and Zn), which are leached out of the alloys in caustic potash. Due to

loss of the non-noble component the precursor alloy is converted under partial shrinking and recrystallisation into a metal sponge of 20% porosity with an inner surface of 30–70 m^2 g^{-1}. Preferred precursor alloys are NiAl$_3$ and Ni$_2$Al$_3$ and the γ- and δ-phase of Ni/Zn which contain 75–60 mol% Al and 80–85 mol% Zn respectively.

Although at least four different technologies (cold rolling, flame spraying, Zn and Al-melt dipping, cathodic deposition of Ni/Zn-precursor alloys) have been described for applying the precursor alloy on the electrode surface, only cold rolling and cathodic deposition of precursor alloys are exercised for commercial production of Raney-nickel coated cathodes. Cold rolling of the powdered precursor NiAl$_3$ (10–20 μm particle size), mixed with granules of Mond nickel (carbonyl nickel, which is very ductile) on stainless steel or nickel toils creates so-called composite coatings of 20–30% coarse porosity (d$_p$>10 μm) which firmly adhere to the supporting toil. Due to the excessive sheer and stress exerted during rolling the toils are deformed and have to be shaped according to demands after this process has been finished. Compare also Chapters 4.8 and 7.2.3.2.

Choosing Ni/Zn (γ- and δ-phase) as a precursor allows to galvanically codeposit nickel and zinc together according to anomalous codeposition of these two metals from acidic electrolytes in the form of layered coatings containing the γ- and δ-phase. The cathodic deposition of NiZn-precursor has the advantage to be applicable to metal electrodes of any size. The coatings are usually roughly 100 μm thick and have a relatively rough, cauliflower-like surface (Fig. 10.7 a). During caustic leaching the Raney-nickel precursor alloy which contains 50 mol% or more of leachable metals (Al or Zn) the precursor shrinks and cracks are formed which subdivide the coating penetrating it through its entire extension (Fig. 10.7 b,c).

10.7.2.3.2
Utilisation of the Catalyst in Raney Nickel Coatings

The limited utilisation of nanoporous Raney-nickel coatings had been treated in Chap. 7 and must be accounted for in designing the coating thickness. Nickel is, compared to platinum metals (e.g. Pt, proper, or Ru) which exhibit exchange current densities of the hydrogen evolution reaction in caustic electrolytes at 30 °C of the order of 10^{-3}–10^{-2} A cm^{-2}, only a moderately active electrocatalyst with i_0 equalling 5–7×10^{-6} A cm^{-2} at ambient temperature. Therefore considering a ratio of the exchange current densities of Pt and Ni of 10^2–10^3 a naive estimation would predict comparable catalytic activities of smooth Pt-metal cathodes and Raney-nickel cathodes, provided the catalytically active inner surface of the microporous Raney-nickel coating would amount to 100–1000 cm^2/cm^2 electrode surface. It is discussed in Chap. 7, that because of mass transport limitations the utilisation of the inner surface of Raney-nickel coatings might be much less than 100%. But at higher current densities Raney-nickel coatings with a thickness of less than 100 μm become almost fully utilised.

10.7.2.3.3
Performance and Ageing of Raney-Nickel Coatings

Figure 10.8 compares current voltage curves for cathodic hydrogen evolution in 30 wt% KOH at 90 °C at smooth nickel and two different Raney-nickel cathodes obtained by caustic leaching of cathodically deposited NiZn and by cathodic reduction of smooth nickel sulfide coatings on gas phase sulfidized nickel electrodes. Both Raney nickel electrodes are comparable in performance immediately at the beginning of a long term experiment but they are ageing at a different rate. The Raney nickel coating obtained by relatively slow cathodic reduction of nickel sulfide, which lasts for several hundred hours while hydrogen is being evolved, shows almost no deterioration over more than 3000 h. But the Raney-nickel coating formed by caustic leaching of a Ni/Zn precursor coating whose formation is already finished within a few hours deteriorates sizeably during long term operation. This difference is very likely due to the more open and perturbed structure of the quickly formed catalyst, which causes restructuring and recrystallisation accompanied by surface losses with time. Recrystallisation may be retarded by dispersing TiO_2 or other metal oxides throughout the porous structure. This measure can only be used for cold-rolled or plasma sprayed, not for cathodically deposited precursor coatings.

An important loss mechanism is oxidation of the highly dispersed, reactive Raney nickel by reaction with water (Ni + $2H_2O \rightarrow$ Ni $(OH)_2$ + 1/2 O_2) under depolarised condition, that means during off-times. In contact with the hot electrolyte after complete release of the hydrogen stored in the pores by diffusion into the electrolyte the electrode potential shifts by 80 mV to the NiO/Ni potential and slow corrosion of Raney nickel can no longer be prevented. Prevention of corrosive degradation during off-times is most effectively achieved by keeping the electrolyzer polarised.

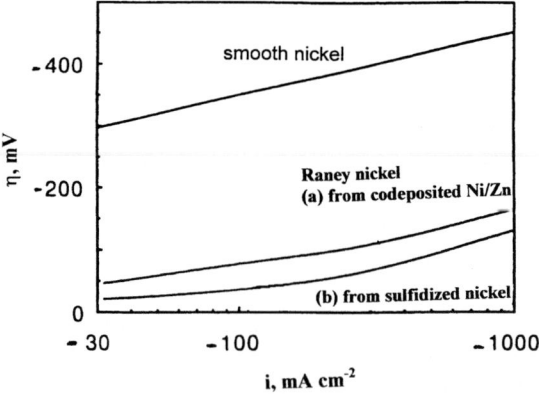

Fig. 10.8. Comparison of current-voltage correlation of cathodic hydrogen evolution from smooth nanoporous nickel and differently activated nickel cathodes

10.7.3
Coatings of Platinum-Metal Oxides

Dimensionally stable RuO_2-anodes have been successfully used also for cathodic hydrogen evolution from acid solutions. One observes low Tafel slopes, low overpotential and they are particularly insensitive to iron poisoning. But they are also electrocatalytically active in alkaline electrolytes The oxide is thermodynamically unstable at the hydrogen potential and should be reduced to ruthenium metal. It can be shown by cyclovoltammetry, that after short exposure of the oxidic coating to hydrogen evolution under technically relevant conditions (30 wt% KOH, 80 °C) that the surface of the coating does not persist as RuO_2 and is converted to the metal which becomes the catalyst proper although the underlying metal oxide layer keeps unchanged. It is also the insensitivity of ruthenium metal to iron poisoning (see below) and the metallic reservoir stored in the bulk of ruthenium oxide which makes DSAs uniquely effective hydrogen evolving cathodes.

Platinum or platinum-metal activated iron, steel or nickel electrodes had never been used for hydrogen evolution from alkaline electrolytes in technical electrolyses. The main reason was certainly the high costs. Such coatings – as concerns platinum proper – show however also detrimental fast deactivation by iron poisoning. The iron contents of the electrolyte in conventional diaphragm chloralkali-electrolyzers and old fashioned water electrolyzers – though relatively low – was always high enough to deactivate the platinum surface by adsorption of ferrous-cations and subsequent deposition of metallic iron – after sufficiently high cathodic overpotential had been obtained due to Fe II adsorption. Quite different from platinum, ruthenium does not adsorb ferrous ions to a comparable extent and its catalytic activity is not impaired by traces of dissolved iron.

10.7.4
Active Coatings of Flame Sprayed, Doped Nickel Oxide

Asahi Glass reports on the development and commercial use of electrodes covered by catalytically active nickel oxide coatings which are said to be fabricated by a proprietary flame spraying process. No detailed information of the composition, morphology and phase content of the coating are communicated. Therefore it can only be assumed that the nickel oxide, which in the form of stoichiometric NiO is a non-conductor is doped by some additives transforming it to sufficiently conductive semiconductor. The electrode coating is said to deteriorate slowly by progressive reduction to metallic nickel. Lifetimes of three years in chloralkali electrolysis membrane cells are reported [1].

10.7.5
Platinum and Platinum Metal Cathodes in Membrane Water Electrolyzers

In acidic electrolytes only noble metals, and in particular the platinum metals, persist under unpolarised conditions and can be used as electrode materials for hydrogen evolution. Platinum in particular is a unique electrocatalyst for H_2 evolution in strongly acidic environment as prevailing in perfluorinated sulfonated polymer membranes used as solid-polymer electrolyte in membrane electrolyzers which had initially been developed by General Electric and later on had been formed into an established technology by ABB and other, smaller companies.

It is state of the art to deposit the platinum catalyst on the membrane surface by a diffusion process in which platinum salt solutions (from the cathode side of the membrane) and a reductant as e.g. hydrazine are counterdiffusing causing reductive precipitation of dispersed platinum close to, and on the surface of the membrane. The ABB MEMBREL cell is reported to contain a catalyst load of only 0.2 mg Pt cm^{-2}. The cathode is reported to exhibit only 50–70 mV overpotential at current densities of 1 A cm^{-2} and 80 °C.

10.8
Fuel Cell Electrodes

10.8.1
Low- and High-Temperature Fuel Cells

According to the electrolyte and working temperature one distinguishes the low temperature fuel cell technologies
(i)   alkaline fuel cell, AFC (70–80 °C),
(ii) proton-exchange-membrane fuel cell, PEMFC (70–80 °C) and
(iii) phosphoric acid cell, PAFC (200 °C) from the high-temperature technologies
(iv) molten-carbonate fuel cell, MCFC (650–700 °C) and
(v)   solid-oxide fuel cell, SOFC (800–1000 °C).

Historically alkaline cells became important as power sources for space vehicles, but they are of no relevance for terrestrial application. All fuel cells are anodically combusting hydrogen, though MCFCs and SOFCs may by supplied with methane or carbon monoxide from which by internal steam reforming and shift reaction within the cell the hydrogen may by generated in situ, Chap. 12. Fuel-cell systems comprising low temperature cells as PAFCs and PEMFCs are also fueled with natural gas. But they need separate steam reforming and shift reactors as steam reforming and shift converting are too slow at temperature around 100 °C. Anodic hydrogen oxidation and cathodic oxygen reduction is kinetically hampered at low temperature, so that anodic hydrogen oxidation in AFCs, PEMFCs and PAFCs demands catalysts of highest activity, that means platinum metals and platinum in particular. Also Raney nickel had been used in AFCs, but this technology had been abolished because Raney-nickel anodes are always under thread of failure due to nickel oxidation. Working temperatures of MCFCs and SOFCs, however, are high enough that the cheaper nickel metal in the form of sintered nickel – a much poorer catalyst than Pt – can be used.

For cathodic oxygen reduction in low-temperature fuel cells platinum is indispensable, whereas the cathodic electrocatalysts in MCFCs and SOFCs are

10.8 Fuel Cell Electrodes

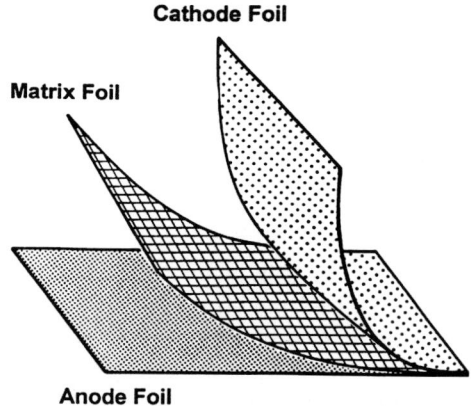

Fig. 10.9. Composing fuel cells from prefabricated cell components in the form of flexible foils or from component precursors formed by foil casting

lithiated nickel oxide and lanthanum – manganese perovskite respectively, which both are p-type semiconductors and much cheaper. Fuel-cell electrodes – different from electrolyzer electrodes – are not produced piecewise as self-supporting parts of the cell. Instead the whole fuel cell is usually composed as one unit comprising anode, electrolyte matrix and cathode together. Figure 10.9 demonstrates schematically how the cell is assembled from anode, matrix and cathode precursor foils in the thin layer technology, which is typical for fuel cell production. A final heat treatment in the cell stack generates the fuel cell in-situ.

10.8.2
Structural Design of Gas-Diffusion Electrodes in Low-Temperature Fuel Cells

Electrodes of low-temperature fuel cell with liquid electrolytes, e.g. phosphoric acid cells are usually constituted of PTFE-bonded nanoporous catalyst particles. These catalyst particles are most often formed of Pt-doped active-carbon agglomerates. Typically, as had been discussed in Chap. 7 and depicted schematically in Figs. 7.2 a and 8.7, these electrodes contain a coherent system of hydrophilic electrolyte flooded nano- and micropores and a hydrophobic, gas filled, micropore system. The hydrophilic pores extent on nanometer scale within the catalyst particles whose outer surface is covered by a thin electrolyte film and on micrometer scale between the catalyst particles. The hydrophobic gas conducting pore system extends on micrometer scale throughout the working-layer of the electrode between the catalyst particles and is constituted by hydrophobic PTFE-fibres or partially PTFE-clad catalyst particles. Both micropore systems are thoroughly interwoven. The hydrophilic pores establish the electrolytic connection between the electrode and the electrolyte-filled interelectrodic gap or matrix and the counter electrode and the hydrophobic pores are connected to the gas lumen through which the working gases are supplied to the electrodes. Figure 7.2 a,b explains the dual pore system in an Raney-nickel anode for outdated alkaline cells, in which the catalyst particles measuring 5–10 μm, are much coarser than Pt-loaded active carbon agglomerates ($d_p < 0.1$ μm) in PTFE-bonded carbon electrodes.

10.8.3
Oxygen Reduction Catalysts in Low-Temperature Cells

Electrocatalysis of cathodic oxygen reduction proceeds in many cases of practical importance on metallic surfaces, not at oxidised metalcatalyst surfaces. A comparison of the cyclic voltammograms of gold and platinum in phosphoric acid shows that gold – the most noble of all noble metals – is anodically oxidised at potentials which exceed the oxygen equilibrium potential by approximately 100 mV, whereas platinum like all other noble metals with the exception of gold is less noble than oxygen, becomes already oxidised at potentials which are 0.2 mV more cathodic than the oxygen potential. Other platinum metals, Rh or Ru and also silver are even less noble. They do not corrode at more anodic potentials because dense, passivating and, sometimes electron conducting oxide layers are formed. Provided such layers are non-conducting, as is PtO, it is essential that a thickness of only several atomic layers is maintained so that the electrons are able to cross this layer by tunnelling. Pt-metal at cathodic overpotentials in excess of –200 mV is the electrocatalysts proper for oxygen reduction. Silver cathodes and silver containing catalysts, which sometimes are used in alkaline electrolytes, have to be protected from oxidation and dissolution of the metal by keeping them permanently in particular in off-time at a cathodic polarisation of at least – 200 mV vs. O_2-equilibrium potential. Summarising these facts, one can state that:

(1) the most important catalyst material for O_2-reduction in acid as well as in alkaline solution at low temperature cells is platinum. It is the material of choice for O_2 reduction in acid electrolytes with respect to catalytic activity as well as with respect to longevity;
(2) in alkaline solution silver is a second option, but silver must be protected against anodic oxidation and partial dissolution by safely polarizing it to at least to –200 mV vs reversible oxygen electrode potential.

10.8.4
Catalysts for Anodic Hydrogen Oxidation

The mechanism of anodic hydrogen oxidation is much simpler than that of oxygen reduction.

Reaction pathways would be the reverse of the Volmer–Heyrovsky or the Volmer–Tafel mechanism that means with anodic adsorption

$$H_2 \rightarrow H_{ad} + H^+ + e^- \tag{10.12}$$

or dissociative chemisorption

$$H_2 \rightleftarrows 2H_{ad} \tag{10.13}$$

as the initial steps.

The electrocatalysts has to be stable under open-circuit conditions. Therefore the material of choice is carbon-supported Pt, Pd and other platinum metals for fuel cells using acid electrolytes as phosphoric acid or proton exchange membranes. Also tungsten carbide (WC) and platinized high surface WC had been proposed and used in acid solution. The mechanism of anodic hydrogen oxidation at tungsten carbide is little understood. Electrocatalysis might be due not to WC, proper, but to tungsten bronze generated in-situ as the material in contact with aqueous electrolyte is thermodynamically unstable and should be oxidised at any electrode potential, so that the formation of surface layers of WO_x are not improbable.

10.8.5
Properties, Preparation and Improvement of Electrocatalysts in Gas Diffusion Electrodes for Low Temperature Cells

10.8.5.1
Pt-Activated Active Carbon

Soot and other active carbons because of their high internal surface, amounting typically to 100 $m^2 g^{-1}$, are the most important catalyst supports for low temperature fuel cell electrodes. Platinum can be utilised on active carbon to a higher extent than in the form of dispersed platinum black. Carbon supported platinum is the fuel cell catalyst of choice as well for the oxygen cathode as for the hydrogen anode.

There exist three different ways to dope the active carbon with platinum or other catalytically active noble metals:

(i) flooding of active carbon with the help of wetting agents by aqueous or organic solvents containing Pt-salts, drying of the flooded material and reduction of the salts deposited within the active carbon particles at low temperatures by hydrogen or other reductants;

(ii) creating acidic groups (carboxylic acids and phenol groups) on the internal active carbon surface by chemical oxidation with e.g. chromic acid, fixation of cationic complexes of the catalyst (e.g. amino complexes of Pt II or Pt IV) at the inner surface of the active carbon particles by ion exchange, followed by chemical or electrochemical reduction of these complexes;

(iii) preparation of stabilised highly dispersed colloidal platinum which means Pt-particles of 1–10 nm size in aqueous solution or non-aqueous solvents like THF by chemical reduction of dissolved Pt-salts followed by adsorptive precipitation of the dispersed Pt-nanoparticles on the outer and inner surface of active carbon particles.

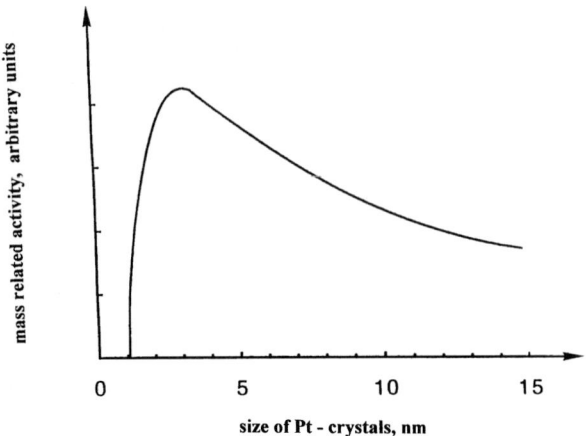

Fig. 10.10. Specific activity against particle size of nanocrystals of platinum [3]

The ultimate goal of catalyst preparation is to obtain very small platinum particles in order to increase the surface to volume ratio ($S/V=3/r$ for a spherical particle). Platinization by ion exchange and impregnation with colloidal platinum, yield the best results in this respect. More recently the transformation of carbon supported Pt colloids of approximately 1 nm diameter into Pt alloys had been reported which seems to yield an even better catalyst, since the alloy particles though coarser than the initial Pt particles show improved catalytic activity and stability [2].

10.8.5.2
Particle Size of Pt Nanocrystals on Active Carbon and Their Effective Catalytic Activity

Obviously it is of the highest commercial importance to utilise the dispersed platinum to the highest possible degree by decreasing the particle size. However, there are limits in increasing Pt utilisation by decreasing particle size. For Pt crystals of variable size the mass – related activity – which is economically spoken the figure of merit – passes through a maximum at a crystal size of 3 nm (Fig. 10.10). Obviously it makes little sense to decrease the crystal size below 2–3 nm.

10.8.5.3
Pt-Alloy Catalysts

Already for more than ten years it has been known that by reaction of dispersed Pt on active carbon with non-noble metals of group IV B and V B dispersed alloys are formed. Treating Pt-impregnated carbons, to which salts of these metals had been added, at 900° C in inert atmosphere (Ar) leads to formation of e.g. highly dispersed Pt-Cr or Pt-V alloys. The metal salts are reduced to the metals by the active carbon and form the respective platinum alloy in situ. These alloy crystallites are highly active cathode catalysts for phosphoric acid fuel cells. With a Pt/metal ratio from 1:1 to 5:1 V, Hf, Zr, Nb and Ta had been tried. All these metals are non-noble and are expected to be dissolved in phosphoric acid in the fuel cell under operating conditions. The initially used binary alloys are indeed not stable enough, and the binary alloy catalysts loose their enhanced catalytic activity during several thousand hours of operation. Ternary and quaternary alloys which contain chromium are remarkably much more stable than the binary alloys, so that the aim of 40,000 h of operation, which is the usually postulated lifetime for phosphoric acid fuel cells, can be achieved. Stabilised-alloy catalysts are used today in commercial fuel cell electrodes of phosphoric acid cells and contribute significantly to the technical success of these electricity generators which today have been produced in units from 50 kW to 11 MW in size [2]. The observed enhancement of the catalytic activity of Pt alloys is not at all understood – a common situation in catalysis research where in general the practitioners use to be far ahead of the theorists.

10.8.6
Morphology and Structure of Complete PTFE-Bonded Active-Carbon Electrodes

Figure 10.11 demonstrates by transmission electron microscopy the distribution of platinum microcrystals on (apparently transparent) graphitized active carbon, a type of soot. Active-carbon particles form agglomerates which measure around 0.1 µm in diameter. These agglomerates, after activation by dispersed platinum, are bonded and externally hydrophilised by an appropriate amount of PTFE. This bonding is achieved by mixing a PTFE-emulsion (particle diameter of dispersed PTFE approximately 0.3–0.5 µm) with active carbon, tape casting this mixture on a coarse porous carbonaceous support and annealing the combined electrode structure at 200° C, at which temperature the PTFE begins to melt. The melting

Fig. 10.11. TEM (Transmission electron microscopy) microgram of platinum dispersed on active carbon. Courtesy Stonehard Assoc.

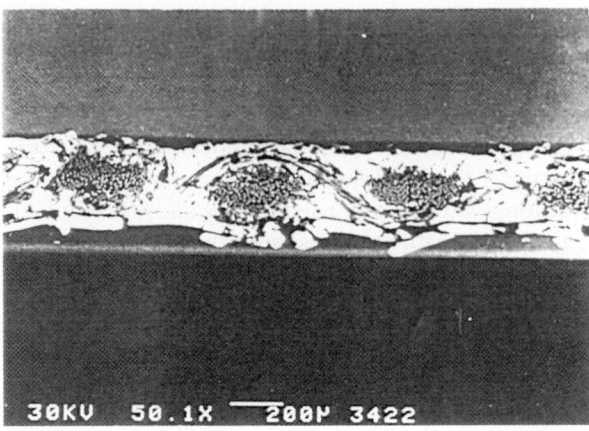

Fig. 10.12. Cross section of a commercial (E-TEK) electrode used for phosphoric-acid fuel cells. The catalyst consists of Pt on Vulcan XC 72 and is supported by and contained in the voids of a carbon fibre cloth. The *light-grey coloured layer at the lower edge* is the SiC-matrix, which contains the electrolyte

PTFE spreads and forms a thin, porous PTFE skin on the soot particles establishing the hydrophobic, gas conducting, pore system in the active layer of the electrode. Figure 10.12 shows the cross section of a commercial electrode, which is supported by carbon cloth. The narrow layer of only 20–30 μm thickness which appears with a light colour is the porous SiC matrix of the cell, which is soaked with the electrolyte. The porous mass between the coarse carbon fibers in the cloth contains 30 wt% PTFE and platinized carbon with 20 wt% platinum.

10.8.7
Ageing of Pt Catalysts

Every catalyst is ageing and losing activity with time. Pt-activated carbon cathodes are ageing faster than anodes. Apart from catalyst deterioration by gaseous poisons the following ageing mechanisms have to be taken into account:
(i) anodic dissolution of Pt in the form of sparely soluble Pt II and diffusion of the dissolved species to the anode where Pt is deposited as metal;
(ii) agglomeration of Pt particles by stochastic movement on the active carbon particles and
(iii) crystal growth and Ostwald ripening of Pt particles due to surface diffusion of Pt atoms.

Catalyst deterioration due to gas poisoning is only avoided by careful gas cleaning. Anodic oxidation followed by dissolution of Pt and transfer to the cathode is a serious cause for loss of Pt. It is potential dependent and accelerates as the cathode potential becomes more anodic, for instance under partial load or in off-time, when the cathode potential drifts towards the oxygen equilibrium potential. Therefore it is of utmost importance that, whenever the cell is switched off, the oxygen in the cathode lumen is rapidly exchanged by inert nitrogen, and that the cell voltage under operation does not surmount 0.8 V. Agglomeration of Pt crystallites due to Brownian motion can really be observed and it can also be shown that, indeed, the interaction between the Pt-particles and the supporting active carbon in presence of the electrolyte, phosphoric acid, is weak enough to allow for relatively free movement of the Pt particles. This fast process obviously is also the reason for the non-observability of slower surface diffusion induced Ostwald ripening. Fortunately alloy catalysts composed of platinum and non-noble metals seem to show a reduced tendency to agglomeration as their deterioration and activity loss is much slower than that of the pure platinum catalyst.

10.8.8
Electrocatalysis of Anodic Methanol Oxidation

10.8.8.1
Technoeconomic Significance of the Process

Membrane fuel cells are considered promising power sources for electrotraction, provided the difficult storage problem of the fuel, hydrogen, can be solved. Cryogenic and pressurised hydrogen is too expensive yet and is unlikely to become much cheaper and therefore storage of hydrogen in form of methanol with on-board steam reforming of methanol is one of the options. Even more advantageous would be the direct anodic oxidation of methanol in a fuel cell, as the complicated integration of the steam reforming process into a car-system could be avoided and – in principle – higher energy conversion efficiencies of the whole system could be expected.

10.8.8.2
Self-Poisoning of Methanol Oxidising Pt-Catalyst by Oxidation Products of Methanol

Although platinum strongly catalyzes methanol decomposition by dissociative chemisorption yielding adsorbed H atoms and adsorbed methoxy species (Pt–CH_2OH), the initially high anodic current density which is observed when a platinum electrode is dipped into a methanol solution decreases within seconds and minutes by orders of magnitude reaching a, technically spoken, intolerably low value of only several mA cm^{-2} at + 500 mV vs RHE.

10.8.8.3
Anodic Methanol Oxidation at Alloy Catalysts

According to long lasting experimental efforts, the use of alloy catalysts which contain a less noble metal whose oxide exhibits low solubilities in acid electrolytes – in particular Ru and further Sn, and W are effective in this respect – enhances the catalytic activity of platinum for anodic methanol oxidation. The rationale of this effect had been, that the oxide of the non-noble component at close atomic distance from the Pt surface atoms, supplies by "spill over" the oxygen which is necessary to oxidise the adsorbed CO-species. A Pt/Ru catalyst of 50/50 mol/mol composition is most effective in methanol oxidation at temperatures around 100 °C. But the catalytic activity is still too low at catalyst loadings of fractions of mg cm^{-2} which would be necessary for mobile electricity generation in passenger cars. Therefore R and D concentrates on developing more elaborate ternary and quaternary electrocatalyst formulations for direct anodic methanol oxidation.

10.8.9
Gas-Diffusion Electrodes in Membrane (PEM) Fuel Cells

10.8.9.1
Rationale of Developing a Method of Internal Wetting for Membrane Fuel-Cell Electrodes

As long as fuel cells are using liquid electrolytes like phosphoric acid, the catalyst utilization is usually not limited by incomplete wetting of the catalyst. Provided the amount of electrolyte is sufficiently high, the hydrophilic nanoporous particles are not only completely flooded but due to their expressed hydrophilicity are wetted also externally by an electrolyte film which together with the whole electrolyte flooded hydrophilic pore system establishes the ionic contact of an electrode to the respective counter electrode. Using proton exchange membranes as electrolytes which are quasi-solid may cause a problem with respect to the complete wetting of the internal surface of the catalyst particles. In spite of this difficulty of developing solid-polymer-membrane fuel cells water-swollen perfluorinated sulfonic acid polymers such as the commercial Nafion have been used for fuel cells very early as they offer the following advantages.

(i) The electrode kinetics of cathodic oxygen reduction at Pt in contact with the acidic polymer is enhanced by a factor of at least 10 compared to aqueous sulfuric or phosphoric acid at temperatures of 50–80 °C because the sulfonic acid groups interact adsorptively less with Pt than SO_4^{-2} or PO_4^{2-}-ions, leaving the greater part of the Pt surface free for adsorption of O_2.

(ii) Transport properties of oxygen are better in the fluorinated polymer than in aqueous electrolytes. Although the diffusion coefficient is lower than in water, the oxygen solubility is largely enhanced so that the transport parameter Dc_{O_2} is higher than in water.

(iii) Although the electrolyte is strongly acidic, cell construction is less demanding as the solid electrolyte does not cause corrosion problems.

10.8.9.2
Improving Catalyst Utilisation by Ionomer Impregnation of Gas-Diffusion Electrodes

The ionomer which forms the proton exchange membrane (PEM) can be dissolved in isopropanol or other organic solvents. This opens the way for improving the ionic contact between the catalyst particles of a gas diffusion electrode and the proton conducting membrane and electrolyte, as it is now possible to impregnate the Pt-activated active carbon with the Nafion ionomer by imbibing it with the Nafion solution in isopropanol. After evaporation of the solvent the ionomer is left attached to the inner surface of the active carbon and the Pt catalyst.

10.8.9.3
The Preparation of Membrane Electrode Assemblies (MEAs) for Membrane Fuel Cells

The proton exchange membrane is sandwiched between the two gas diffusion electrodes. State of the art are thin film electrodes, whose thicknesses do not exceed 10 μm with commercially available platinized active carbon as catalyst. A slurry of this catalyst in a 5 wt% solution of Nafion in *t*-butanol is prepared and sieve-printed or foil cast at first on one side, and after drying on the other side of the membrane. Curing the MEA precursor at 200 °C is the final step, which yields the MEA ready for incorporation into the fuel cell stack. Figure 10.13 a shows schematically a cut through a typical electrode membrane assembly. As shown in Fig. 10.13 b by the micrograph of a cut through a real MEA in membrane fuel cells with a membrane thickness of 150 μm the electrodes with a thickness of 5–10 μm are remarkably thinner than the interelectrode gap.

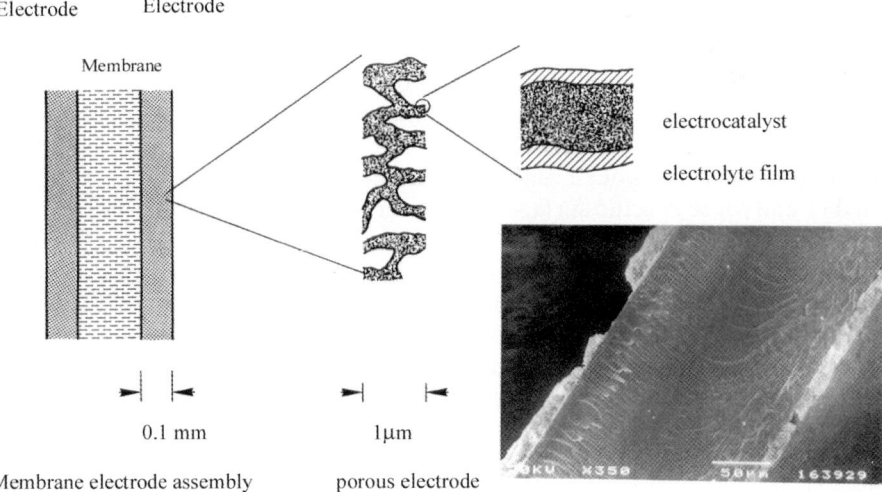

Fig. 10.13. a Schematic presentation of a cut through a membrane-electrode assembly (MEA) of a membrane fuel cell. **b** Micrograph of a cut through a PEMFC-MEA

10.8.10
Electrodes for High-Temperature Fuel Cells

10.8.10.1
Stability of Electrode Structures at High Temperatures

An increase in process temperature changes the reaction rates and hence the overall kinetics of more evolved electrochemical reactions which are composed of charge transfer and chemical steps to a sizeable extent. For hydrogen evolution from aqueous solutions, for instance, effective activation energies of 40–60 kJ mol^{-1} are reported. Increasing the temperature from ambient temperature to 600 °C would be expected to increase the rate of such reactions by a factor of at least 2×10^4. Therefore the electrochemical evolution and oxidation of hydrogen which exhibits at room temperature on nickel exchange-current densities of the order of 10^{-6} A cm^{-2} would proceed at 600 °C with exchange-current densities of 10^{-1} A cm^{-2} with almost vanishing overpotential even at high current densities. Also the much slower oxygen electrochemistry with somewhat higher activation energies would become relatively fast at 600 °C. Therefore in high temperature fuel cells the demand for highly efficient and expensive noble metal catalysts simply does not exist. Finding appropriate electrode materials for high temperature fuel cells is rather a question of finding materials which are resistant against chemical attack of the working gas and, in the case of molten carbonate fuel cells, an aggressive electrolyte and which keep dimensionally stable over long times and which are not subject to creep or shrinking by progressive sintering. As highly dispersed catalyst materials only oxide ceramics and refractory metals with melting points above 1800 °C would be expected to persist in the dispersed state at temperatures exceeding 600 °C. Nickel with a melting point of 1400 °C, which is the preferred anode metal of both high temperature fuel cell technologies (molten carbonate and solid oxide cells) is not stable above 150 °C in the form of the nanoporous Raney nickel. At temperatures above 600 °C highly dispersed Raney-nickel particles sinter to dense granules of micrometer size within fractions of an hour. Even with micrometer size the particles tend to grow further and loose specific surface at these temperatures. Therefore the sintered nickel anodes of high temperature fuel cells have to be stabilised against sintering and creep. Sintered nickel anodes of molten carbonate fuel cells, for instance, are stabilised by so called dispersion hardening achieved by adding a highly dispersed oxide (LiAlO$_2$) to sintered nickel. Cathode materials of both high temperature cells are composed of chemically and morphologically relatively stable oxide ceramics. For molten carbonate cells lithiated NiO is used and the cathode of an oxide-ceramic cell usually is made of porous LaMnO$_3$.

10.8.11
Electrode Kinetics and Electrocatalysis in Molten-Carbonate Fuel Cells

10.8.11.1
Anodic Hydrogen Oxidation

As one would assume, the electrode kinetics of the electrochemical hydrogen reaction in molten alkali carbonates on nickel is fast at the process temperature of 650 °C. The reaction order 0.5 of hydrogen is understood in terms of a fast established dissociative adsorption equilibrium at vanishing coverage. As hydrogen possesses in the molten electrolyte at 600 °C a solubility comparable to that in water close to ambient temperature and the electrode particles are coarse (of µm size as shown in Fig. 10.14 a) it is essential to establish a morphology and structure of the porous anode with larger gas conducting pores which must not be flooded and also are providing a thin electrolyte film covering the sintered nickel structure. Nickel is not well wetted by the melt as it exhibits a wetting angle of approximately 30°. Dispersing oxide ceramic materials as $LiAlO_2$, or Li_2TiO_3, which are almost insoluble in the melt, induces an improved wetting of the sintered nickel together with improved creep resistance, Fig. 10.14 b

10.8.11.2
Cathodic Oxygen Reduction

The cathodic reduction of oxygen in carbonate melts involves carbon dioxide, since carbonate ions are formed cathodically, decreasing the steady state concentration of oxygen anions at the cathode surface and depolarizing the cathode.

$$\frac{1}{2}O_2 + CO_2 + 2e^- \rightarrow CO_3^{2-} \tag{10.12}$$

The carbonate anion migrates from the cathode to the anode by this way serving as a shuttle for the oxygen dianion which is released as water vapour at the anode after combining with protons:

$$H_2 + CO_3^{2-} \rightarrow H_2O + CO_2 + 2e^-. \tag{10.13}$$

The cathode consists of a porous layer of lithiated nickel oxide $Li_xNi_{1-x}O$ which, being a p-type semiconductor of moderate conductivity ($\sim 10\ \Omega^{-1}\ cm^{-1}$) provides the necessary electronic conductivity and an internal cathode surface of approximately 2000–3000 $cm^2\ cm^{-3}$. Lithiated nickel oxide is a good catalyst for cathodic oxygen reduction with exchange current densities of the order of $10^{-2}\ A\ cm^{-2}$ at operating conditions.

Oxygen dissolves in the melt by reaction with carbonate, forming hyperoxide and peroxide anion, Eqs. (10.14) and (10.15):

$$2CO_3^{2-} + 3O_2 \rightarrow 4O_2^- + 2CO_2 \qquad \text{hyperoxide} \tag{10.14}$$

Fig. 10.14a–c. Microstructure of sintered-nickel anodes of molten-carbonate fuel cells (MCFC): **a** non-stabilised sintered nickel subject to rapid shrinking; **b** sintered nickel stabilised against compaction by so called external dispersion hardening – the dispersed oxide ceramic material is Li_2TiO_3 generated in-situ by reaction of TiO_2 with Li_2CO_3; **c** microstructure of the cathode of a molten carbonate fuel cell – the cathode is prepared from sintered nickel similar to that shown in a – the nickel sinter is oxidised in the cell by oxidation in contact with air

$$2CO_3^{2-} + O_2 \rightarrow 2CO_2 + 2O_2^{2-} \qquad \text{peroxide} \qquad (10.15)$$

These equilibria are responsible for a relatively complicated electrode kinetics. The typical cathode structure is depicted in the micrograph of Fig. 10.14 c. The nickel oxide clusters stem from a porous sintered nickel matrix, similar to

10.8 Fuel Cell Electrodes

that shown in Fig. 10.14 a. The sintered-nickel precursor is oxidised in-situ by reacting it with air, so that from metallic nickel granules agglomerates of lithiated nickel oxide are formed, which are composed of crystallites, whose diameter measure in fractions of micrometers. Only the outer parts of these agglomerates accounting for no more than a quarter of the agglomerates' volume are utilised because electrode kinetics are fast. In so far the cathode structure of MCFCs is still far from being ideal.

10.8.12
Electrodes in Solid-Oxide Fuel Cells (SOFC)

10.8.12.1
Electrodes and Electrode Structure

Figure 10.15 shows the enlarged cross section of a solid oxide fuel cell composed of a 50–100 μm thick ZrO_2 (more precisely yttria stabilised zirconia, YSZ) membrane, a relatively thick (≈100 μm) porous cathode and a porous Ni/ZrO_2-cermet anode, whose thickness is only approximately 10 μm.

10.8.12.2
The SOFC-Anode

The anodes and cathodes of solid oxide fuel cells exhibit quite different morphologies. The anode structure is characterised by a finely dispersed cermet composed of YSZ and nickel. The electrode reaction (Eq. 10.16) takes place near the phase boundary of the Ni/YSZ-grains. There the actively working anode is not nickel but the surface of the zirconia in the immediate neighbourhood of a Ni/ZrO_2 contact point. The coherent particulate-nickel matrix serves mainly as

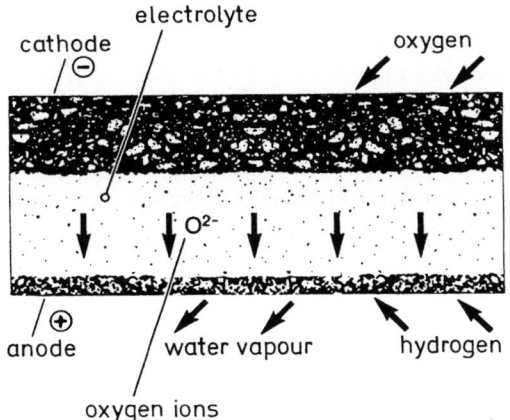

Fig. 10.15. Micrograph of the cross section through a solid oxide fuel cell (SOFC)

a current collector and is not the anode proper. The main purpose of using a porous cermet is to stabilise a high specific surface area for the interface ZrO_2/fuel gas where the equilibrium

$$O^{2-}_{ad} + H_2 \rightarrow H_2O + 2e^- \tag{10.16}$$

is established. As the mobility of electrons in the zirconia surface is relatively high, whereas their concentration is relatively low, the effective mean migration length for electrons in zirconia has to be kept as short as possible and the volume specific interfacial area of metal/solid electrolyte interface (Ni/ZrO_2) has to be high enough in order to avoid charge transfer limitations. Therefore, the morphology of the SOFC-anode resembles in some sense that of Raney nickel anodes of alkaline fuel cells with liquid electrolytes, although the limiting factors for electrode performance are very different: low solubility and diffusivity of gases in the case of low temperature alkaline fuel cells, low minority charge carrier concentrations in zirconia in the case of SOFCs. The cermet anode is therefore composed of small zirconia particles of 0.5–1 µm diameter and 5–10 µm nickel grains, the small dimensions of the zirconia grains rendering short transport distances for electrons in the ceramic material.

10.8.12.3
The SOFC-Cathode

The SOFC-cathode is usually composed of mixed oxides (La(Sr)$_x$MnO$_{3+x}$ in particular) which possess good oxygen ion mobilities together with moderate electron conductivity. Such electrode materials could, in principle, work as flat, two-dimensional films. However, their limited electronic conductivity demands an enhanced, not too small thickness of the electrode layer in order to provide a reasonably low lateral resistivity of the electrode for current collection.

Only this is the reason why SOFC-cathodes are formed as porous and relatively thick layers of LaMnO$_3$. The cathode thickness for cells with current collector spacings of approx. 1 cm length is about 200 µm with a mean porosity of 40% and a mean particle and pore diameter of 5 µm.

Current densities in the cathode are mainly determined by the respective value of oxide anion conductivity compared to the electronic conductivity ($\kappa_{O^{2-}}$ and κ_{e^-} both are coupled to each other in Wagner diffusion). Equation (10.17) describes the current density limit for coupled transport of oxygen anions and electrons:

$$i_l = \left(\frac{RT}{Fd_{LaMnO_3}} \right) \kappa \tag{10.17}$$

d_{LaMnO_3} = particle diameter of electrode material. Therefore, the thickness of closed cathode layers or coatings, d, would have to be kept below a value of several 10 µm. Coatings of this thickness, however, would provide too low lateral electronic conductivities for current collection along the cell surface. Therefore, a porous cathode is constructed with approx. 50% porosity and at least 200 μm thickness, and with pore- and particle size of approx. 10 μm. The main part of this porous layer serves simply as current collector, and only the lowest part, i.e. those crystallites which contact the zirconia electrolyte, are acting as cathode surface across distances of the order of 10 μm.

References

1. E. Endoh, H. Otouma, T. Morimoto, Y. Oda, Int. J. Hydrogen Energy, *12*, 473, (1987)
2. P. Stonehart, Ber. Bunsenges. Phys. Chem., *94, 913, (1990)*
3. P. Stonehart, K. Kinoshita, J. A. S. Bette, Electrocatalysis *79*, (1976) 275

Further Readings

F. Hine, Electrode Processes and Electrochemical Engineering, Plenum Press, New York 1985
H. Wendt, S. Rausch, Th. Borucinsky, Advances in Applied Catalysis, *40*, 87 (1994)

CHAPTER 11

Industrial Processes

11.1
Introductory Remarks

Electrolysis is since more than hundred years applied on large scale as a unique chemical process technology for chlorine and aluminium production. Both processes are strongly endothermic and since they are associated with a relatively high positive Gibb's energy, electrolysis has been found to be particularly suited for their production. But electrolyses consume expensive electrical energy, are performed in relatively expensive electrolysis cells which are reactors with relatively low space time yields and demand additionally relatively expensive purification steps for preparing purified raw materials from relatively impure precursors or ores. (For instance rock salt or brine for chloralkali and bauxite for aluminium electrolysis, respectively.)

Since electrolysis processes are capital and energy intensive they have always to strive for economic superiority in competition with thermo-chemical process technology. Indeed, also for electrolytic chlorine and aluminium production there had been developed alternative thermochemical processes which, however, could not compete with, and beat electrolysis. The persistent success of electrochemical technology in these two cases lasting for more than a century is not the least due to a steadily improving electrolysis technology which led to dramatically reduced energy consumption and to generally reduced costs by also reducing investment costs due to improved process engineering and cell design. Although the electrolysis cell can be assumed to be the heart of an electrochemical process the additional process equipment for raw materials purification and product conditioning forms an essential part of the whole process and demands an essential fraction of the invested capital. Therefore the electrochemical engineer dealing with development and improvement of electrolysis processes cannot dispense with general chemical engineering science and technology and the tools and methods which are used by the normal chemical engineer. For this reason in this chapter not only the electrochemical process steps and electrolysis technologies are described, but the processes are treated as a whole including unit operations of raw material purification and product conditioning.

11.2
Inorganic Electrolysis and Electrosynthesis

By far the greatest part of commercial electrochemical processes and the respective energy consumption of the chemical process industries is connected with chloralkali electrolysis, that means coupled production of chlorine, hydrogen and alkali hydroxides. Chloro-oxoacids and their salts come next, all other electrochemical processes in the chemical industries running on much smaller scale. Metal winning is another segment, which is traditionally not allocated to the chemical but the metallurgical industries. But this division is rather given by convention, than founded on serious differences. The electrochemical engineering principles prevailing are, of course, the same. Therefore there is not made a fundamental distinction between these two different types of processes in this chapter. Inorganic electrolysis processes deal mainly either with chlorine production or production of compounds derived from elemental chlorine or they deal with electrolytic metal winning either from aqueous electrolytes – as part of general hydrometallurgy – or from melts as a unique technique for generation of expressly non-noble metals which cannot be comparably easily be won in the necessary purity by pyrometallurgical processes. Hydrogen production by water electrolysis is an old, not so well developed technology. Its commercial role is nearly insignificant because hydrogen production by steam reforming of methane is by far cheaper. Nonetheless this technology is worth not only of mentioning but also of broader treatment since remarkable improvements had been achieved recently with respect to its potential role for a hydrogen economy.

Accordingly the inorganic electrolytic processes will be ordered according to

(i) production of chlorine and chlorine-derived compounds,
(ii) water electrolysis,
(iii) electrolytic metal winning and refining in hydrometallurgy, and
(iv) metal winning by high temperature melt electrolyses.

11.3
Chloralkali Electrolysis

Worldwide electrolytic generation of chlorine and caustic soda stays in the second place of electrochemical electricity consumption following aluminium electrolysis. Electrolytic chlorine generation forms the basis for production of vinyl and polyvinyl chloride, of chlorinated hydrocarbons and it is an important oxidant in industrial organic syntheses e.g. in the production of propylene oxide and other epoxides and it cogenerates caustic soda as one of the most important reactants consumed in the chemical industries on very large scale for base catalysed reactions and even more frequently for acid neutralisation.

11.3.1
The Electrochemical Reaction

The electrochemical cogeneration of caustic soda, chlorine and hydrogen

$$NaCl + H_2O \rightarrow NaOH + 1/2\,Cl_2 + 1/2\,H_2 \tag{11.1}$$

is performed in three different variants namely as the (i) diaphragm process, (ii) the amalgam and (iii) the membrane process. Diaphragm and membrane process comprise anodic chlorine evolution together with cathodic water decomposition and hydrogen evolution:

$$(Cl)^-_{aqu} \rightarrow 1/2\,Cl_2 + e^- \qquad \text{anode reaction} \tag{11.2}$$

$$H_2O + e^- \rightarrow 1/2\,H_2 + (OH)^-_{aqu} \qquad \text{cathode reaction} \tag{11.3}$$

The cathodic reaction in the mercury process is the cathodic deposition of sodium dissolved in mercury forming an amalgam with maximally 0.4 wt% of sodium. In the amalgam process cathodic amalgam formation is followed by heterogeneously catalysed decomposition of the amalgam with water

$$(Na)_{Hg} + H_2O \rightarrow NaOH + 1/2\,H_2 \qquad \text{decomposition reaction} \tag{11.4}$$

which is accomplished in a separate reactor – the so called decomposer or pile which is filled either with a carbon comb or with carbon particles. In this reactor the corrosion reaction at Eq. (11.4), which is strongly hindered on mercury and also on carbon, is sizeably accelerated by heterogeneous electrocatalysts, for instance by MoO_2, which accelerate the cathodic hydrogen evolution. Reaction (11.4) is to be understood as composed of the anodic sodium ionisation and dissolution ($(Na)_{Hg} \rightarrow Na^+ + e^-$) and the cathodic hydrogen evolution (11.3).

11.3.2
Thermodynamics and Energy Demands

In the diaphragm and the membrane process the anolyte contains 260–300 g NaCl dm^{-3} (4.4–5.1 mol dm^{-3}) whereas the catholyte contains either a mixture of 15 wt% NaOH and 13 wt% NaCl (or of roughly 2.7 mol dm^{-3} NaCl and 3.5 mol dm^{-3} NaOH) in the diaphragm process or 30 wt% NaOH which corresponds to approximately 8 mol dm^{-3} NaOH in the membrane process. As had been pointed out in Chap. 3 the equilibrium cell potential due to a rough estimation neglecting activity corrections can be calculated at 80 °C to be

$$U_0 \text{ (diaphragm process)} = U_0 \text{ (membrane process)} = 2.3 \text{ V} \tag{11.5}$$

For the mercury process with more than 3 V a remarkably higher equilibrium cell potential is calculated by accounting for a shift of the equilibrium potential of the $(Na)_{Hg}/Na^+$ amalgam electrode which is by 0.9 mV more positive than the potential of the Na/Na$^+$ electrode due to the negative free enthalpy of sodium amalgam formation.

$$U_0 \text{ (amalgam process)} = 3.2 \text{ V} \tag{11.6}$$

From these data one calculates a minimum expenditure of electrical energy of approximately 444 kJ/mol Cl_2 or 1.810 kWh/to Cl_2 for the diaphragm and membrane process.

The minimal electrical energy demand of the mercury process amounts to 618 kJ/mol Cl_2 or 2.523 kWh/to Cl_2. Under practical operating conditions, however, the effective energy consumption is 30–50% higher. Furthermore the diaphragm process consumes a substantial amount of thermal energy for water evaporation as the obtained caustic soda contains only 15 wt% NaOH and 13 wt% NaCl so that it must be concentrated by a factor of approximately 3 to obtain 50 wt% NaOH containing a remainder of roughly 1 wt% of NaCl.

11.3.3
Anodic Chlorine Evolution

For more than 80 years graphite (artificial Acheson graphite) had been used for chlorine evolving anodes as only this material – though being thermodynamically unstable – was sufficiently inert enough to persist for reasonable time in chloralkali electrolyzers. These anodes "burnt" away within several months by anodic oxidation and chlorination of the graphite and had to be replaced by fresh ones every three to four months. Since the beginning of the 1970s, however, graphite anodes had been substituted more and more by ruthenium-dioxide coated titanium anodes. These so called dimensionally stable anodes (DSA) are almost perfectly stable with life times extending from 3–5 years. The lifetime is limited by the stability of the coating which is determined by the ampere-hours consumed per unit electrode surface. But even more important is that the RuO_2 coatings constitute an anode surface of high catalytic activity. After the life cycle of an anode the expensive titanium support is not spoiled but recovered and can be reactivated by applying fresh catalytic coatings.

Figure 4.15 compared the current voltage curves for anodic chlorine evolution at graphite and RuO_2-coated titanium anodes. The very low anodic overpotential at chlorine evolving RuO_2-coated anodes is mainly due to concentration polarisation as the anodically generated chlorine is dissolved at relatively high supersaturation in immediate neighbourhood of the electrode surface where chlorine bubbles form continuously under steady state conditions releasing chlorine gas. The form of the anodes depends on the process technology. For mercury cells working with horizontal electrodes wires or rods are applied (Fig. 10.5) in order to assure unimpeded gas release. For diaphragm cells, extended metal electrodes are used and for the membrane technology perforated plates, louvred and extended metal electrodes are usual (Fig. 10.1). RuO_2 coatings are formed on the titanium substrates after a special etching procedure by the spray–sinter procedure described in Chap. 10.

11.3.4
The Cathodic Reaction

11.3.4.1
Cathodic Sodium Deposition in the Mercury Process

According to the scheme Fig. 11.1 b which shows the principle of the mercury or amalgam process, the cathode of the mercury cell is formed by the surface of a freely moving mercury film flowing along a steel plate, the cell bottom, which is inclined by 2–3 degrees. Although at a pH of 2–3, which is the adjusted brine-pH, the equilibrium potential of 0.2 wt% Na-amalgam in the cell electrolyte containing roughly 280 g NaCl/l is by more than 0.9 V more negative than the equilibrium potential of the hydrogen electrode (–100 to –200 mV vs NHE), hydrogen evolution at the mercury surface is almost completely supressed. The cathodic sodium deposition proceeds with negligible overpotential at high current densities (1 A cm^{-2}) with 95–97% current efficiency. The mercury enters the cell with less than 0.003 wt% sodium. As the current density varies but little along the cell, the sodium content increases linearly with increasing flow length and reaches 0.3–0.4 wt% at the lower end of the cell. The electrode reaction is very fast and activation overpotential is negligible.

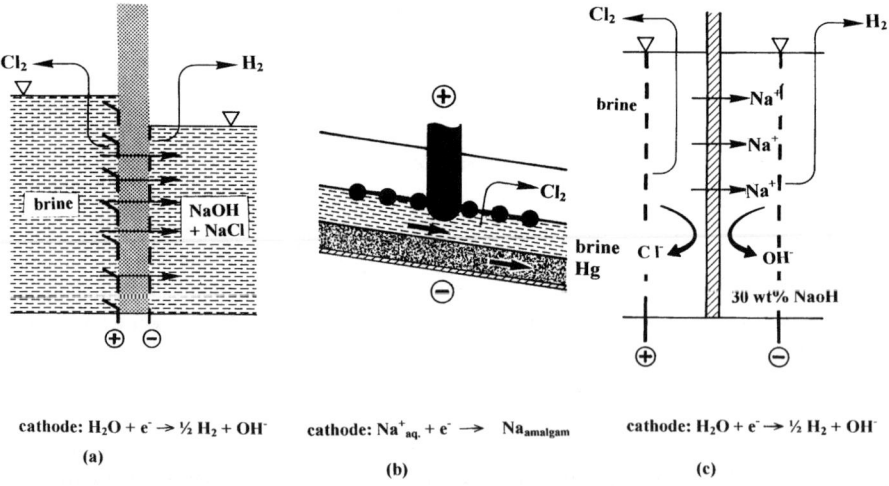

Fig. 11.1 a–c. Principle cell design and functioning of the three types of chloroalkali electrolysis: **a** diaphragm cell; **b** cell of the mercury process; **c** membrane electrolyzer

11.3.4.2
Cathodic Hydrogen Evolution in the Diaphragm and Membrane Process

Figures 11.1 a and c demonstrate the principle cell designs of the diaphragm and the membrane process. Old fashioned diaphragm electrolyzers used steel or nickel-coated steel cathodes. In more sophisticated membrane electrolysis cells steel electrodes could not be used because corrosion of these cathodes in off-time produced enough – though only slightly soluble – Fe II ions which lead to irreversible membrane deterioration by blocking the anionic groups of the cation exchange membrane. Therefore nickel cathodes became indispensable. Stainless steel cathodes exhibit considerable cathodic hydrogen evolution overpotential. But also at smooth carbon steel and nickel cathodes sizeable overvoltages are experienced, 200–300 mV being usual. Therefore up to date electrolyzers use catalytically activated cathodes in order to take advantage of very low hydrogen overpotentials and corresponding energy savings. As catalytic coating either high-surface Raney nickel is used or the surface of the nickel cathodes is doped with platinum, palladium, ruthenium of low loadings or ruthenium oxide whose surface becomes reduced to the metal, which in acidic as well as in alkaline solutions is a good cathodic electrocatalyst with an activity, which is comparable to that of the other metals of the platinum group and which is particularly insensitive against iron poisoning. Compare Chaps. 4 and 10.

Raney-nickel coatings deteriorate by slow recrystallization of the highly disperse Raney metal which may be retarded by addition of titanium to the precursor alloy. Since recrystallization as well as corrosion hazards during off-times are less severe under the conditions of the diaphragm process (lower caustic concentration of the catholyte) one prefers for diaphragm cells the cheaper Raney nickel activation. Nickel cathodes with proprietary coatings containing platinum group metals are coming into use more and more for membrane cells.

11.4
Process Technologies

11.4.1
The Amalgam Process

According to Fig. 11.1 b, mercury forming the cathode flows down the inclined steel base which typically has a width of from 1.50 m to 2 m and a length of 10 m. Above the flowing mercury cathode which forms a film of several millimetres thickness the brine forms a slowly moving layer of several centimetres in height. The purified brine enters the cell at 80 °C (or even several degrees more) and contains approximately 310 g dm^{-3} NaCl. The brine concentration is depleted by 10% as it leaves the cell due to simultaneous anodic chlorine evolution and cathodic formation of sodium amalgam. The sodium concentration in the mercury is initially almost nil (less than 0.003%) and increases to between 0.2 and 0.4%. As shown in Fig. 8.10 the RuO_2-coated titanium grid or rod electrode al-

Table 11.1. Typical operational data of modern mercury cells

Cathode area, m^2	15 to 30
Cathode dimensions /m × m	9.6 × 1.6 to 14.6 × 2.1
Slope of cell base, %	1.5 to 2.0
Rated current, kA	160 to 350
max. current density, kA m^{-2}	10 to 12.5
cell voltage at 10 kA m^{-2}	4 to 4.3
numbers of anodes	24 to 54
quantity of mercury per cell, kg	1650 to 5000
Energy requirement per ton of chlorine, kWh (d.e.)	3200 to 3400

Manufacturers: Uhde, De Nora, Olin-Mathiesen, Solvay, Krebs Paris

most touches the surface of the mercury so that one could speak in this case of a "zero gap" cell and the corresponding cell resistance (k value) is the smallest of the three different chloralkali electrolysis technologies – amounting to well less than $1 \times 10^{-4}\,\Omega\,m^2$ or $1\,\Omega\,cm^2$.

Usual current densities amount to 1 A cm^{-2} the ohmic potential drop which adds to the decomposition potential (3.2 V) is approximately 0.3 V and the summed overpotential and ohmic losses do not surmount 800–900 mV. One measures accordingly cell voltages above 4 V (4.3–4.6 V). The anodically evolved chlorine and depleted brine leave the cell either as two separate streams or as a two phase mixture. The amalgam flows across a weir into an amalgam decomposer. This is either a trickle-tower filled with a packed bed of graphite particles (tower decomposer or pile) or a flat bed arranged below the cell which contains on its bottom graphite combs, which float on the mercury and protrude with their fins into the caustic soda electrolyte above (vertical decomposers). At the carbon particles or comb surfaces a molybdenum-containing catalyst is applied which allows cathodic hydrogen evolution at largely enhanced exchange current densities and correspondingly reduced overpotential so that the amalgam corrosion Eq. (11.7)

$$(Na)_{Hg} + H_2O \rightarrow NaOH + 1/2\,H_2 \tag{11.7}$$

is accomplished by fast hydrogen evolution at the surface of the catalytically activated carbon, Eq. (11.8)

$$H_2O + e^- \rightarrow 1/2\,H_2 + OH^- \tag{11.8}$$

coupled to the practically unhindered anodic dissolution of sodium, Eq. (11.9)

$$(Na)_{Hg} \rightarrow (Na)^+{}_{aq.} + e^- \tag{11.9}$$

In this way a very pure 50 wt% solution of caustic soda is continuously produced and the mercury almost free of sodium is recycled. Conditioning and purification of chlorine which is particularly clean in the amalgam process as it contains less than 0.5% oxygen consists of cooling and drying. The hydrogen needs a particular treatment freeing it of residual mercury vapour. Table 11.1 collects characteristic data of a number of different types of mercury cells.

11.4.2
The Diaphragm Process

The diaphragm process uses a cell divided by a polymer-bonded asbestos diaphragm of low hydraulic resistance through which the electrolyte flows with controlled flow rate from the anode to the cathode side. As shown in Fig. 11.1, a continuous consumption of chloride anions at the anode and migration of sodium ions from the anolyte into the catholyte leads to a depletion of the brine concentration in the anolyte to approximately half of the initial value in the feed stream (280 g dm^{-3}) so that due to the influx of the anolyte and superimposed sodium ion migration – which matches closely cathodic hydrogen evolution and hydroxyl ion generation, the catholyte consists of a mixed solution of NaCl and NaOH with the sodium concentration amounting to that of the feed brine.

$$H_2O + e^- \rightarrow 1/2\,H_2 + OH^- \tag{11.10}$$

The catholyte leaving the cell contains approximately 18 wt% NaCl and 11% caustic soda. In order to free the product of sodium chloride the caustic concentration is increased by multieffect evaporation to 50 wt% with the consequence of precipitating almost all NaCl in the form of very pure salt producing a 50 wt% caustic soda with approximately 1 wt% dissolved NaCl. In monopolar diaphragm cells with hollow cathodes submersed in the anolyte due to a relatively low hydrostatic head of several centimetres a continuous flow of the electrolyte through the cell is maintained. The hollow cathode whose exterior is formed by extended metal covered by asbestos, is flanged to the side wall of the electrolyzer. NaCl/NaOH containing electrolyte leaves the hollow cathode together with hydrogen through a pipe on the upper end of the electrode flange. The catholyte together with the hydrogen is removed from the cell by pumping them out of the cathode chamber. The chlorine is collected under a hood covering the cell from where it is pumped to the chlorine purification step.

Comparing the three process designs with respect to the product quality, the diaphragm process yields the lowest purity of caustic soda and chlorine. The chlorine may contain up to 4% of oxygen because any hydroxyl ions which reach the anolyte by back-migration or backflow are immediately oxidised to oxygen if not a continuous addition of hydrochloric acid neutralises these hydroxyl ions keeping the pH under steady state conditions below a value of 3. If this is not done the steady state pH increases to 5 and oxygen is cogenerated together with chlorine. As already pointed out above, 50 wt% NaOH obtained in the diaphragm process by evaporating two thirds of the water contents of the anolyte contains 1 wt% NaCl, which diminishes its value to a certain degree because it

Table 11.2. Typical operational data of diaphragm cells

Electrode gap, mm	6 to 10
Current density, kA m^{-2}	1.5 to 3
cell voltage, V	2.8 to 3.4
Current efficiency, %	95 to 96
Energy consumption, kWh (d.c.) per to Cl$_2$	2000 to 2500
O$_2$ contents of anode gas, %	1.5 to 2
Anode life, years	8 to 10
Cathode life, years	10 to 15

Manufacturers: Hooker, Glanor, OxyTech Systems, Uhde, Diamond-Shamrock

cannot be used for every purpose. The cell liquor also contains up to 1 wt% KClO$_3$ which is formed in the anolyte by disproportionation of hypochlorite.

$$3ClO^- \rightarrow ClO_3^- + 2Cl^- \tag{11.11}$$

Caustic soda of this quality is, for instance, not appropriate for the copperamine process producing cellulose fibres (rayon process). Today there exist as well monopolar as bipolar diaphragm electrolyzers. Table 11.2 collects the design data of a number of modern diaphragm cells. Cell voltages of 3.4–3.5 V are usual at designed current densities between 0.14 and 0.24 A cm^{-2}.

11.4.3
The Membrane Process

The development of perfluorinated ion exchange membranes, compare Chap. 8.4, which withstand the aggressive action of chlorine and hot concentrated caustic soda opened the way for the membrane process; see Fig. 11.1 c. Today this process is able to produce caustic soda and hydrogen and chlorine gases of a quality comparable to that of the amalgam process with lower investment and energy costs than the diaphragm or the amalgam process.

Today a composite membrane made of a very thin layer of a highly permselective perfluorinated carboxylate polymer membrane (Fig. 11.2) and a less selective but better conducting perfluorinated sulfonate polymer membrane is used, which allows the production of 30 wt% caustic soda with a cell voltage of 3.2 V at current densities around 0.3 A cm^{-2} (3 kA m^{-2}) and current efficiencies of 98%. Figure (11.1 c) demonstrates the principle of the cell with recirculated highly purified brine, which is depleted by from 20–30% of its initial content. At the cathode water is decomposed delivering hydrogen and hydroxyl ions the latter forming together with solvated sodium ions (Na$^+ \cdot$ 4 H$_2$O) which migrate through the membrane 30 wt% NaOH. At higher NaOH concentrations current efficiencies for NaOH would drop below 98%. Due to residual hydroxyl-ion per-

11.4 Process Technologies

Table 11.3. Characteristics of different membrane electrolyzers

	Bipolar type						Monopolar type			
	Asahi Chemical Industry		PPG	Tokuyama Soda	Hoechst -Uhde	Krebs kosmo	Asahi Glass	De Nora	Oxy- Tech	ICI
	Standard	Super	BIZEC	TSE-270	BM	MZB	AZEC	K-40	MGC	FM-21
Effective membrane area, m²	2.7	5.08	3.83	2.7	1.2–3	2.5	0.2	0.64	1.5	0.21
Cells per electrolyzer	80–110	80–110	20–50	30–120	up to 100	4×18	30–540	20–60	2–30	1–120
Current load, kA	10.8	20.3	15.3	8.1– 10.8	3–15	variable	18–340	40–150	6–225	1–100
Current density, kA m⁻²	4	4	4	3–4	2–5	variable	3–4	3–4	2–5	1.5–4.1

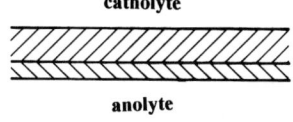

catholyte

anolyte

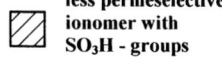

less permselective ionomer with SO₃H - groups

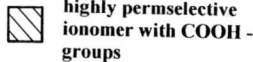

highly permselective ionomer with COOH - groups

Fig. 11.2. Schematic of laminated ion exchange membranes with highest permselectivity

meability, traces of hydroxyl ions may enter from the catholyte into the anolyte giving rise to anodic oxygen production unless the anolyte is kept at a pH of 3 by continuous addition of hydrochloric acid. Therefore chlorine and caustic current efficiencies differ from each other for membrane cells. As shown in Table 11.3 there exist today quite a number of producers of cells for the membrane process. The cells are working typically at nominal current densities between 3 and 4 kA m⁻² with cell voltages around 3.2 V. The cross section of the cells today is 0.2–5 m². They are most often constructed according to the zero-gap principle which demands high precision metal working with highest possible flatness of cell components. Figure 11.3 a shows the bipolar Hoechst-Uhde cell composed of stacked single cells made of separate stainless steel shells housing the cathodes and titanium shells for the anode side. Figure 11.3 b shows the monopolar ASAHI glass cell.

Fig. 11.3 a,b. Membrane cells: **a** bipolar Hoechst-Uhde electrolyzer composed of steel and titanium shells; **b** monopolar ASAHI glass electrolyzer

11.4.3.1
Process-Flow Sheets

Although the electrolysis cell is the centrepiece of the three chloralkali-electrolysis processes the initial preparation and purification of brine and the subse-

11.4 Process Technologies

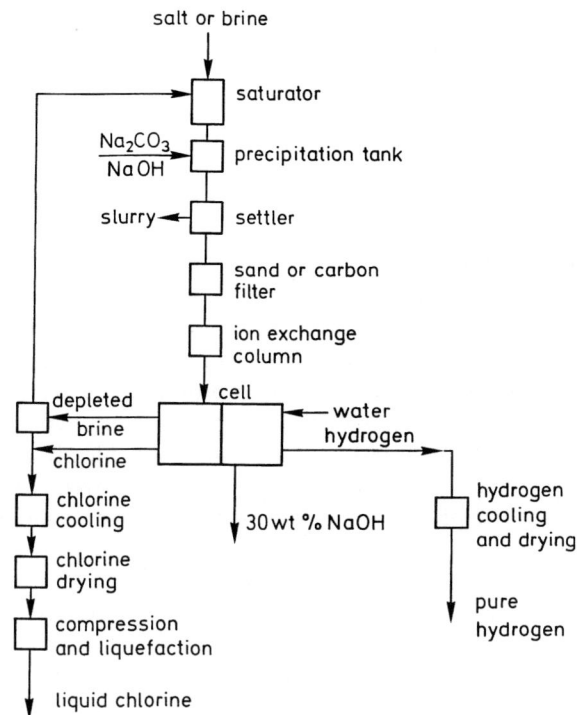

Fig. 11.4. Flow sheet of complete membrane process

Table 11.4. Impurities in rock salt and sea salt

	Rock salt	Sea salt
insolubles	<2%	0.1–0.3%
water	<3%	2–6%
calcium	0.2–0.3%	0.1–0.3%
magnesium	0.03–0.1%	0.08–0.3%
sulfate	<0.8%	0.3–1.2%
potassium	<0.04%	0.02–0.12%

quent conditioning of the product gases and produced caustic soda must be considered essential parts of chloralkali-electrolysis plants which are representing a considerable part of the invested capital. Figure 11.4 shows as an example the flow sheets of the membrane process. All three processes, irrespective whether their raw material is rock salt or natural brine, demand brine purification. Table 11.4 collects the contents of impurities of rock salt and sea salt respectively.

Magnesium and calcium (and traces of iron) have to be removed as cathodic calcium and magnesium deposition on mercury would cause the formation of solid amalgams. These two ions are also precipitated in the diaphragm causing diaphragm clogging in the diaphragm process. In the membrane process they would deteriorate the cation-exchange membrane by binding to the anionic sulfonate groups whose ability to exchange hydrated cations freely is essential for the high ionic conductivity and permselectivity of the membrane.

Therefore the first process step following the preparation of almost saturated brine containing approximately 300 g dm^{-3} NaCl is precipitating the alkali earth metal cations by precipitation with NaOH, (Mg(OH)$_2$), and Na$_2$CO$_3$, (CaCO$_3$). Sulfate is usually removed by addition of BaCO$_3$, (BaSO$_4$). Only the diaphragm process allows to keep sulfate in solution because a special procedure uses the relatively low solubility of the double-salt NaCl · Na$_2$SO$_4$ for continuous removal of sulfate. Its precipitation and generation of purified Glauber salt (Na$_2$SO$_4$·10 H$_2$O) in the final brine-purification step is an elegant method for profitably separating sulfate. The precipitated mixture of calcium carbonate and magnesium hydroxide is sedimented in big settler tanks during residence times of typically two hours. A small amount of finely dispersed solid matter is finally removed from the brine in sand filters. The chemical precipitation decreases the concentrations of calcium, magnesium and sulfate to < 2 mg dm^{-3}, < 1 mg dm^{-3} and < 10 mg dm^{-3} respectively. For the membrane process the concentration of the divalent cations must be decreased to lower than 0.05 ppm which is accomplished by passing the brine purified by precipitation through a bed of a sodium loaded cation exchange resin.

11.4.3.2
Brine Recycling

In the amalgam and the membrane process the highly purified brine is depleted in the anode chamber and the brine is recycled for resaturation. Leaving the cell it contains dissolved chlorine, hypochlorite, hypochloric acid and chlorate. All chlorine contained in the brine in form of any of these species has to be removed from the recycled brine stream. This is accomplished by acidifying the relatively hot effluent brine with enough HCl to maintain a pH of 2 and subjecting the acidified electrolyte to a vacuum dechlorination step. Hypochlorite and chlorate are reconverted according to Eqs. (11.12) and (11.13)

$$ClO^- + 2HCl \rightarrow H_2O + 3/2\,Cl_2 \tag{11.12}$$

$$ClO_3^- + 6HCl \rightarrow 3H_2O + 7/2\,Cl_2 \tag{11.13}$$

The dechlorinated, recycled brine is resaturated and purified as described above. Since in the membrane process the brine needs further purification by passing it over an ion exchange bed the trend today is to remove also sulfate ions by anion exchange rather than to stay with BaSO$_4$ precipitation since due to the relatively high costs of barium carbonate, ion exchange would help to reduce operating costs. Regenerating sulfate loaded anion-exchange resins with highly concentrated NaOH or NaCl allows to recover Na_2SO_4 in relatively pure form. Although the efforts to purify the brine are higher for the membrane than for the diaphragm process the investment for brine purification is comparable because the volumetric flow rate of fresh brine is by a factor of two smaller for the membrane process as the brine utilisation is almost 100% and by that factor higher in the membrane process than in the diaphragm process in which the brine is only converted to 50%.

11.4.4
Gas Purification and Conditioning

11.4.4.1
Chlorine

The chlorine which is recovered from the recycled brine by vacuum dechlorination is added to the chlorine main stream leaving the gas fully saturated with wa-

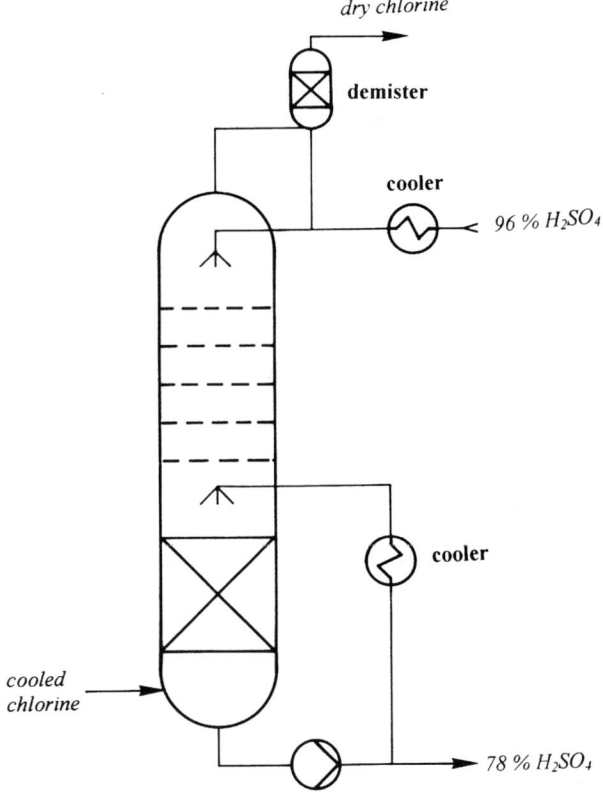

Fig. 11.5. Schematic of the counter current drying process of chlorine by concentrated sulfuric acid

ter vapour at the operating temperature. This gas contains – depending on temperature and brine concentration – from 40–60 vol.% water vapour which is separated in a tube and shell cooler made of titanium. For further purification the chlorine is dried with concentrated sulfuric acid. Figure 11.5 explains this counter current drying process which is accomplished stepwise by at first drying the chlorine with recycled 78 wt% H_2SO_4 and subjecting it to a final drying process with 98% H_2SO_4. The dry chlorine leaving the drying towers can be distributed and handled in carbon steel tubes and containments. But strict temperature control must assure that 180 °C is never exceeded in order to prevent chlorine-ignition of the steel. Chlorine from the mercury cells would now be appropriately pure for distribution and consumption. Chlorine from diaphragm and membrane cells contains 2–3% oxygen and must be purified for many purposes by chlorine liquefaction.

11.4.4.2
Hydrogen

Hydrogen from diaphragm and membrane cells is 99.9% pure or purer and needs only drying for conditioning which is most effectively performed by condensing the larger part of the water vapour by injection of cool water, for instance in counter current absorption towers. The water vapour contents of the gas is further decreased to some percent by indirect cooling with refrigerated water. Mercury saturation of hydrogen leaving amalgam decomposers of the amalgam process gives rise to substantial mercury losses if not properly mitigated. Hydrogen from mercury cells is always indirectly cooled in order to keep the amount of mercury-contaminated water low since together with water the mercury vapour is also precipitated. Cooling at 2 °C followed by adsorption of the last traces of mercury on iodine-loaded active carbon leaves only 1 to 15 µg Hg per m^3 H_2 in the gas and allows to bind and recycle evaporated mercury to an extent that today mercury losses can be kept safely below 1 g per ton of chlorine.

11.4.5
Comparison of the Three Processes

With respect to product quality the mercury process has particular advantages as the purity of chlorine and caustic soda and the immediately obtained high caustic soda concentration are not met by the diaphragm and the membrane process. But mercury retention and recovery from the product streams – from hydrogen in particular – is a serious problem causing extended additional investment. The purity of the caustic soda from membrane cells is quite comparable to that obtained by the mercury cell. But as its concentration still does not exceed 30 wt% an evaporation step is often needed. Chlorine from membrane cells usually contains oxygen and necessitates a liquefaction step for chlorine purification. In this respect the diaphragm process is particularly disadvantageous delivering impure, diluted caustic soda and impure chlorine. Nonetheless this process is still operational –

11.4 Process Technologies

Table 11.5. Energy consumption of chloralkali electrolyzer per to of chlorine

Type of technology	Diaphragm	Mercury	Membrane
electric energy (kWh)	2800–3000	3200–3600	2600–2800
steam equivalent (kWh)	800–1000	0	100–200
total kWh	3600–4000	3200–3600	2700–3000

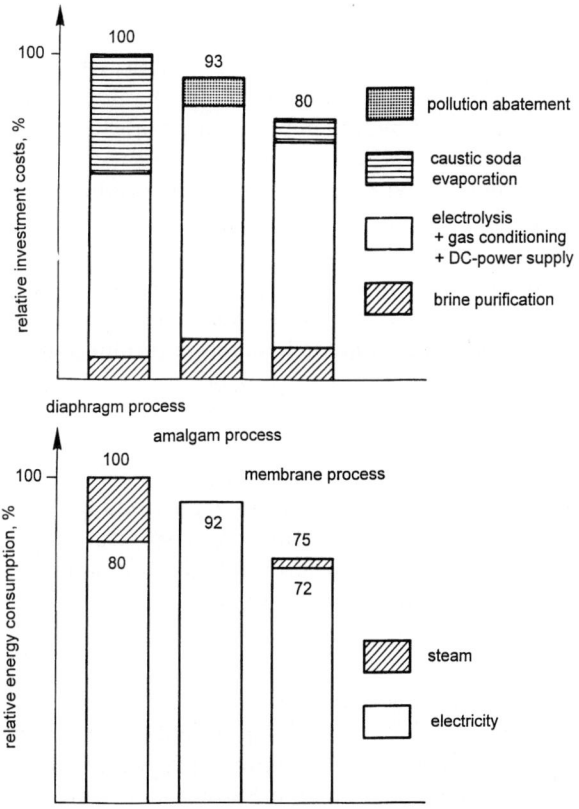

Fig. 11.6. Relative costs and energy consumption of the three chloralkali electrolysis technologies according to D. Bergner Chem. Ing. Technik, *66* (1994) 1026

very often combined to an amalgam or a membrane process as it delivers roughly half of the initially dissolved salt in form of very pure sodium chloride which can profitably be used as raw material for either of the two other processes. Concerning the economics today the membrane process is undoubtedly the most advantageous. Not only the energy consumption is lowest for this process but also the investment costs cannot be met by the other two processes. Table 11.5 collects the energy consumption for the three processes showing that the membrane process is the most energy efficient. Figure 11.6 gives an impression of the relative energy consumptions and costs and cost break-up of the three different electrolysis processes showing that according to total costs the membrane process is around 20% cheaper and the amalgam process around 5–7% cheaper than the diaphragm process.

11.5
Hypochlorite, Chlorate and Chlorine Dioxide

Chlorine and the products hypochlorous acid and chlorine dioxide, which are obtained from chlorine, are still the most important bleaching chemicals in the paper and pulp industry, though the use of hydrogen peroxide avoids the generation of chlorinated aromatic compounds and will be likely to supercede the use of chlorine-derived bleaches during the next decade. But still the production of sodium chlorate and its consumption for production of chlorine dioxide in the pulp industry is steadily increasing. Hypochlorite and chlorate production start from electrolysis of aqueous solutions of sodium chloride in undivided cells. The recent introduction of RuO_2-coated titanium anodes has revolutionised the cell design and electrolysis technology, in particular, in chlorate production.

11.5.1
Production of Sodium Hypochlorite

Solutions of hypochlorites are only stable for a limited time, provided the solutions are kept definitely alkaline (pH >9). Therefore the most appropriate procedure is to absorb chlorine in absorption towers in sodium hydroxide solutions. The absorption proceeds according to the disproportionation reaction

$$Cl_2 + 2OH^- \rightarrow ClO^- + Cl^- + H_2O \tag{11.14}$$

Keeping the process temperature low by cooling of the absorbent to 30–35 °C and regulating the contents of active chlorine (which is equivalent to the twofold concentration of ClO^-) at 150 g dm^{-3} or less keeps losses due to hypochlorite disintegration ($ClO^- \rightarrow Cl^- + 1/2 O_2$) low. The final stage of exhaust-gas treatment after chlorine liquefaction in the chloralkali-electrolysis industry in which chlorine is removed almost completely from the gas by absorption in sodium hydroxide solutions is today the most important source for hypochlorite solutions.

11.5.1.1
Electrolytic Generation of Hypochlorite

Production of sodium hypochlorite solutions on-site for immediate use in the paper and pulp industry and on board seagoing vessels for disinfecting purpos-

11.5 Hypochlorite, Chlorate and Chlorine Dioxide

es is performed in undivided flow through cells. Turbulent mixing of the electrolyte at current densities low enough to avoid evolution of gaseous chlorine allows for fast hydrolytic disproportionation of dissolved chlorine by reaction with cathodically dissolved hydroxyl anions according to Eq. (11.14). As the brine is unbuffered and not alkaline, electrolysis and storage of the product solution demands lowered temperatures not exceeding 25 °C in order to suppress chlorate formation (see below). For storage on long term it is essential to keep the solution strongly alkaline (pH >12.5), which suppresses chlorate formation safely.

11.5.1.2
Current Efficiency Losses

In the main, three electrochemical reactions can be identified to cause current efficiency losses: (i) cathodic hypochlorite reduction (Eq. 11.15), (ii) anodic hypochlorite oxidation (Eq. 11.16) and (iii) anodic oxygen evolution.

The first two reactions are mass transfer controlled, that means that their rate increases linearly with the hypochlorite concentration. The contribution of hypochlorite reduction can be diminished by increasing the current density at the cathode by choosing smaller cathode surfaces (grids or wires vs. flat anodes) and by additionally applying diffusion barriers in the form of porous cathode coatings. On board ships, where sea water is used instead of pure NaCl solutions MgO and $CaCO_3$ deposits are formed on the cathode which form such porous coatings though at the expense of enhanced cell resistances and cell voltages. Anodic hypochlorite oxidation and water electrolysis can be suppressed by use of RuO_2-coated anodes as they reduce the anodic chlorine overvoltage.

$$ClO^- + H_2O + 2e^- \rightarrow Cl^- + 2OH^- \tag{11.15}$$

$$3ClO^- + 1.5H_2O \rightarrow ClO_3^- + 3H^+ + 2Cl^- + 0.75O_2 + 3e^- \tag{11.16}$$

11.5.2
Production of Sodium Chlorate

Worldwide more than 10 million tons per year of sodium chlorate are produced. Sodium chlorate and – to a much smaller extent – potassium chlorate are produced by chloralkali electrolysis in undivided cells. The primary reaction is the formation of hypochlorite anions according to Eq. (11.14). At higher temperatures and hypochlorite concentrations chlorate is formed from hypochlorous acid which exists in hydrolytic equilibrium in unbuffered solution according to the homogeneous disproportionation reaction Eq. (11.18).

$$ClO^- + H_2O \leftrightarrow HClO + H^+ \qquad \text{hydrolysis} \tag{11.17}$$

$$2HClO + OCl^- \rightarrow ClO_3^- + 2Cl^- + 2H^+ \qquad \text{disproportionation} \tag{11.18}$$

Table 11.6. Typical operational data of sodium chlorate cells

Current per cell, kA	15–100	(480*)
Cell type	unipolar	(multipolar*)
Anode surface, m^2	7–92	
Interelectrodic gap, mm	3.5–6	
Current density, kA m^{-2}	1.5–3.0	
Operating voltage, min, V	2.7–3.0	
Operating voltage, max, V	3–3.7	
Energy consumption, kWh/t NaClO$_3$	4300–5500	
Current efficiency,%	94–96	
Operating temperature, °C	70–80	
pH of cell liquor	5.5–6.5	
Kind of coating	Pt/IrO+RuO$_2$	
cathode materials	in most cases steel	
Cell liquor composition		
Min. NaCl conc., g dm^{-3}	70–110	
Max. NaClO$_4$ conc., g dm^{-3}	600–700	
NaOCl conc., g dm^{-3}	1–3	

Manufacturers: Krebs, Paris; Chemetics, Vancouver; Peroxid Chemie, Zürich; De Nora – Pestalozza; PC Ugine Kuhlmann; Solvay, Bruxelles; Krebs Kosmo, Berlin; Fröhler – Lurgi – Uhde; Pennwaldt, Philadelphia; Huron Chemicals, Kingston*.

The cathodic hypochlorite reduction is almost completely suppressed in chlorate cells by adding chromate in 4–5 g dm^{-3} concentration to the cell liquor. Chromate is reduced cathodically to hydrous chromium oxides forming a porous layer on the cathode which is an effective diffusion barrier for hypochlorite without hindering cathodic hydrogen evolution which still proceeds with almost unaffected overpotential. Anodic oxygen evolution is usually negligible provided sodium chloride concentrations are kept above 100 g dm^{-3} and RuO$_2$ coated anodes or Ti anodes coated by oxides of other platinum metals are used which establish low anodic overpotentials.

Electrical efficiency losses of the process are mainly due to anodic hypochlorite oxidation according to Eq. (11.16). The electricity consumed by this undesired process is completely lost as it is absorbed by oxygen evolution. The rate of the cathodic reduction of hypochlorite, Eq. (11.15), is mass transfer controlled. To keep its contribution low, the hypochlorite concentration in the electrolyzer must be as low as possible. Care must also be taken to keep the concentration of dissolved, unreacted chlorine low as high chlorine concentrations would cause chlorine losses due to gas stripping by the cathodically evolved hydrogen. Therefore continuously operated chlorate electrolysis plants are composed of the undivided electrolysis cell proper and a chemical reactor of much larger volume in series. The electrolyte is recycled with less than 10% NaCl conversion per single

11.5 Hypochlorite, Chlorate and Chlorine Dioxide

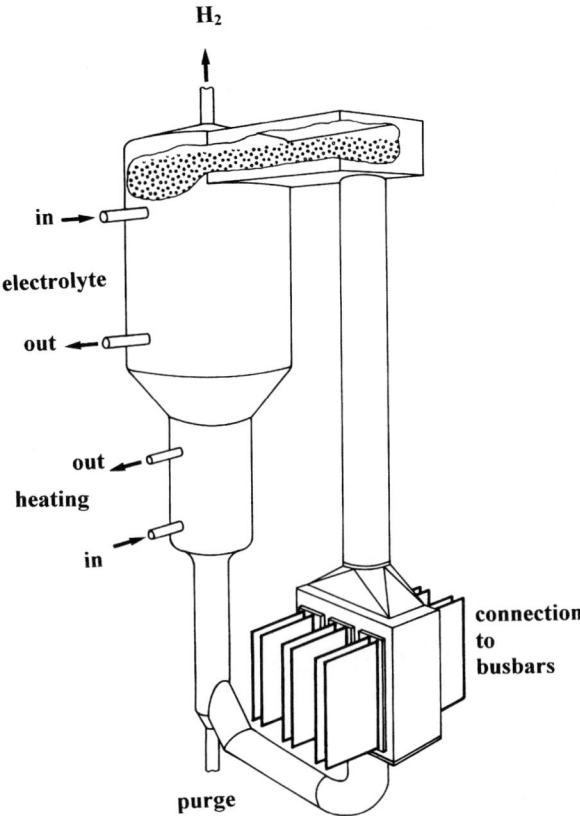

Fig. 11.7. Design of the Krebs chlorate cell

pass and the electrolyzer plus reaction vessel are operated according to the continuous stirred tank reactor (STR) mode. A steady state concentration of chlorate is maintained in the loop, while adding at the inlet the saturated brine and extracting at the outlet a product electrolyte stream of identical volumetric flow rate.

The homogeneous reaction at Eq. (11.18) proceeds mainly in the tank reactor with low steady state hypochlorite concentration. The degree of anodic conversion of chloride in the recycled electrolyte is kept relatively low (10%) in order to keep the hypochlorite and chlorine concentration at the outlet of the flow cell low. The usual current density is between 200 and 300 mA cm^{-2}. Today most cells are operated not on forced convection by pumping the electrolyte but using natural convection driven by the buoyancy of the released hydrogen. The cells of chlorate electrolyzers are forming a stack of anodes and cathodes with 3–4 mm distance connected electrically in parallel or they are realised in the form of a flow-through bipolar cell stack. Table 11.6 collects typical operational

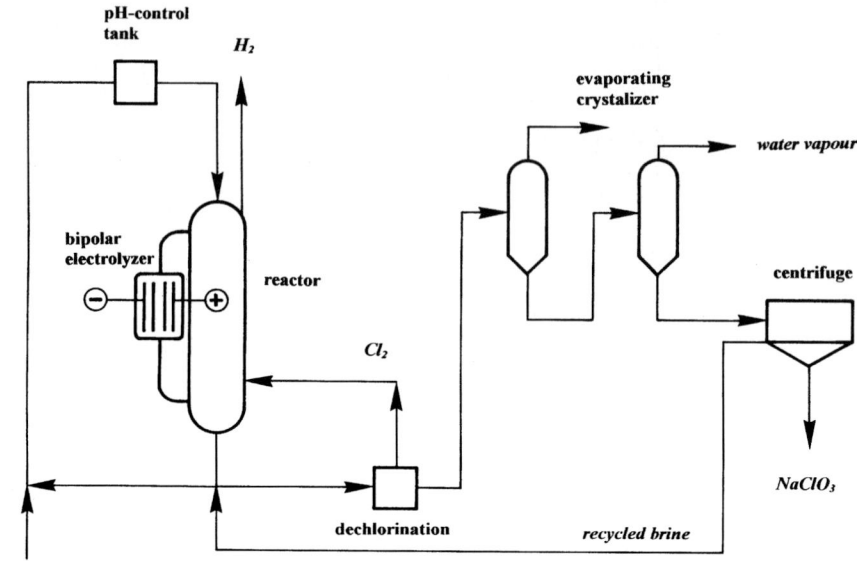

Fig. 11.8. Process scheme of electrolytic NaClO$_3$ production

conditions of commercial chlorate electrolyzers. The quoted low steady state NaOCl concentration underlines its importance for obtaining sufficiently high current efficiency. All chlorate electrolyzers use today noble metal oxide coated titanium anodes as the precondition for enhanced process temperatures (65–80 °C) speeding up the rate of the chemical conversion of hypochlorite to chlorate and thus allowing for a relatively small reactor volume. Figure 11.7 depicts as an example the design of the Krebs cell for chlorate production. In the Krebs cell the reactor is rather a tube than a stirred tank reactor. Its well defined and narrow residence time distribution comes close to that of an ideal tube reactor and allows to obtain the desired hypochlorite conversion in a smaller tank volume than would be needed with a stirred tank. Most often the Krebs electrolyzer comes with three tubes containing the electrolysis cells, coupled to one reactor tube.

11.5.2.1
Balance of Plant of Chlorate Electrosynthesis

Solar-heat-dried salt from sea water, mined rock salt or brine are used as raw material for NaClO$_3$ production. Magnesium and calcium are removed from the brine by precipitating MgO and CaCO$_3$ with NaOH or Na$_2$CO$_3$ in order to prevent fowling of the cathode by CaCO$_3$ deposition. Hydrochloric acid is then added in order to adjust the pH to approximately 6.5. The product solution containing approximately 100 g dm^{-3} NaCl and 600 gdm^{-3} NaClO$_3$ which according to the process schematic of Fig. 11.8 is continuously taken from the electrolyzer-reactor loop is stripped by air in order to remove unreacted chlorine. After passing through a filter the electrolyte is concentrated in a multieffect evaporation in which crystallisation of sodium perchlorate takes place while the solution cools from 85 to 10 °C. One obtains a slurry with 15–20 wt% of crystals. The salt is separated from the electrolyte in a centrifuge or hydrocyclone. After passing a rotary dryer the conditioning of the product salt is finished. The electrolyte is recycled and replenished with salt so that the residual, purified, sodium chloride is not lost but NaCl utilisation approaches 100%.

11.5.2.2
Construction Materials

The anode is made of RuO$_2$-coated titanium and the cathode material is stainless steel. In the electrolyzer/reactor loop where the electrolyte is slightly acidic and active chlorine is present the usual construction material today is PVC, glass fiber reinforced PVC in particular. For the brine purification step the electrolyte is kept slightly alkaline. Therefore for these parts of the process steel may be used. Flash evaporators, hydrocyclones and filters are usually PTFE-clad steel or titanium-clad steel. The salt dryer is made of stainless steel as it is imperative to avoid contamination of the product salt by iron. More recently the use of titanium for all the equipment is reported to be favoured. One third of the production costs is usually due to electricity consumption.

11.5.3
Chlorine Dioxide from Sodium Chlorate

Sodium chlorate is converted in the paper and pulp industry to chlorine dioxide – a most effective bleach. It is generated by reacting dissolved NaClO$_3$ with hydrochloric acid according to Eq. (11.19)

$$2HClO_3 + 2HCl \rightarrow 2ClO_2 + Cl_2 + 2H_2O \tag{11.19}$$

Chlorine evolution according to Eq. (11.20)

$$NaClO_3 + 6HCl \rightarrow 3Cl_2 + NaCl + 3H_2O \tag{11.20}$$

has to be suppressed by maintaining a high sodium chlorate concentration but a low hydrochloric acid concentration in the reactor. According to Lurgi and other firms, a process integrates chlorate electrolysis and HCl production by combusting chlorine from a chloralkali electrolysis with hydrogen from the chlorate cell generating ClO$_2$ according to reaction Eq. (11.19). The gas leaving the reactor is reabsorbed in purified water in an absorption column.

The Mathieson and the Solvay processes dispenses with hydrochloric acid as a reductant. It uses concentrated sulfuric acid as proton donor and sulfur dioxide (Mathieson) or methanol (Solvay), as reductants, facilitating the whole process as chlorine cogeneration is safely avoided according to Eqs. (11.21) and (11.22).

$$2NaClO_3 + SO_2 + H_2SO_4 \rightarrow 2ClO_2 + 2NaHSO_4 \tag{11.21}$$

$$2NaClO_3 + CH_3OH + H_2SO_4 \rightarrow 2ClO_2 + HCHO + Na_2SO_4 + 2H_2O \tag{11.22}$$

11.6
Perchloric Acid, Perchlorates, Peroxidsulfates

11.6.1
Perchloric Acid

Although $NaClO_3$ can be converted electrochemically to $NaClO_4$ it is also desirable to produce pure perchloric acid as it is particularly easy to prepare any desired perchlorate salt $M(ClO_4)_x$ neutralising $HClO_4$ with stoichiometric amounts of metal hydroxides $M(OH)_x$ or oxides respectively. According to a process patented by Merck company (E. Merck AG, DE 1,031,288, 1956) it is possible to oxidise chlorine dissolved in chilled 40% $HClO_4$ at Pt anodes according to Eq. (11.23)

$$1/2Cl_2 + 4H_2O \rightarrow HClO_4 + 7H^+ + 7e^- \tag{11.23}$$

The production volume is relatively small and therefore the cell design is little elaborated. The electrolyzer is a divided cell composed of PVC frames and the anodes are Pt foils spot welded to corrosion resistant tantalum rods through which the anodes are connected to the current source. At Pt anodes the oxygen overvoltage is sufficiently high to avoid excessive current efficiency losses. The working temperature is maintained between −5 and +3 °C by external cooling of the circulated electrolyte. The current density is moderately high with 250–500 mA cm^{-2} achieving a current efficiency of 60% (balance: O_2 evolution) at a cell voltage of approximately 4.4 V. The chlorine concentration at the cell inlet is approximately 3 g dm^{-3}. The product solution is distilled to remove the by-product HCl and remaining chlorine to yield 70% pure perchloric acid.

11.6.2
Sodium Perchlorate

The standard potential of the anodic chlorate oxidation to form perchlorate, Eq. (11.24),

$$ClO_3^- + H_2O \rightarrow ClO_4^- + 2H^+ + 2e^-; \quad E^0 = 1.19 V \tag{11.24}$$

Table 11.7. Typical technical data of perchlorate cells

Current density, A cm^{-2}	0.15–0.5
Cell potential, V	5–6.5
Current efficiency (Pt anodes)	90–97%
Energy consumption, kWh per–NaClO$_4$	2500–3000
Process temperature, °C	35–50
Electrolyte, concentrations in g dm^{-3}	
Na$_2$Cr$_2$O$_7$	0–5
Cell inlet	
NaClO$_3$	400–700
NaClO$_4$	0–100
Cell outlet	
NaClO$_3$	3–50
NaClO$_4$	800–1000

is very close to the oxygen equilibrium potential (1.23 V). It is therefore essential to find anode materials which are corrosion resistant at E >1.2 V and exhibit a sufficiently high oxygen overpotential. Smooth platinum anodes are optimal in this respect provided sufficiently high current densities are applied. Commercial perchlorate electrolyzers are operated at 0.15–0.5 A cm^{-2}. With smooth Pt anodes 90–97% current efficiency are obtained whereas at second best lead dioxide anodes only 85% current efficiency is achieved the balance being due to oxygen evolution. Cathodes are made from steel, stainless steel, nickel or bronze. The conventional cell schematic – as far as it is disclosed – is relatively simple – not to say primitive. The Cardox cell contains cylindrical anodes (copper rods of 1.3 cm diameter clad by Pt-foil) surrounded by a steel tube of 7.6 cm diameter serving as cathode. The tubes are perforated at the bottom and the top allowing for free circulation of the electrolyte driven by ascending hydrogen bubbles. The Pechiney cell is composed of Pt-foil anodes confronted from either side with cathode plates made from bronze. Table 11.7 gives an overview of typical operating data of a sodium perchlorate generating plant.

11.6.3
Peroxidisulfates

Comparable to anodic chlorine oxidation to perchloric acid is the anodic generation of peroxidisulfates by anodic oxidation of sulfuric acid according to Eq. (11.25).

$$2\text{HSO}_4^- \rightarrow \text{S}_2\text{O}_8^{2-} + 2\text{H}^+ + 2e^- \tag{11.25}$$

Ammonium peroxidisulfate which is used as bleach, as additive to explosives and propellants and as radical starter for acrylonitrile polymerisation is the only peroxo compound which today is produced electrochemically. – Peroxidisulfates can only be produced by anodic oxidation of sulfuric acid whereas anodic oxidation of alkali sulfate solutions produces only oxygen. A typical electrolyte for peroxidisulfate production contains at least 100 g dm^{-3} H$_2$SO$_4$ in some cases of the different technologies more – up to 500 g dm^{-3} H$_2$SO$_4$. The temperature gradient of the solubility of ammonium peroxidisulfate in water is by far greater than of any of the alkali peroxidisulfates. This facilitates separation of the salt from the electrolyte a lot and therefore it is the ammonium salt which is produced electrochemically by oxidation of an anolyte containing sulfuric acid and ammonium sulfate. Divided cells are used as peroxidisulfate would be reduced cathodically.

From the engineering point of view the most elegant design is that of the so called München process as it works at a relatively high temperature in an undivided cell. It solves the problem how to prohibit reduction of the product by wrapping thick sheets of porous materials (formerly asbestos, today porous hydrophilic polymers or cation exchange materials) around the cathode, thus establishing an effective mass transport barrier. All other processes use true divided cells. As high current densities at the anode are indispensible (ranging from 0.5 to more than 1 A cm^{-2}) and simultaneous fast and efficient heat transfer must be established in order to prevent overheating of the electrolyte at the anode surface all processes – with the exception of the München process – choose a concentric cell arrangement. Around a central rod-shaped-electrode a cylindrical narrow porous tube 4–6 mm in diameter contains the anolyte which is circulated in cascading it from anode tube to anode tube. The anode tubes are suspended from a ceiling plate into a common trough containing the catholyte and the cathode rods made of lead or of graphite. Only these materials are sufficiently corrosion resistant, also in off times, in the sulfuric acid electrolyte. The anode is made of Pt-foil-clad material – for instance a silver rod, covered by a tantalum coating and this is clad by Pt foil. Another solution of how to apply smooth Pt on a central rod is that of the München process where around a central rubber coated aluminium rod a Pt wire is wound.

The München process, in particular, may be operated as well continuously as batchwise. The effluent is concentrated by evaporation, the product crystalized by cooling and the anolyte is recycled. Table 11.8 collects the relevant technical data of four different ammonium peroxidisulfate processes. The first two processes (Weissensteiner and Degussa) oxidize essentially sulfuric acid to peroxidisulfuric acid eventually converting it to the ammonium salt by adding ammonium sulfate.

11.7 Fluorine

Table 11.8. Technical data of four different process versions for anodic electrosynthesis of ammonium peroxodisulfate

	Weißensteiner–Teichner process	Degussa process	Münch process	Riedel–Löwenstein process
current density anode, A cm^{-2}	0.8 to 0.9	0.5 to 0.7	1.0 to 1.2	0.4 to 1.1
current density cathode, A cm^{-2}	0.1	0.1	0.03 to 0.05	–
current yield %	70 to 73	70 to 75	82 to 84	85
Energy (d.c.) consumption/kWh kg^{-1}	12.4	9.4	10.7	9.3
process temp., °C	20 to 22	–	≈40	28 to 33
anode material	Pt or Pt/Ta	Pt/Ta/Ag	Pt	Pt
cathode material	Pb	Pb	C	Pb
electrolyte (inlet) H$_2$SO$_4$, g dm^{-3}	550 to 590	550 to 590	100	260
(NH$_4$)$_2$SO$_4$, g dm^{-3}	–	–	300	210 to 220
electrolyte (outlet) H$_2$S$_2$O$_8$, g dm^{-3}	250 to 280	330 to 340	–	–
(NH$_4$)$_2$S$_2$O$_8$, g dm^{-3}	–	–	160	230 to 240
platinum demand g per t	2.2	1.6	2.0	2.0

11.7
Fluorine

Elemental fluorine, being by far the strongest elemental oxidant, can only be produced electrochemically – namely by the anodic oxidation of fluoride anions in HF/alkali fluoride electrolytes at amorphous carbon electrodes. Fluorine is produced commercially since 1946 in the USA and since 1948 in Europe. Initially the incentive was the fluorination of UF$_4$ to UF$_6$ which is used for diffusional separation of the uranium isotopes. Today fluorine is used also by the chemical industry for production of SF$_6$ and for fluorinated organic compounds and polymers. The production of freons had meanwhile been stopped in the US and Europe because of their detrimental effect on the ozone shield in the stratosphere.

The electrolysis decomposes hydrogen fluoride according to reaction Eq. (11.26)

$$HF \rightarrow 1/2\,H_2 + 1/2\,F_2 \tag{11.26}$$

The composition of the usual electrolyte is close to the eutectic KF · H$_2$F$_2$ with a melting point of 82 °C. As fluorine electrolyzers are intermittently resupplied with HF, the KF concentration changes in the course of the electrolysis between 38 and 42 mol%. In order to avoid crystallisation of the electrolyte and to increase the relatively low ionic conductivity of the electrolyte and to reduce the cell voltage correspondingly the process temperature is raised to between 90 and 110 °C. Because of the aggressive electrolyte and the even higher reactivity of the product gas a simple cell design, avoiding as far as possible pumps and gas gaskets, is imperative. The cell is usually a tank reactor with a volume of approx. 1 m^3, whose inner wall very often serves as cathode. As mild steel or nickel is passive against the electrolyte and fluorine, these materials can be used as construction and also as cathode material. The anodes are carbon blocks. It is essential to keep the product gases separate from each other. A nickel shirt, which dips into the electrolyte separates the ascending gas–electrolyte emulsion and diverts the two different product gases into two different collector vessels. A cell may have 20–40 anode blocks being operated in parallel with a total current of 1000 to 10,000 A.

A particular problem is the relatively high anodic overpotential which is the main reason for the excessively high cell voltages of 10–15 V at current densities of 100–200 mA cm^{-2} – compared to the equilibrium potential of $E_0 = 1.8$ V. Poor wetting of the anode by the electrolyte is supposed to cause gas blanketing up to the generation of "anode effects", i.e. complete gas coverage of the anode accompanied by sparking with an intolerable increase of the cell voltage to several tens of volts. The fluorine leaving the cell is saturated with hydrogen fluoride which is removed and recovered by passing the gas through a packed bed of highly porous KF pellets. Heating this bed releases the hydrogen fluoride which is recycled into the cell.

11.8
Hydrogen by Water Electrolysis

11.8.1
Technoeconomic Environment

Water electrolysis had been performed on industrial scale since more than 100 years. The market for water electrolyzers, however, had always been small and stagnant and therefore innovation had been notoriously sluggish. Hydrogen has the singular capability of storing energy of high quality on a very large scale. Therefore it has been visualised to become the cornerstone of future energy systems which have been expected to be based, to a great deal, on solar energy and other renewable but typically unsteady energy sources. Hydrogen has been considered as a means for storage and transport of solar energy across very long distances in time and space. But energy transmission and transport by hydrogen became superseded by electricity transmission with high-voltage-dc technology. But still water electrolysis is singular since it is the only established technology for hydrogen production from high-graded, non-fossil energy, e.g., photovoltaic and nuclear electricity. Today its energy storage capability is assumed to become of importance rather for smaller self-sustaining energy systems than for large integrated energy grids. In this context the future of water electrolysis

11.8 Hydrogen by Water Electrolysis

will depend essentially on the development of cheap and highly efficient electrolyzers of several tens of kW and not on larger units consuming electric power in the MW-range. Load levelling of nuclear electricity generation is also supposed to be an attractive option for water electrolysis in particular because short time storage of hydrogen and oxygen is not very expensive and offers the possibility to regenerate electricity in peak-demand times by either high-value steam generation in hydrogen burners or electricity generation in hydrogen/oxygen fuel cells.

It is unlikely that hydrogen from electrolysis will ever become competitive with hydrogen from methane. Nonetheless, as a response to the first and second oil crisis, publicly financed research and development programs of the years 1975–1985 (Commission of the European Communities, International Energy Agency, Canada, Japan) led to a remarkable improvement of alkaline water electrolysis – the current technology – and the development of completely new electrolysis technologies, like membrane and high temperature steam electrolysis, all with improved energy efficiency. But the incentive to go ahead with the development and to develop small cheap units – perhaps of membrane electrolyzers – is definitely not yet accepted. On the contrary: large companies producing MW-electrolyzers, like Lurgi, have abandoned these activities and left the field to their competitors.

11.8.2
Thermodynamics and Technological Principles of Electrolytic Water Splitting

For water electrolysis (see Eqs. (11.27–11.30) for the alkaline system), the minimum amount of electrical energy is given by ΔG^0 of the water splitting reaction

$$H_2O_{(g\ or\ l)} \rightarrow H_{2(g)} + 1/2 O_{2(g)} \tag{11.27}$$

According to Eq. (11.30) the theoretical decomposition voltage of (liquid) water electrolysis at 25 °C and 1 bar, with $\Delta G^0_{298K} = 237.2$ kJ/mol, is only $U^0 = 1.23$ V. Referring to the lower heating value under standard conditions (1 bar, 25 °C), water vapour is split with a heat consumption ΔH^0 of 241.8 kJ/mol

$$2H_2O_{(g)} + 2e^- \rightarrow H_2 + 2OH^- \quad \text{(cathode)} \tag{11.28 a}$$

$$2OH^- \rightarrow 1/2 O_2 + H_2O + 2e^- \quad \text{(anode)} \tag{11.28 b}$$

$$H_2O_l \rightarrow H_{2(g)} + 1/2 O_{2(g)} \tag{11.29}$$

$$\Delta G^0 = 2FU^0 \tag{11.30}$$

Figure 3.5 indicates the temperature dependence of ΔG^0, ΔH^0 and the equilibrium cell voltage, U^0, at normal pressure (below 100 °C for electrolysis of liquid water and above 100 °C for water vapour decomposition). U^0 drops from 1.23 V at 100 °C to approximately 0.9 V at 900 °C or 1200 K. The cell voltages that are achieved in practice with present commercial alkaline electrolyzer technology are considerably higher than the equilibrium voltage of 1.25 V. With roughly 1.65–1.8 V (attributed to the best and most advanced commercial electrolyzers), they correspond to an energy efficiency of only 70–79%, if the lower heating value ΔH^0 of hydrogen (which under normal conditions is only slightly – by a factor of 1.07 – larger than $\Delta G^0{}_{298K}$) is used as reference for the energy yield.

11.8.3
Process Technologies

Three process versions have been developed for electrolytic water splitting.
- Water electrolysis with alkaline aqueous electrolytes, employing a porous diaphragm to separate the cathode and anode in order to avoid remixing of hydrogen and oxygen (Fig. 11.9 a). Alkaline-water electrolyzers are available in the power range of 10, 100 and several 1000 kW. The usual process temperature is 70–80 °C for ambient pressure processes and 90–100 °C for pressurised electrolyzers.
- Membrane or Solid-Polymer-Electrolyte (SPE) water electrolysis, employing a proton-conducting ion exchange membrane, which serves as electrolyte as well as a separator between the two porous electrodes. Highly purified water is added to the anode side. Membrane electrolyzers had been developed by General Electric and ABB but are today not available commercially (Fig. 11.9 b). The process temperature is 70–80 °C.

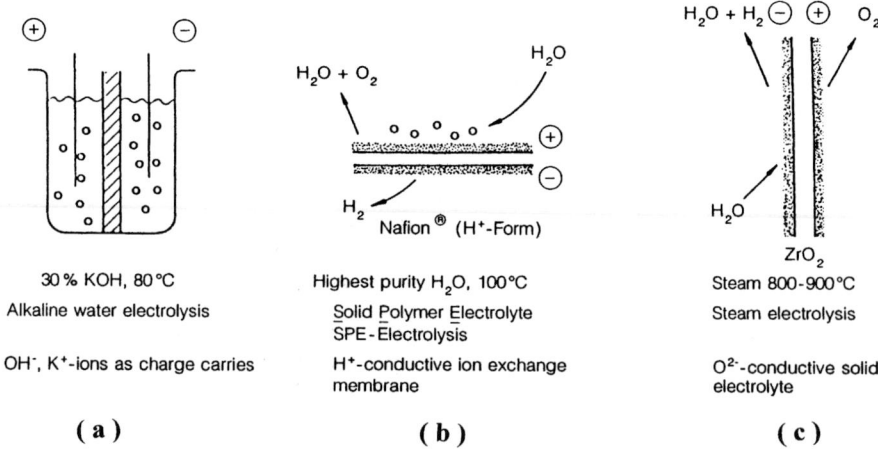

Fig. 11.9. Principles of the three existing water-electrolysis technologies. **a** alkaline, **b** membrane and **c** High temperature steam electrolyzer

11.8 Hydrogen by Water Electrolysis

Table 11.9. Key data for different water electrolyzers

Manufacturer	Electrolyzer Corp. Ltd.	ABB	Norsk Hydro A.S.	de Nora S.p.A.	(Lurgi Giovanola)[a]	ABB	MTU[b]
Cell Type	Monopolar	Bipolar	Bipolar	Bipolar	Bipolar	Bipolar	Bipolar
Operating pressure	normal	normal	normal	normal	30 bar	20 bar	30 bar
Electrolyte	alkaline	alkaline	alkaline	alkaline	alkaline	Nafion	alkaline
Operating temperature, °C	70	80	80	80	90	80	130
Electrolyte	28% KOH	25% KOH	25% KOH	29% KOH	25% KOH	Nafion	30 wt% KOH
Current density, kA m^{-2}	1.34	2.0	1.78	1.5	2.2	7	7–10
Cell voltage, V	1.9	2.05	1.75	1.85	1.80	1.65	1.65–1.8
Current yield	99.9	99.9	98	98.5	98.75	98	99
O$_2$ purity	>99.7	>99.6	>99.3	>99.6	>99.3	>99.5	>99.5
H$_2$ purity	>99.9	>99.8	>98.8	>99.9	>99.8	>99.8	>99.8
Required energy[c], kWh m^{-3}	4.9	4.9	4.3	4.6	4.5	4.0	4–4.4

[a] technology transferred from Lurgi to Gianovola
[b] MTU Motoren und Turbinen Union Friedrichshafen GmbH, Energy Conversion, New Technologies
[c] Per m^3 of hydrogen (S.T.P.)

– High-temperature steam electrolysis (850–1000 °C), employing oxygen ion-conducting ceramics (cubic ZrO_2 stabilized by Y_2O_3, MgO or CaO) as electrolyte. With the water entering on the cathode side as steam, a steam–hydrogen mixture is formed during electrolytic water splitting. O^{2-} ions are transported through the ceramic electrolyte to the anode where they are discharged as oxygen (Fig. 11.9 c). Steam electrolysis is still far from technical maturity, and it is questionable, whether it will ever become an established technology. A success in the development of solid-oxide fuel cells (SOFC) may eventually bear positively on the development of this type of electrolyzers, as the cell design of electrolyzers and solid-oxide fuel cells would be identical.

11.8.4
Conventional Alkaline Water Electrolysis

Potassium hydroxide (KOH) is used as electrolyte in conventional alkaline water electrolysis, mostly in 20–30 wt% solutions because of the conductivity optimum and the remarkable corrosion resistance of stainless steel in caustic solutions of this concentration. Hydrogen production via electrolysis with alkaline electrolytes is a well-established technology, but for cost reasons it is mainly used in medium-sized plants (0.5–5 MW, corresponding to roughly 100–1000 $m^3 h^{-1}$ of hydrogen (S.T.P.). Only very few large-scale plants with capacities of up to 30,000 $m^3 h^{-1}$ (in Aswan, Egypt, for example) used this technology to make hydrogen for ammonia synthesis on a very large scale.

11.8.4.1
Monopolar Technology

With one exception (Electrolyzer Corp. Ltd., Canada technology) most commercial electrolyzers are bipolar (see Table 11.9). So-called electrode supports, attached to the supporting plate, carry the electrodes. They are either louvered metal sheets or two-dimensional perforated surfaces (perforated plates, metal mesh or expanded metal). They serve as "fore-electrodes", forming a hollow space between the electrode and the supporting plate in which the two-phase mixture of electrolyte and hydrogen or oxygen is collected and drawn off – compare Section 8.3.

11.8.4.2
Bipolar Technology

Figure 11.10 depicts the bipolar pressurised cell stack of the Lurgi electrolyzer. Figure 11.11 gives the technical details of the layout of the bipolar filter press arrangement of the Lurgi pressure electrolyzer designed to operate at 30 bar. The electrolyzer stack has a diameter of approx. 1.5 m and might extend to 5 or, in a duplex arrangement, even to 10 m in length. Figs. 11.11 and 10.4 c show how the bipolar separation plates take the shape of waffled sheets that carry the metal net electrodes (nickel or nickel-plated iron), pressing them tightly against the

11.8 Hydrogen by Water Electrolysis

Fig. 11.10. Pressurised 30 bar Lurgi electrolyzer; the technology is now continued by Giovanola S.A., CH – 1870 Monthey Courtesy Lurgi GmbH

2–3 mm thick, very flexible asbestos diaphragms (so called distance free or zero gap geometry). Since asbestos will be phased out, very different porous materials as oxide ceramic materials like nickel oxide or polysulfon-bounded ZrO_2, Zirfar will be used in the future. The product gas/electrolyte emulsions are collected in gas ducts and transported to so called separators (long, pressurised tubes, arranged on top of the electrolyzer), in which the electrolyte emulsion can segregate due to sufficiently long residence times (about 1 min) to release the hydrogen or oxygen. The separators are equipped with water coolers and function as heat exchangers to remove excess heat. The effort required to condition the very pure gases is low: a gas scrubber removes KOH aerosol carried in the gas streams, in some cases followed by a gas drying step which frequently can be omitted in pressurised electrolysis systems. Oxygen in hydrogen and hydrogen in oxygen are safely kept well below 1 vol.%. If needed, a Pt catalyst takes care of the removal of remaining impurities (O_2 in H_2 or H_2 in O_2) between the gas-washing and drying steps by catalytic combustion.

Table 11.9 summarises typical operating data for a number of commercial water electrolyzers. In the last two columns the data of an improved alkaline technology and of an ion-conducting membrane electrolyzer are given, which have not yet advanced beyond the demonstration stage. Today's state of art is characterised for the conventional technologies by following key data: input-specific investment costs of 1500–2000 DM kW^{-1} (1988: 870 DM kW^{-1}) for very large plants, working temperature 80 °C, current density 2 kA m^{-2}, cell voltage about 1.8 V (corresponding to an energy consumption of about 4.5 kWh m^{-3} of hydrogen, including all losses and additional energy requirements for pumps and other auxiliary equipment). These data not only result in very high energy costs

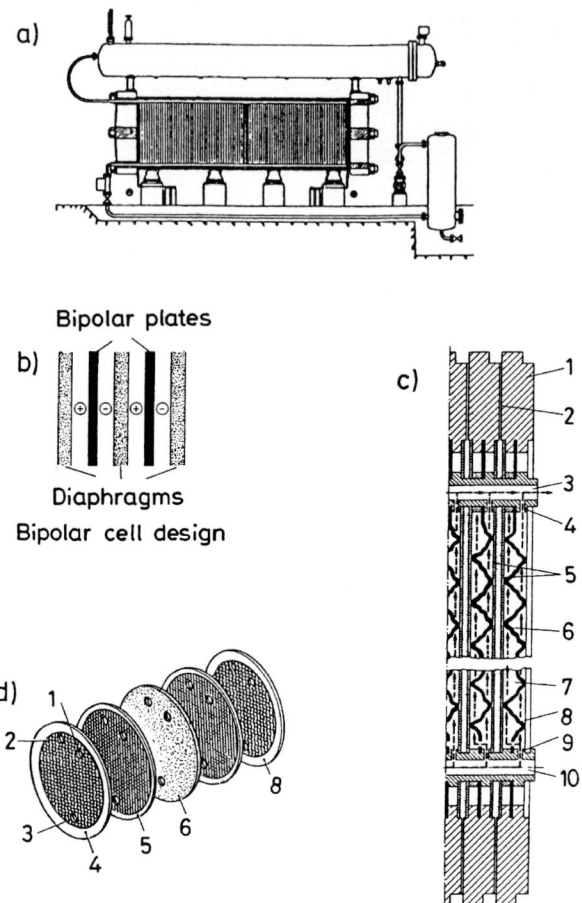

Fig. 11.11 a–d. Lurgi electrolyzer: **a** bipolar cell stack; **b** schematic of stacked cells with diaphragm (*grey*) and bipolar plate (*black*); **c** basic cell design; **d** exploded view of the cell; C:1 cell frame, 2 seal, 3 electrolyte/gas collecting conduit, separate for H_2 and O_2, 4 electrolyte outlet, 5 wire-mesh electrode, 6 diaphragm, 7 flow direction of the electrolyte mixture, 8 bipolar waffled metal sheet, 9 electrolyte inlet, 10 electrolyte-feed manifold. Courtesy Lurgi GmbH

(60–80% of total costs under Central European economic conditions), but also in relatively low space/time yield caused by low current densities which, if increased, would considerably increase the cell voltage and energy losses and costs (compare 9.5). Low temperature operation (< 80 °C) under ambient pressure permits the use of cheaper materials and light equipment such as normal steel for electrodes, cells, pipes, pumps, etc., and helps to keep investment costs low. Higher temperature and pressure demand heavier construction and far more care in particular to selection of steels with reliable corrosion stability and insensitivity towards stress corrosion cracking.

11.8 Hydrogen by Water Electrolysis

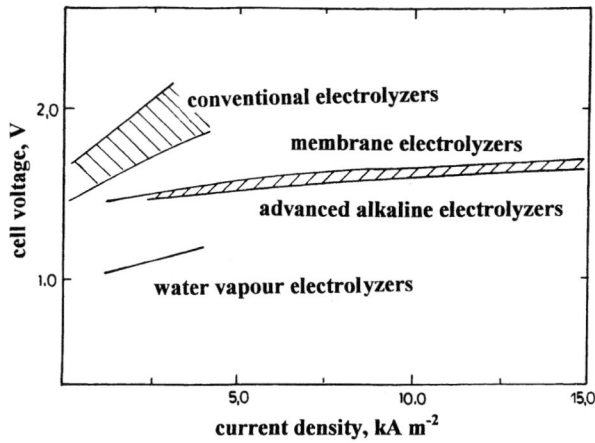

Fig. 11.12. Current voltage curves of various water electrolysis technologies

The advantages of higher temperatures and pressures are, however, convincing with respect to power consumption. Higher temperatures reduce the cell resistance because of improved electrolyte conductivity and reduce the electrode overpotentials due to thermal activation of the electrode process. High pressure reduces the volume of the evolved gases and gas holdup in the cell and thus reduces the effective electric resistance of electrolyte – compare Chap. 7.3.

Figure 11.12 which compares current voltage curves of different electrolyzer technologies depicts as the steepest current voltage curve one that is typical for the present conventional electrolyzer generation having effective surface specific resistances around 1 Ω cm^2. For improved pressurised electrolyzers this resistance is some 70–80% less. The steep increase of cell voltage with current density – mainly due to the relatively high cell resistance – is a good reason to keep current densities in conventional electrolyzers low, 0.2–0.3 A cm^{-2}, thus conserving expensive electrical energy. Only the Gianovola (former Lurgi) electrolyzer in its most modern version, the electrolyzer of MTU and the ABB membrane electrolyzer aim at current densities upwards of 0.5 A cm^{-2} and can do so because of their relatively low surface specific resistances. They allow hydrogen production with an energy consumption of 4 kWh m^{-3} at current densities as high as 0.8 A cm^{-2}.

11.8.4.3
Improved Alkaline Technologies

The main improvements which had been achieved in alkaline water electrolysis are as follows.
- Changing the cell configuration and geometry in particular reducing the interelectrodic gap with the aim of reducing the surface-specific cell resistance by a factor of at least three in order to reduce the ohmic voltage losses despite

considerably increased current densities. This included the development of a new diaphragm concepts which allows to dispense with the conventionally applied thick asbestos tissue. Interelectrodic gaps are now of the order of fractions of millimetres instead of several millimetres. In the MTU electrolyzer porous electrodes and diaphragm form a cosintered unit according to the true zero-gap concept. Compare Fig. 8.12 b
- Increasing the process temperature (possibly to an upper limit of 150 °C) with the aim of exploiting the increasing electric conductivity of the electrolyte in order to reduce the electric cell resistance further and to take advantage of the thermal activation of the electrode processes, i.e., to reduce overvoltage at increased temperature.
- Developing new electrocatalysts able to reduce the sum of the anodic and cathodic overpotential to about 0.3 V or less at current densities increased by a factor of 3 (0.5–1 A cm^{-2}).

11.8.5
New Technologies

11.8.5.1
Membrane Water Electrolysis

In membrane water electrolysis the high electrolytic conductivity of proton-loaded Nafion membranes is exploited. The electrode materials (ruthenium–iridium oxide–oxo–hydrates for the anode, finely dispersed Pt metals for the cathode) are deposited on both membrane surfaces via a diffusion precipitation method. The dispersed electrodes are brought into contact across the entire surface with a porous current collector (carbon felt at the cathode, platinized sintered titanium at the anode). Figure 11.12 compares the typical current voltage characteristics of membrane electrolyzers with those of other electrolyzers. The consequence of using proton-conducting Nafion membranes can be clearly seen. Cell voltages of the membrane technology is not any better than advanced alkaline electrolysis. This technology, however, dispensing with the uncomfortable and dangerous alkaline electrolyte, seems to be particularly suitable for smaller units.

11.8.5.2
Steam Electrolysis

Because of the negative temperature coefficient of the free enthalpy, high-temperature steam electrolysis requires the lowest cell voltages. However, there are many material and fabrication problems connected with the high operating temperature which are not yet solved. So the practical use of this technology is still well away. Figure 10.15 shows the cut through an oxide ceramic fuel cells which is identical with that of a steam electrolyzer.

Figure (11.13 a) illustrates the design principle of series-stacked cylindrical individual segments, each 1–2 cm in diameter and made of ZrO_2, each representing

11.8 Hydrogen by Water Electrolysis

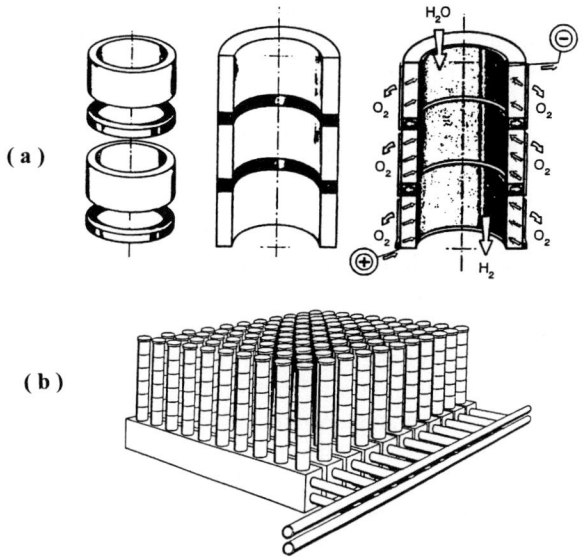

Fig. 11.13. a Design principles of the segmented HOT-ELY steam electrolyzer composed of zirconia rings and LaCrO$_3$-interconnect rings making contact between cathode and anode of subsequent ring-segments. **b** Cell registers forming a complete electrolyzer

one cell according to the HOT-ELLY concept of Dornier-System. Rings made of electrically conductive oxide ceramics (LaCrO$_3$), which connect the inner surface (cathode) of a segment electrically with the outside (anode) of the next segment, serve as seal between the segments. The cathode is made of ZrO$_2$-Ni cermet, while the anode is constructed as a porous layer of an electronically conductive mixed oxide (LaMnO$_3$). Cell voltages between 1 and 1.3 V were obtained in bench scale experiments. Figure 11.13 b demonstrates how stacks of several cylindrical cells are connected in parallel in registers, which are the essential parts which compose the steam electrolyzer. Also flat-plate electrolyzers might be envisaged as flat plate solid oxide fuel cells (Sect. 11.5) are today already in an advanced state of development and these units also might be used as steam electrolyzers.

11.8.6
Economic Implications of Technical Innovations for Alkaline Water Electrolysis

Any attempt to portray the consequences of technological innovation on the production costs of electrolytically produced hydrogen should be made with the assumption of fixed energy prices. The main results of an economic analysis for alkaline water electrolysis are the following.
- At present only conventional and advanced alkaline water electrolysis have a chance of being used in large-scale electrolytic hydrogen production facilities.

- Increasing the process temperature of alkaline water electrolysis beyond 140–150 °C does not produce any advantages; 120–130 °C at moderately enhanced pressure of 10–30 bars is likely to be the preferred process temperature in the future.
- The optimum current density is closely correlated to the price of electricity. It is slightly below 2 kA m^{-2} at a price of 0.12 DM kWh^{-1} and 6–7 k Am^{-2} at a price of 0.05 DM kWh^{-1}.
- At optimum current density the introduction of an improved alkaline water electrolysis technology permits a reduction of the production costs of electrolytically produced hydrogen by 20–25%.
- The data for small advanced alkaline and membrane electrolyzers (100 kW) are comparable with respect to both energy consumption and investment costs. The disadvantage of their small scale and the limited market makes 100 kW units today more expensive than large conventional units by a factor of approx. 6–7 on a power-specific basis. Small units are today offered for roughly 10,000 DM kW^{-1}.
- The situation might change in favour to advanced technologies if load levelling of nuclear power plants, and the possibility to store renewable energy in small isolated stand-alone units, would turn out to be a profitable option for electricity producing utilities.

11.9
Electrowinning and Electrorefining of Metals

11.9.1
Metal Electrowinning and Refining from Aqueous Electrolytes

Pyrometallurgical winning of metals is by far less expensive than electrowinning of metals by cathodic deposition from aqueous solutions of their salts. But the pyrometallurgical processes have slight or only partial purification effects and yield impure metals, which do not satisfy current quality standards. Therefore, electrowinning and electrorefining have become important for the production of high-quality copper, zinc, lead, nickel and cobalt. Electrorefining is based on the ordering of the equilibrium potentials of different metals according to the voltage series (compare Chap. 3). Anodic dissolution of impure metals if the current densities are small and therefore the overpotential negligible leaves the noble component of raw, unrefined metals undissolved. These nobler metals precipitate in the so-called anode slime or sludge, whereas the metals that are less noble than the metal being electrorefined dissolve with the main metal but are not redeposited on the cathode. Therefore, the less noble metals accumulate in the aqueous electrolyte. For any two metals M_1 with the standard electrode potential E_1^0 and M_2 with E_2^0, the voltage difference (E_1^0-E_2^0) establishes the anodic dissolution equilibrium and defines the concentrations c_1 and c_2 of the two metal ions in equilibrium at a given electrode potential according to the Nernst equation:

11.9 Electrowinning and Electrorefining of Metals

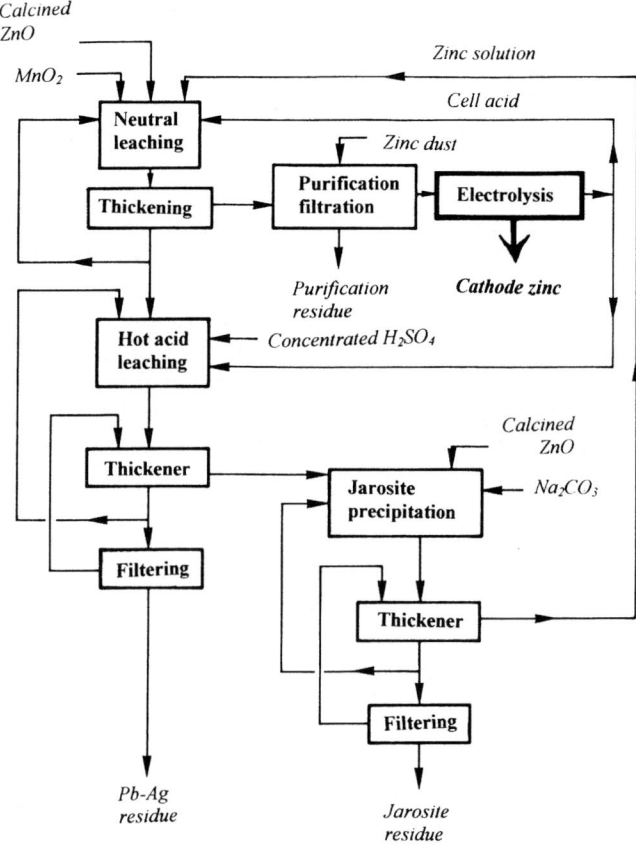

Fig. 11.14. Process scheme of zinc electrowinnng from galena (ZnS); galena is initially "roasted", i.e. oxidised to ZnO

$$E = E_1^0 + \left(\frac{RT}{\nu_{e,1}F}\right)\ln c_1 = E_2^0 + \left(\frac{RT}{\nu_{e,2}F}\right)\ln c_2 \tag{11.31}$$

$$c_1^{1/\nu_{e,1}} / c_2^{1/\nu_{e,2}} = \exp\{F(E_1^0 - E_2^0)/RT\} \tag{11.32}$$

The equations neglect the activity coefficients of the metals and are furthermore applicable only if electrochemical dissolution/deposition is a fast process for both metals. This is the case with the metals Cu, Zn and Pb but not with Ni, Co and Fe. The deposition/dissolution kinetics of nickel is particularly sluggish, which diminishes the efficiency of an electrorefining process involving nickel because enhanced overpotentials are generated at both the anode and the cathode. Alloy formation, which may change the chemical activity of the minor components greatly, is also not allowed for in Eq. (11.31). For instance, zinc alloyed in nickel ultimately codeposits with nickel at higher current densities, although according to the difference in standard electrode potentials, $E^0_{Ni}-E^0_{Zn}=0.53$ V, the equilibrium concentration of the two ions should differ at ambient temperature by a factor of 10^{17}. However, zinc becomes much more noble upon alloying with nickel.

Such effects and the accumulation of impurities in the recirculated electrolyte make chemical treatment and purification of the electrolyte indispensable for all electrowinning and electrorefining processes. Since electrowinning with insoluble oxygen evolving anodes starts from cruder materials than electrorefining, the chemical purification of the electrolyte must be even more elaborate and expensive for electrowinning than for electrorefining. Therefore, the important purification steps for electrowinning of non-noble metals, like zinc, are hydrometallurgical rather than electrochemical. This is exemplified by the flow sheet of the Jarosite zinc electrowinning process in Fig. 11.14. The process begins with roasted galena, that is calcined zinc oxide, which is leached with recirculated electrolyte (cell acid). The silver and lead remain in the sludge. All elements nobler than zinc are then precipitated by cementation with zinc dust, and the cemented sludge is treated further to recover Cd, Ni, Co and Fe. Iron can disturb zinc electrowinning seriously by electrochemical short circuiting:

$$Fe^{3+} + e^- \rightarrow Fe^{2+} \quad \text{cathode} \quad (11.33\ a)$$

$$Fe^{2+} \rightarrow Fe^{3+} + e^- \quad \text{anode} \quad (11.33\ b)$$

Therefore, iron III ions are precipitated as jarosite, a crystalline iron III sulfate oxide hydrate of very low solubility.

In copper refining the electrolyte purification is much easier because copper is a relatively noble metal. Accumulated nickel and other contaminants are removed from the recirculated electrolyte in a bleed stream. From this stream at first copper is completely electrodeposited by electrodeposition, then nickel sulfate is separated from the electrolyte by evaporative crystallisation.

For electrowinning and electrorefining of metals, mass transfer towards the cathode is of supreme importance. Formation of dendrites must be prevented. It is initiated if the applied current density comes close to or exceeds the mass transfer limited value. Convective diffusion driven by free convection caused by density differences of the bulk electrolyte and the electrolyte close to the electrode due to depletion and accumulation of copper in the solution at the cathode and the anode respectively dominates mass transfer in these hydrometallurgical electrolysis processes. Anodes and cathodes, typically 0.8×1.2 m^2, are suspended vertically at a distance 5–10 cm from each other. They are arranged alternately with the plate anodes one after the other in large rectangular tanks or troughs, which contain several hundred electrodes. Although there is a steady flow of electrolyte from one end of these huge tanks to the other, the flow velocity is small and electrolyte flow is not directed parallel to the electrode surface. Therefore forced flow does not contribute to convective mass transfer towards the electrodes to a significant degree. The electrolyte for copper electrorefining contains ca. 40 g dm^{-3} of Cu (a concentration somewhat less than

1 mol dm^{-3}) and total depletion of the copper from the solution decreases the density by several grams per 100 ml, i.e., by several percent. The Grashof number under these conditions – as had been discussed in Chap. 5 – is defined by:

$$\text{Gr} = \frac{x^3 g \Delta \rho / \rho}{v^2} \tag{11.34}$$

and would exceed the critical value of Gr·Sc of 10^{16} for turbulence if the distance x from the leading edge of the plate cathode exceeds several centimetres. Therefore, for the greater part of the cathode length, usually of the order of 1 m, turbulent flow, with correspondingly intense mass transfer, prevails. But since even then mean mass-transfer coefficients do not exceed ca. 10^{-3} cm s^{-1}, mean current densities for cathodic metal deposition in electrowinning and electrorefining processes must be kept low, of the order of 80–200 mA cm^{-2}, to stay well below mass-transfer-limited current densities. Higher current densities cause the quality of the deposited metal to deteriorate, the deposit becomes coarse, forms whiskers and begins to occlude or incorporate electrolyte, floating anode slime, and other impurities. Even at current densities significantly lower than mass-transfer-limited values, the deposit quality must often be improved by addition of surfactants to the electrolyte. These additives increase the overpotential of crystal growth for the deposited metal leading to frequent nucleation, which gives rise to smooth cathode surfaces, which are composed of very tiny crystallites with little inclusion of foreign substances. Such surfactants are absolutely necessary for obtaining highly purified metals and for preventing the formation of dendrites, which can short-circuit the cell. Cheap, widely used additives are animal glue (gelatin), lignin sulfonates and aloe extract; the last mainly contains polyphenols. All these additives are continuously consumed because of hydrolysis, oxidation by air, and adsorptive incorporation into the deposited metal and the anode slime. Oxygen evolving anodes for electrowinning processes are conventionally made of lead with few percents of silver. They have remarkably high oxygen-evolution overpotentials. In the future, activated titanium anodes will certainly become more important for this purpose.

The design of electrorefining and electrowinning cells is very similar. Large rectangular tanks or troughs of up to 2 m width, 10 m length and 2 m depth are arranged in long rows and switched in series in huge tank houses. Into the tanks are immersed the flat anodes and cathodes in vertical position with a relatively wide interelectrodic gap measuring 1–5 cm. The cathodes are thin starter sheets of the respective metal which are produced separately by electrodepositing the metal on aluminium or titanium cathodes. These deposits measuring approximately 1 mm in thickness are detached from the support by peeling, supplied with metal-lashes which are then fixed to a supportive metal rod or tube. Anodes and cathodes are connected to the anodic and cathodic bus bars on opposite sides of the tank simply by gravity-loaded contacts – often on sharp edged profiles of the bus bars or the rods which support the cathodes. The electrolyte is circulated between cells, electrolyte recovery and purification process units, but its volumetric flow rate is relatively slow not contributing significantly to mass transfer enhancement.

Table 11.10. Potential distribution and energy requirements in electrorefining and electrowinning of copper, nickel and zinc

	Copper		Nickel refining		Zinc electro-winning
	Refining	Winning	Metal anode	Sulfide anode	
Current efficiency, %	97	85	96	96	90
Cell voltage, V	0,28	2	1,9	3,7	3,5
Reversible cell voltage, mV	0	900	0	350	2000
Cathodic overpotential, mV	– 80	– 50	– 250	– 1500	– 150
Anodic overpotential, mV	30	600	300	250	600
Ohmic voltage drops, mV	100	400	1050[a]	1250[a]	500
Cell hardware IR, mV	70	500	300	350	250
Electric energy requirement, kWh kg^{-1}	0.25	2	1–9	3.5	3.3

[a] With diaphragm

Unlike all other electrochemical and chemical processes, the cost factor labor is extraordinarily high for electrowinning and electrorefining of metals because of the frequent electrode handling. In a process with anode lives of approx. 30 days (Cu, Ni) or only 10–14 days (Pb), and even shorter life spans for cathodes, there is the constant necessity of anode casting, starter sheet production and the exchange of anodes and cathodes. Continuous progress in automation is vital for the survival of these processes. Aside from the labour costs, the cost structure of these processes are unique because additional to the investment cost for solution processing and the electrolyzer tank house the capital required for the metal inventory in electrodes and circulated electrolyte is a substantial part of the invested capital. Copper, nickel and zinc electrowinning/refining are using very similar technologies. Table 11.10 gives an overview of cell voltages and potential distributions and energy requirements of copper, nickel and zinc electrorefining and winning electrolyses.

11.9.2
Copper Electrowinning and Electrorefining

Copper is almost always electrorefined – less frequently electrowon. The raw material is the raw copper obtained by reduction of the oxidic ores or concentrates by carbon (coke) in a blast furnace. Raw copper is cast into anodes containing roughly 98% Cu, the rest being mainly nickel and lead, with a fraction of one percent of noble metals, especially silver. Sulfuric acid is the electrolyte of choice for electrowinning as well as for electrorefining.

Table 11.11 summarises the operational conditions. Nickel sulfate removal is accomplished in a bleed stream from which copper is electrowon prior to evaporation and crystallisation of $NiSO_4$. The removal of Cu in liberator cells serves to balance the copper content of the bath and offset the chemical dissolution of

11.9 Electrowinning and Electrorefining of Metals

Table 11.11. Operating data of copper electrorefining

Maximal impurities, g dm^{-3}	
As	20
Ni	20
current density, mA cm^{-2}	150–240
electrolysis temperature, °C	55–40
Cu^{2+} concentration, g dm^{-3}	35–40
H$_2$SO$_4$ concentration, g dm^{-3}	150–200
Anode sludge contents of noble metals, %	
Ag	5–50
Au	0.02–0.7
Pt-metals	0.1–0.25

copper due to oxidation by dissolved oxygen, which tends to increase the Cu^{2+} content of the electrolyte steadily. Electrowinning of copper is not fundamentally different from electrorefining, except that the depleted electrolyte is recirculated to leach copper ores. Purification of the electrolyte in electrowinning is today mainly based on solvent extraction. The current densities can be greater for electrowinning than for electrorefining – which is performed with no more than 40–60 mA cm^{-2} – because in electrowinning cells stirring by the evolved oxygen at the anode enhances bath convection and mass transfer also to the cathode.

The cost contribution of electrical energy in refining and electrowinning of copper is an almost insignificant fraction of the whole process cost. Therefore, increasing the space-time yield is of special concern as it allows to decrease capital costs. Increasing the current density by at least 50% by periodic current reversal is one approach being used. The current is reversed for 10 s out of every 200 s, and correspondingly 5% of the deposited copper is redissolved. This causes electropolishing, which produces high-quality, smooth deposits even at higher cathodic current densities. Another approach (Onahama refinery, Japan) consists of casting thinner anodes, which reduces the copper inventory by 50%, and increases the space-time yield by ca. 20% because the electrodes can be spaced closer together without short-circuiting. However, labour costs are increased by this measure, and also increased is the amount of recycled anode scrap.

11.9.3
Nickel Electrowinning

Electrowinning begins with either crude nickel or nickel matte (Ni$_2$S$_3$) anodes. In both cases the metallic contaminants to be removed are the same: mainly copper, cobalt and iron. In both cases the anolyte must not be allowed to enter the

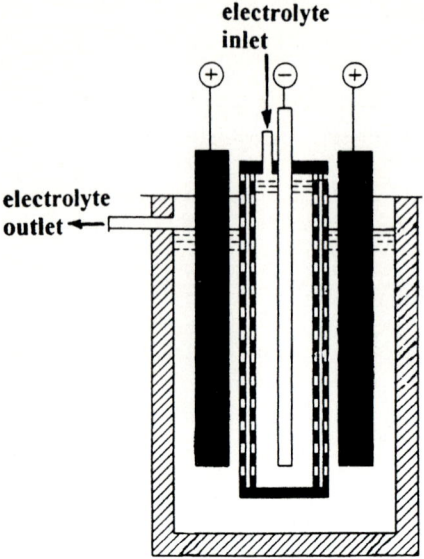

Fig. 11.15. Schematic of the Hybinette cell for nickel electrowinning from nickel matte

cathode chamber prior to purification. In the Hybinette cell, Fig. 11.15, the cathodes are surrounded by a cathode bag. Because always a small hydrostatic head is maintained in the cathode chamber the depleted catholyte flows out of the cathode bag into the anode chamber. At the nickel-matte anode nickel ions pass into the electrolyte according to Eq. (11.35) and (11.36):

$$Ni \rightarrow Ni^{2+} + 2e^- \quad (11.35)$$

or

$$Ni_2S_3 \rightarrow 2Ni^{2+} + 3S + 4e^- \quad (11.36)$$

Then the recycled electrolyte is purified. Copper(II) is extracted from the crude liquor by cementation with nickel powder, and iron and cobalt are precipitated as Fe_2O_3 and $Co(OH)_3$ and Co_3O_4 respectively together with lead by oxidation with chlorine and addition of basic nickel carbonate. If nickel matte is the starting material, sulfur is recovered from the anode sludge. Starting from nickel matte, rather than raw nickel, requires higher cell voltages, much higher (3.7 V, Table 11.10) than would be expected from the difference between the two standard electrode potentials, which differ by only ca. 0.3 V. This much higher cell voltage is caused by electrode kinetic hindrance and additional ohmic potential drop on account of an adherent porous layer of sulfur that covers the anode.

11.9.4
Nickel from the Chloride Leach Process

Falconbridge at Kristiansand, Norway, and Societé Le Nickel (SLN) at Le Havre-Sandonville have developed a chlorine leach process which uses chlorine oxidation of nickel/copper matte at 110 °C, the boiling point of the leaching solution, which is essentially an aqueous $NiCl_2+CuCl_2+HCl$ electrolyte. The chlorine generated anodically in the final nickel-electrowinning step is sparged into the emulsion of nickel/copper matte in acid Ni/Cu chloride solution and the leaching procedure is extended in a train of agitated tanks. By controlling the redox potential in the leaching electrolyte carefully it is possible to convert the matte into dissolved nickel chloride, insoluble copper I chloride, copper II sulfide and elemental sulfur according to Eqs. (11.37)–(11.40):

$$2CuCl + Cl_2 \rightarrow 2CuCl_2 \tag{11.37}$$

$$Ni_3S_2 + 2CuCl_2 \rightarrow 2NiS + NiCl_2 + 2CuCl \tag{11.38}$$

$$NiS + 2CuCl_2 \rightarrow NiCl_2 + 2CuCl + S \tag{11.39}$$

$$2CuCl + S \rightarrow CuS + CuCl_2 \tag{11.40}$$

The solid residue still contains approximately 8 wt% undissolved NiS and appreciably high copper concentrations are found in the leach solution. Autoclaving the slurry to 140–150 °C in two vertical reactors in series promotes the precipitation of copper and the dissolution of nickel sulfide. Treating the effluent solution with fresh pulverized matte reduces the copper concentration from 7 g dm^{-3} to less than 0.5 g dm^{-3}. Nickel utilisation of the copper nickel matte amounts to 90%, the final residue – mainly CuS – containing 6–8% nickel. Iron in the solution is precipitated as iron III hydroxide after chlorine-oxidation and neutralising the solution with nickel carbonate. Cobalt is extracted with trisoctylamine and, after acidic stripping, is recovered by electrowinning. Dissolution of the prepurified leach solution followed by a second chlorine treatment and buffering with nickel carbonate removes the residual traces of cobalt, iron copper, arsenic, lead and manganese. This solution is electrolyzed in undivided cells generating anodically chlorine and cathodically purified nickel. Dimensionally stable – IrO$_2$ coated titanium – anodes are used which are enclosed by a porous polyester bag which allows collection of the chlorine gas which is recycled to the leaching vessels.

The leach residue is roasted to form SO$_2$ and impure copper oxide the latter being dissolved in sulfuric acid. The insoluble residue of this stage contains 50% nickel, 18% Cu and the platinum metals, from which the nickel and copper is dissolved by chlorination leaving the platinum metals in a separately treated residue. The operational concentrations of the electrolyte at different process stages of the Falconbridge process are collected in Table 11.12. The Falconbridge proc-

Table 11.12. Concentration data in the Falconbridge process for nickel electrowinning

	Ni	Co	Cu	Fe	S	Cl
Matte, %	45	1	28	2	22	–
Feed solution, g dm^{-3}	60	–	25	1	–	90
Sulfide leach residue, %	7	0.5	56	2	33	0.3
Pregnant solution, g dm^{-3}	230	5	0.5	6	–	270

Table. 11.13. Operating data of zinc electrowinning

Maximal impurities of cell liquor, g dm^{-3}	
As	0.01 to 1
Sb	0.05 to 0.1
Ge	0.002 to 0.05
Ni	0.05 to 1.0
Fe	20 to 30
current density, mA cm^{-2}	400 to 600
electrolysis temperature, °C	40
Zn^{2+} inlet conc., g dm^{-3}	55 to 70
H$_2$SO$_4$ inlet conc., g dm^{-3}	160 to 180

ess is remarkably more involved than the Jarosite process for Zinc electrowinning. It is again an example for the high importance of hydrometallurgy for purifying the less noble metals in their electrowinning processes.

11.9.5
Nickel Refining

It is a particular property of metallic nickel, that it tends to passivate in aqueous solutions of sulfuric acid. In order to avoid passivation of nickel anodes in the electrorefining process it is therefore necessary to use a mixed sulfate/chloride or a chloride electrolyte – as chloride prevents passivation. Furthermore chloride electrolytes have a higher ionic conductivity than sulfate electrolytes and they permit the use of chlorine treatment for the purification of the electrolyte.

11.9.6
Zinc Electrowinning

The chemical purification procedures involved in electrowinning zinc were discussed, as an example, at the beginning of this section (jarosite precipitation of iron). The peculiar circumstance of zinc electrowinning is the negative standard electrode potential of zinc (–0.76 V), which would make zinc deposition impossible except for the fact that hydrogen evolution at zinc cathodes has a very high overpotential documented by a very low hydrogen exchange current density ($i_0=10^{-11}$ A cm^{-2}).

Nevertheless, care must be taken to keep the current efficiency for zinc deposition as high as approx. 90%. Therefore the process is carried out at the rather low temperature of 35 °C. Ten percent of the current goes into hydrogen evolution, which improves mass transfer enough that current densities can exceed 0.5 A cm^{-2}. The operational conditions for zinc electrowinning are listed in Table 11.13.

11.9.7
Lead Electrorefining

The electrorefining of lead, introduced in 1903 as the Betts process, is especially effective in separating bismuth, which goes into the anode sludge. Chemical treatment of the so called bullion is used to remove copper, and subsequently tin (together with As, Sb and Bi) by precipitation with caustic soda. The crude anodes consist of (wt%) approx. 98% Pb, 0.5% Bi, 0.01% As, 1% Sb, 0.02% Cu and 0.02% Sn. The electrorefining process uses either hexafluorosilicic acid or sulfamic acid electrolyte. Current densities are somewhat higher for the processes working with H_2SiF_6 (160–200 mA cm^{-2}) than for the sulfamic acid process (120–160 mA cm^{-2}). In both cases a relatively low temperature is maintained to retard hydrolysis of the electrolytes.

11.10
Metal Electrowinning from Molten Salt Electrolytes

11.10.1
General Considerations

Manganese ($E^0=-1.05$ V) is the most reactive metal that can be electrodeposited from aqueous solution. More reactive or less noble metals – the alkali and alkaline earth metals, aluminium, magnesium and the lanthanides – are electrodeposited from salt melts. Cathodic hydrogen evolution, due to water decomposition, cannot interfere with cathodic metal deposition in these electrolytes. Normally melt electrolysis is not used as a purification step because of the difficulty and high costs of purifying and recycling contaminated melts or salts respectively. Thoroughly purified oxides or chlorides are the starting materials for electrowinning. The Bayer process, for instance, is an integrated step in aluminium production since it produces aluminium oxide of very high purity (impurities < 0.1 wt%) from crude bauxite. The Bayer process is responsible for 40–50% of the production cost of aluminium. Typical problems posed by melt electrolysis are first of all material problems: the stability of:
(i) cell linings against chemical attack of the aggressive melt;
(ii) anodes against the corrosiveness of the anode gases (Cl_2, O_2); and
(iii) the cathode material against reaction with the liquid metal.

These three problems are especially severe because of the high process temperature, usually above the melting point of the metal: 660 °C (Al), 648 °C (Mg) and 97 °C (Na). A second typical problem is dispersion or dissolution of the cathodically deposited liquid metal into the electrolyte, so called metal mists, which results in partial reoxidation due to convective transport of the metal mists toward the anode. There the metal reacts with evolved gas, or it is directly reoxidized at the anode. This phenomenon is caused by the low surface tension of molten metals in contact with the molten electrolyte, the relatively small density differences between molten metals and molten salt and strong convection of the electrolyte due to magnetohydrodynamic effects and gas stirring. Cell geometry must be designed to suppress excessive convection or to provide a sufficiently large interelectrodic gap of several centimetres to minimise metal reoxidation. A further engineering aspect that is unique and typical for all high-temperature electrolytic processes is the careful calculation of heat balances and heat isolation devices for electrolyzers and potlines, which is indispensable for energy conservation and for steady, reliable operation of molten-salt electrolyzers over months and years.

11.10.2
Aluminium Production – the Hall–Heroult Process

The electrolytic production of aluminium by electrolysis of Al_2O_3 dissolved in cryolite, Eq. (11.41), (Hall–Heroult process) is by far the most important electrochemical process based on the use of a molten-salt electrolyte.

$$Al_2O_3 + 3C \rightarrow 2Al + 3CO \tag{11.41}$$

11.10.2.1
The Melt

Figure 11.16 a depicts the phase diagram of the binary mixture NaF/AlF_3. The composition 25/75 AlF_3/NaF corresponds to the stoichiometry of cryolite Na_3AlF_6, proper, the rounded maximum indicating partial dissociation of the complex anion AlF_6^{3-} at the melting point. Technical electrolytes are made up of cryolyte containing a slight excess of aluminium fluoride, as indicated by the shaded area. The addition of Al_2O_3 to cryolite (mp ca. 1000 °C) decreases the melting point. As the alumina content is increased up to 10 wt%, the eutectic of this composition melts at 960 °C as shown in Fig. 11.16 b. These circumstances limit the variation of the melt temperature to a relatively narrow range taking into account that an aluminium smelter is operated as a discontinuous batch reactor because of the discontinuous addition of alumina maintaining its concentration within the limits of 2–8 wt% (mp-swing: 995–970 °C). Therefore it is desirable to run the electrolysis above 1000 °C, but increased vapour pressure of the melt, decreasing current efficiencies and increasing heat losses make it ad-

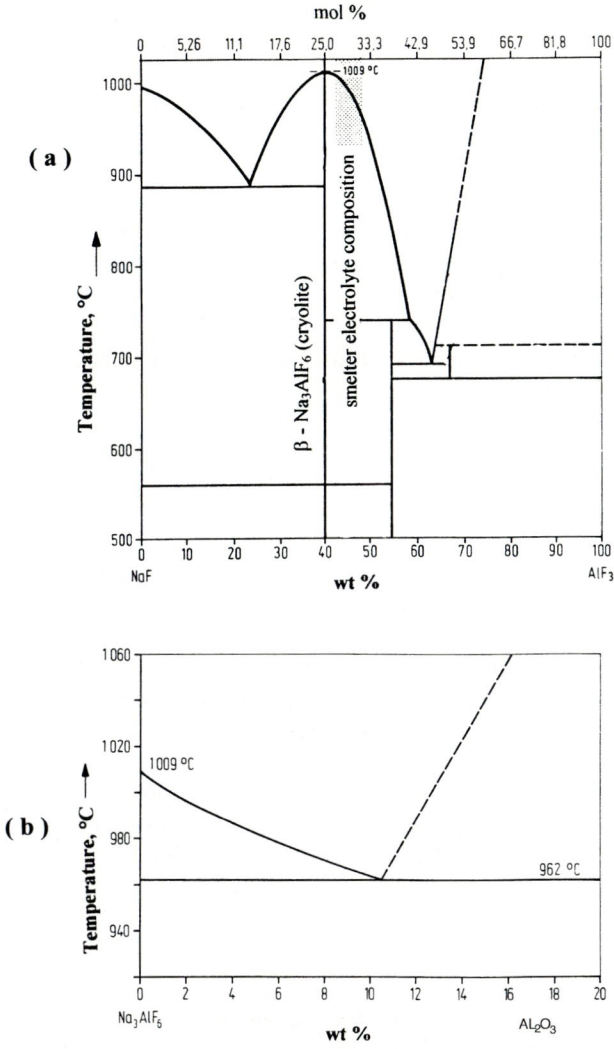

Fig. 11.16. a Melting point diagram of the binary system NaF/AlF$_3$. **b** Melting point diagram of the binary system cryolite/Al$_2$O$_3$ [1]

visable to limit the pot temperature with 1100 °C as a maximum. The conductivity of the melt at 1000 °C is – depending on its composition – and alumina content – between 2.4 and 2.8 Ω^{-1} cm^{-1}. The density of the melt (2.1 g cm^{-3}) is decreased by dissolved Al$_2$O$_3$ so that the small difference in density with respect to liquid aluminium (2.3 g cm^{-3}) is favourably affected, diminishing the danger of dispersing the liquid metal in the melt.

11.10.2.2
Electrode Reactions

For the decomposition of one mol of Al_2O_3 6 F (579 kAs) are required. The mechanism of the cathodic deposition of aluminium on a liquid aluminium cathode is still under debate. But it is important to state, that the reaction is fast and that cathodic overpotentials are mainly due to concentration polarisation and almost negligible not exceeding 0.05 V. More complicated and still less understood is the anode process, the anodic release of oxygen at carbon anodes leading to release of a mixture of carbon monoxide and carbon dioxide according to Eqs. (11.42) and (11.43). This gas mixture is not in accordance with the Boudouard reaction ($CO_2+C \rightleftharpoons 2CO$) which would render almost pure (99%) carbon monoxide at 1000 °C and 1 bar total pressure.

$$(O^{2-})_{melt} + C \rightarrow CO + 2e^- \tag{11.42}$$

$$2(O^{2-})_{melt} + C \rightarrow CO_2 + 4e^- \tag{11.43}$$

The theoretical decomposition potential of Al_2O_3 (activity=1) at 1000 °C of 2.2 V is decreased by reaction of oxygen with carbon to 1.18 V (CO_2 formation, Eq. (11.43)) and 1.08 V (CO formation, Eq. (11.42)) respectively. The actual cell voltage is almost four times higher.

Figure 11.17 which divides the total cell voltage of 3.9–4.6 V into the ohmic potential drops in the cathode (0.3–0.6 V) the anode (also 0.3–0.6 V) and the cell voltage (3.1–3.5 V) allows determination of the IR-corrected cell voltage under operating conditions. $E_{cell\ operating}$, is 1.6–1.8 V and remarkably higher than the 1.08 or 1.18 V respectively mentioned before, because there exists an anodic overpotential which amounts to at least 0.4 V at current densities which are usually

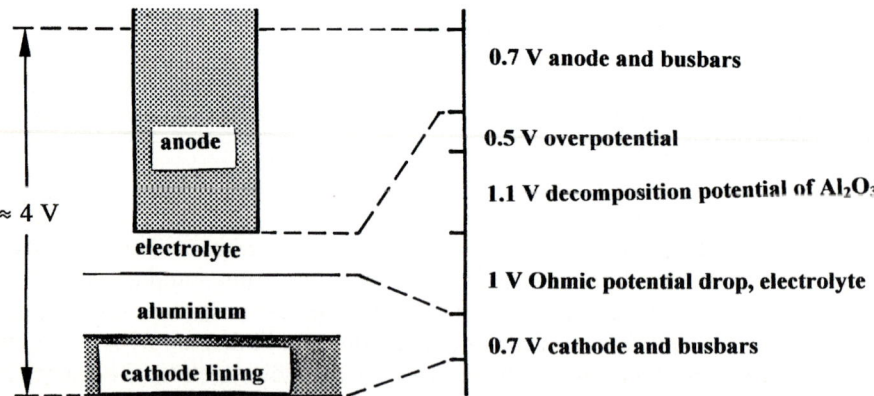

Fig. 11.17. Division of cell voltage in aluminium electrolysis

around 1 A cm^{-2}. There are almost no means to decrease the overpotential at the anode except by the quality of the carbonaceous anode material. Today prebaked anodes are more and more superseding the so-called Söderberg anodes. Söderberg anodes undergo their formation from a physically and chemically ill defined mixture of petrol coke and tar to solid carbon in the cell under the influence of the heat generated by the cell as they move slowly downward towards the melt surface replacing the volume of the anodically consumed part of the anode.

The charge transfer overpotential depends on the anode material being approximately 100 mV higher on prebaked than on Söderberg anodes. On the other hand prebaked anodes cause an increase of the CO_2 content of the anode gas, compared to Söderberg anodes, 60–80%. This means a saving in petrol coke consumption by 15% and as petrol coke is an essential cost factor in aluminium electrowinning the decreased carbon consumption trades well off against a slight increase of the anodic overpotential. Although the horizontal anode surface pointing downwards evolves continuously anode gas the additional ohmic resistance of the gas bubble curtain under the anode is relatively low producing an additional ohmic potential drop of well less than 0.1 V. The anode is consumed quite evenly because rigorous stirring of melt and liquid metal below the anode prevents gas accumulation. Any part of the anode surface which would protrude would also be consumed faster as enhanced current densities provide a faster burnoff.

11.10.3
The Cell

Figure 11.18 depicts the construction of a modern Hall–Heroult cell schematically. The cell is composed of an outer carbon or graphite lining, surrounded by a heat insulation made of heat-resistant ceramics and a row of adjustable prebaked carbon anodes. At the bottom of the pot there is the cathode, a pool of liquid aluminium, which is emptied periodically by suction through intermittently inserted steel pipes and which is covered by the molten electrolyte into which the gas-evolving anodes are suspended. The melt is covered by a crust of solidified electrolyte and a supply of fresh, loose Al_2O_3 powder. Periodically crushing the crust serves to introduce alumina and to replenish the Al_2O_3 content of the bath. There is violent stirring of the molten metal and the electrolyte due to magnetohydrodynamic effects and anodic gas evolution. Stirring may cause severe losses in current yields and energy efficiency due to enhancing reoxidation of metal mists at the anode or the reoxidation of the dispersed metal by the anode gases. These energy losses are avoided to a certain extent by increasing the anode–cathode distance but this measure, due to an increase of ohmic losses, increases the cell voltage and hence energy consumption. Today electrolyte and metal motion can be modelled numerically and – based on this modelling – also controlled with considerable precision. In particular the influence of cell geometry and geometric arrangement of power lines leading to and away from the cell and the influence of the situation of the cell pots in the line with respect to each

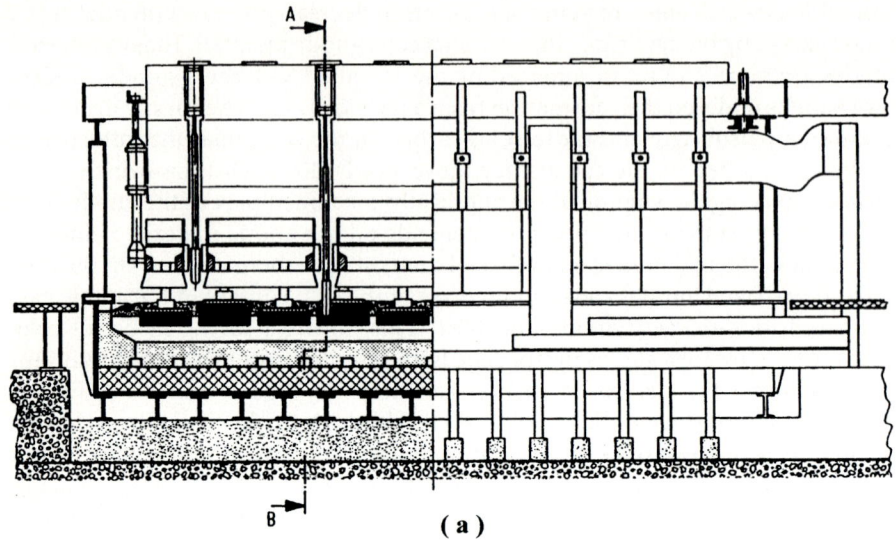

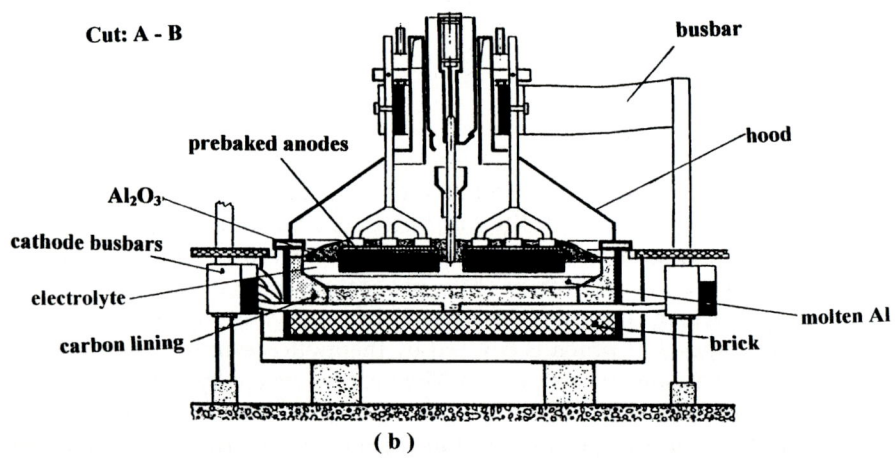

Fig. 11.18 a,b. Schematic of a modern Hall–Heroult cell: **a** side view; **b** cross-cut

other on magnetohydrodynamic stirring can be optimised. Carefully designed aluminium smelters can therefore be operated with optimal interelectrode distances of a little less than 6 cm (in earlier times 10 cm) achieving current efficiencies exceeding 90% and electrical energy consumption of optimally 13 kWh kg^{-1} (theory: 6.34 kWh kg^{-1}). Table 11.14 collects the relevant data concerning energy and raw materials of the Hall–Heroult process.

Table 11.14. Demand of raw materials and electrical energy per 1000 kg of aluminium

Material	wt (kg)/elect. energy
aluminium oxide	1920 to 1950 kg
aluminium fluoride	10 to 40 kg
cryolite	10 to 30 kg
sodium carbonate	1 to 3 kg
prebaked anodes	420 to 470 kg
Söderberg anode mass	500 to 570 kg
electrical energy (d.c.)	13,000 to 14,500 kWh
current density	>10 kA m^{-2}

Due to the chemical composition of the electrolyte Hall–Heroult cells emit remarkable amounts of fluoride-containing dusts and fumes. All modern Hall–Heroult cells are today encapsulated by a hood which permits to collect all emissions of the cell – gaseous or dusts – by suction and to subject the effluent gas stream to a vigorous cleaning. Cleaning might be wet by using washing towers or – today more and more popular – it might be dry. Dry cleaning is performed by passing the contaminated air through a fluidized bed of adsorption active alumina which after being loaded with fluorides is recirculated to the cell.

The relatively low energy efficiency of the Hall–Heroult process is the reason that more energy-efficient electrolysis processes have been sought. Twenty years ago, Alcoa resumed the work of BUNSEN, who was first to produce aluminium metal by electrolysis of alkali chloride-aluminium chloride melts. The Alcoa electrolyzer used bipolar stacks of carbon electrodes. The lower, grooved faces of these carbon electrodes pointing downward served as gas evolving anodes. The Alcoa electrolyzer was reported to produce aluminium metal with a specific energy of 10 kWh kg^{-1}, offering a 30% saving in electrical energy over the Hall–Heroult process. R and D on the Alcoa process had, however, been terminated. The reason is the unavoidable generation of noxious perchlorinated and notoriously persistent organic chloro compounds like hexachlorobenzene, octachlorostyrene and related materials whose removal from the recycled chlorine and effluent gases could not be accomplished cost effectively.

11.10.4
Alkali Metals from Chloride Melts

Magnesium chloride, with a decomposition voltage of 2.6 V at 700 °C is the most easily decomposed chloride among the group 1 and 2 chlorides. The light alkali metals lithium, sodium and potassium and the alkaline-earth metal magnesium, all have lower densities than the melts of their chlorides. Therefore, the elec-

Table 11.15. Working temperatures, cell voltages and current efficiencies in the production of light alkali metals and alkaline-earth metals

Cell operating characteristics	Metal		
	Lithium	Sodium	Magnesium
Working temperature, °C	400–420	590–600	660–670
Cell voltage, V	8–9	6.5–7	5.5–6
Current efficiency,%	85–90	90–95	78–80

trodeposited metals ascend through and collect on top of the chloride melts. They must be prevented from coming into contact with the electrogenerated chlorine. Interelectrodic gaps of several centimetres together with steel diaphragms between the electrodes and steel collectors that gather the ascending metal droplets above the cathode allow an effective separation of chlorine and liquid metal – but at the expense of relatively high ohmic losses and low energy efficiencies. Metal dispersion and the solubility of metal in the melt are reasons for low current efficiency (Table 11.15). The trade-off between high current efficiency and low cell voltage results in energy yields between 30 and 40%, which are typical for salt-melt electrolysis but which are low compared to large-scale electrolysis processes with aqueous solutions.

11.10.5
Magnesium Electrolysis

The knowledge disclosed on magnesium electrolysis is remarkably sparse compared to that of aluminium smelters. The reason is obvious. Magnesium production amounts to only a small fraction of aluminium production and the number of producing companies and smelters is small and most technical know-how is therefore kept secret. The respective electrolysis technology is determined by the chosen feed. It may be either pure, dry $MgCl_2$ – which today is most common – or magnesium chloride dihydrate $MgCl_2 \cdot 2\,H_2O$ for the Dow process. In both cases the actually used electrolyte in the cell is molten dry $MgCl_2$ with some $CaCl_2$ added in order to increase the density to facilitate separation of magnesium from the melt. The magnesium chloride dihydrate is dehydrated on adding the feed to the melt at 750 °C but some fraction of magnesium chloride undergoes hydrolysis, Eq. (11.44),

$$MgCl_2 \cdot 2H_2O \rightarrow MgO + 2HCl + H_2O \qquad (11.44)$$

so that the chlorine released anodically contains water vapour, and hydrochloric acid. Traces of water, dissolved in the melt as well as dissolved MgO cause slow consumption of the carbon anodes due to anodic carbon monoxide formation – compare Eqs. (11.42) and (11.43).

11.10.5.1
Production of the Feed Salt

As shown in Fig. 11.19 a the Dow process is based on the use of $MgCl_2$ from sea water. In this process the anode gas mixture is reacted together with methane and steam in a burner to form gaseous hydrochloric acid, which is absorbed from the gas mixture and recycled into the magnesium chloride dihydrate production. The essential difficulty in producing dry magnesium chloride is the partial hydrolysis of hydrated magnesium chloride in the last stage of dehydration. The Dow process avoids just this last, critical step at the expense of an increased consumption of the anodes, because traces of water vapour and magnesium oxide in the melt cause a burn off of the anodes by CO and CO_2 formation.

The Norsk-Hydro process, starting from dolomite ($CaMg(CO_3)_2$) and magnesium chloride from sea water produces calcified magnesia, MgO, by precipitation and Ca/Mg exchange. In a dry carbochlorination step the MgO is converted according to the exothermal reaction at Eq. (11.45) to anhydrous magnesium chloride in a type of moving bed reactor at temperatures of 800–900 °C.

$$MgO + C + Cl_2 \rightarrow MgCl_2 + CO \tag{11.45}$$

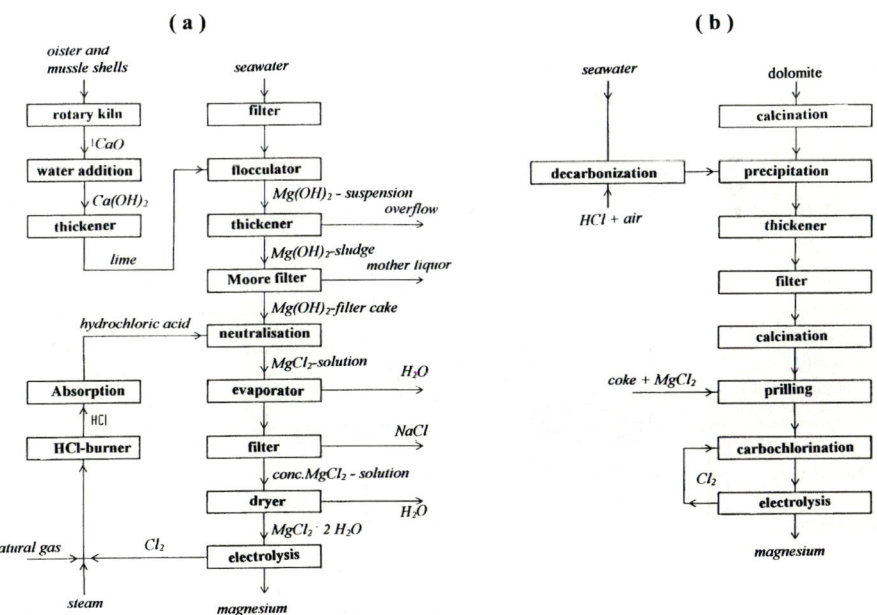

Fig. 11.19 a,b. Process schemes for magnesium electrowinning; **a** Dow process; **b** formation of anhydrous $MgCl_2$ by carbochlorination of MgO according to the Norsk–Hydro process

Figure 11.19 b depicts the process scheme. In case of "anhydrous" $MgCl_2$ which contains no more than 0.1% MgO and less than 0.1% water the anodically generated chlorine is recycled immediately to the carbochlorination reactor. In this process the consumption of carbon anodes by CO-formation is almost negligible.

11.10.5.2
Magnesium Electrolysis Cells

A cut of the Dow cell is depicted in Fig. 11.20. The cell consists of a steel vessel to which the conical iron cathodes are attached. The bottom of the cell forms the ceiling of a gas fired furnace. Into the conical cathode foils protrude the continuously readjusted cylindrical prebaked carbon anodes whose slow consumption ($O^{2-}+C \rightarrow CO+2e^-$), accompanying the anodic chlorine evolution shapes the lower part of the anode conically following the shape of the surrounding cathodes. Liquid magnesium is collected on a side trough from which it is removed discontinuously. The chlorine ascends along the anode and is collected at the cell

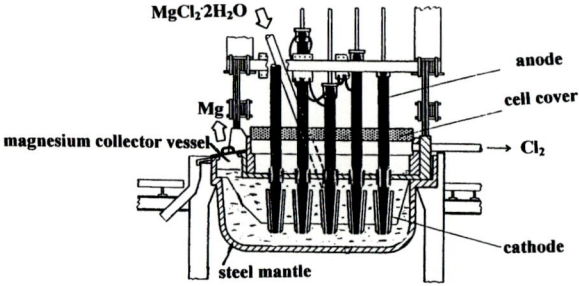

Fig. 11.20. Schematic of the Dow cell for electrowinning of magnesium

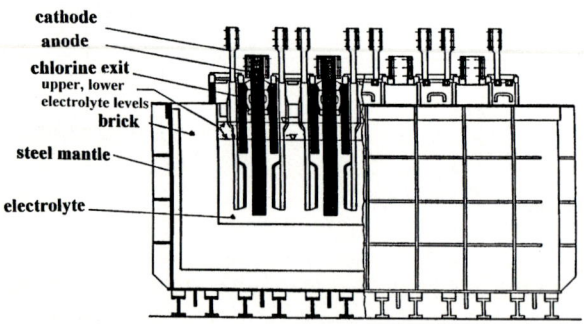

Fig. 11.21. IG cell for Mg production: side view

ceiling and extracted under reduced pressure so that a mixture of air and chlorine which contains 11.5% Cl_2 and around 1.5% HCl is eventually combusted with methane and water vapour in the HCl burner. The $MgCl_2 \cdot H_2O$ feed is added through a tube from above. External heating from below allows to operate the cell over a broad range of current densities. No chlorinated organic compounds are generated anodically at the chlorine (+CO) evolving anodes.

Figure 11.21 shows a cut through the IG cell developed in the 1930s at Bitterfeld. A steel trough contains the cell in a heat insulating melt-resistant MgO-lining. Steel plates, forming the cathodes, and carbon anode plates are suspended from the cell top into the cell. The ascending chlorine is directed by a concrete plate-skirt toward the chlorine exit slit to be collected in the hollow cell lid from which it is collected by suction. The anode gas containing 85–90% chlorine (balance air) is recycled to the carbochlorination reactor. Coextracted anode dust is contaminated by pentachloro and hexachlorobenzene and other noxious perchlorinated compounds which must be recovered and disposed of with particular care. Liquid magnesium gathers on the electrolyte surface protected against reaction with nitrogen, oxygen and moisture by a thin layer of melt which wets and covers the surface of the molten metal completely.

11.11
Organic Electrosynthesis Processes

11.11.1
General Overview

Organic electrosynthesis has been known for at least 140 years. In 1849 Kolbe discovered the reaction named after him, the Kolbe synthesis, shown at Eq. (11.46), which is essentially an anodic decarboxylating dimerization of carboxylate anions:

$$2RCOO^- \rightarrow R-R + 2CO_2 + 2e^- \quad (11.46)$$

After two phases of intensive experimental investigation (1920–1940 and 1970–1980) one understands much better the reaction mechanisms underlying the different types of electrosynthesis reactions and one gains improved insight into the possibilities and limits of organic electrosynthesis processes. Industrial electroorganic synthesis began on a large scale with the Nalco process in which lead anodes were converted to tetraethyl lead by anodic dissolution of lead in presence of the Grignard's reagent of the ethylbromide.

$$Pb + 4ClMgEt \rightarrow Pb(Et)_4 + 4MgCl^+ + 4e^- \quad (11.47)$$

This very successful process, was singular in design and process techniques as it used a shell (cathode) and tube (anodes filled with consumable lead shot) reactor so well known for heat exchangers as an electrolyzer. This cell design did

Table 11.16a. Selected organoelectrosyntheses: Cathodic hydrogenations

Group	Reactant	Product	Status	Company
C = C	1) maleic acid	succinic acid	bread board	CECRI[a], India
C = N	2) 2-methylindene	2-methylindoline	bread board	BASF
	3) 2.2 dinaphthyl ether	1.4-dihydronaphthylether	pilot plant	Hoechst
C = O	4) glucose	sorbitol/mannitol	commercial	CECRI[a]
	5) p-methoxy benzaldehyde	p-methoxy benzylalcohol	commercial	BASF
	6) oxalic acid	glyoxylic acid	commercial	Rhone Poulenc
COOH	8) salicylic acid	salicyl aldehyde	former commercial	CECRI[a]
C ≡ N	9) benzonitrile	benzylamine	pilot plant	CECRI[a]
NO_2	10) nitrobenzene	aniline	bread board	CECRI[a]

[a] CECRI=Central Electrochemical Research Institute, Karaikudi – 623,006, India

Table 11.16b. Selected organoelectrosyntheses: Anodic oxidations

Group	Reactant	Product	Status	Developer
C = C	1) butene	methylethyketone	commercial	Exxon
COO^-	2) ethyloxalate	ethylisocyanate	bench scale	Royal Dutch Shell
COH	3) diacetone-L sorbose	diaceton-2 keto gulonic acid	commercial (terminated)	Merck GmbH
	4) propargyl-alcohol	propiolic acid	commercial	BASF
	5) 2 butyne-1.4-diol	acetylen-dicarboxylic acid	commercial	BASF
$ArCH_3$	6) p-methoxy toluene	p-methoxy-benzaldehyde	commercial	BASF
	7) 4-*tert*-butyltoluene	4-*tert*-butylbenzaldehyde	commercial	BASF
Arenes	8) naphthalene	naphtoquinone	commercial	Holliday
	9) anthracene	anthraquinone	commercial	Holliday
	10) toluene-o-sulfonamide	saccharin	commercial (terminated)	Boots, Holliday

not stimulate any further engineering development in electroorganic synthesis and it certainly will disappear during the next decade as lead tetraethyl will disappear as antiknock agent for gasoline.

Quite different, the technical and commercial success of the Baizer–Monsanto process for producing adipodinitrile from acrylonitrile which used a relatively conventional cell design triggered a host of inventions and revived organic electrochemistry after a long period of stagnation. Table 11.16 copies selected data on electroorganosynthesis processes compiled recently by F. Beck. There is no doubt that there is a number of unpublished productions which concern the synthesis on smaller production capacities that means electrochemically synthesised products in particular in the pharmaceutical industries. These products are often produced in multipurpose electrolyzers of the bipolar plate and

Table 11.16c. Selected organoelectrosyntheses: Electrochemical coupling

educt	product	status	company
Cathodic coupling			
1) acrylonitrile	adipodinitrile	big commercial	Monsanto, BASF, Asahi Chemicals
2) formaldehyde	ethylene glycol	bread board	Electrosynthesis Co.
3) acetone	pinacol	commercial	BASF, Bayer, Diamond Shamrock
Anodic coupling (Kolbe reaction)			
4) adipic-hemiester	sebaic acid	commercial	BASF, Asahi Chemicals
5) monomethyl suberate	tetradecanioc diester	commercial	Soda Aromatics, JP
6) malonic-diester	ethane tetra carboxylate	bench scale	Monsanto

Table 11.16 is far from giving a complete survey. It is only supposed to give some hints to the reader not-experienced in the field. For further details see the article of F. Beck, Organic Electrochemistry in Ullmann's Encyclopedia of Industrial Chemistry, Vol. A9, VCH Verlagsgesellschaft mbH, Weinheim 1987 and N.L. Weinberg, Introduction to industrial electrosynthesis in Electrosynthesis from laboratory to pilot plant production, J.D. Genders, D. Pletcher eds., Published by the Electrosynthesis Company Inc., P.O.Box 430, E. Amherst, New York 1990

frame type which are easily converted or adapted to different processes. In general all commercially performed electrolysis processes have to strive hard to become and stay competitive with catalytic process routes. The bigger the process and the more integrated the chemical plant is, the harder it turns out to make an electrochemical process a success – mainly because catalytic processes utilising hydrogen or air instead of cathodes and anodes as reductants or oxidants are easier to perform and mostly demand lower investment than electrolytic processes.

11.11.2
Cell Types Used in Commercial Electroorganic Synthesis

For the larger, more conventional inorganic electrolysis processes exclusively highly specialised cells had been developed, whose design had been optimised to the purpose of the respective electrochemical production. The situation is different in commercial organoelectrochemical synthesis. Production capacity is mostly relatively small not measured in tons, but kilograms per hour and frequently these processes are not operated continuously and are rather operated for limited times on the prevailing demand. Therefore composing anode, cathode and separator (in the form of membranes and diaphragms) in stacked, multipurpose filter press cells according to the demands of the respective electrosynthesis process is quite usual and is the most convenient and often the cheapest way to find and construct the cell in most cost effective way and possibly shortest time. Figure 11.22 a,b shows two different multipurpose cells. Today several companies are offering these multipurpose cells among which Svenska utvecklingsaktiebolaget, SU-AB, and ICI should be mentioned as pioneers in the field. Interesting enough the monopolar FM 21 multipurpose electrolyzer of ICI had been developed starting with the design of a bipolar chloralkali electrolyzer.

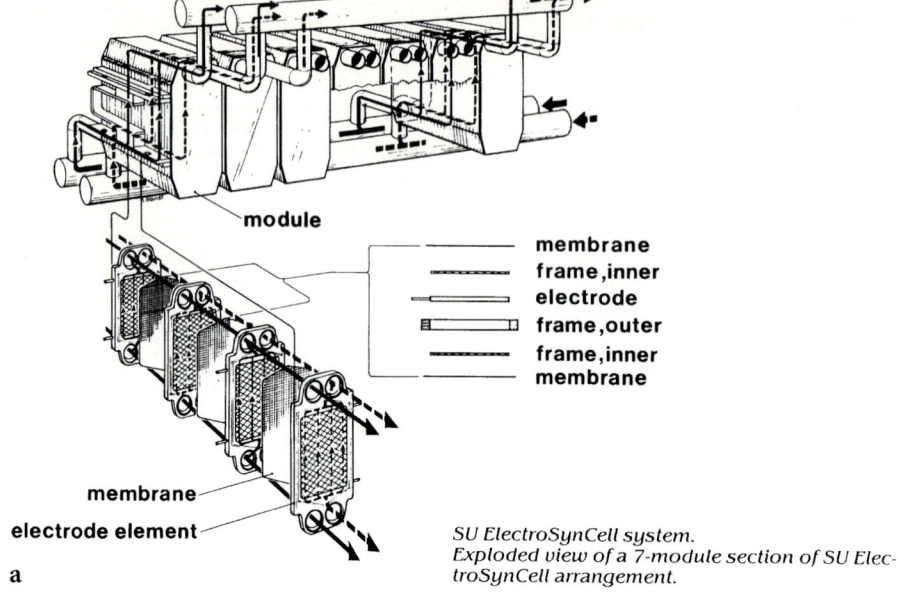

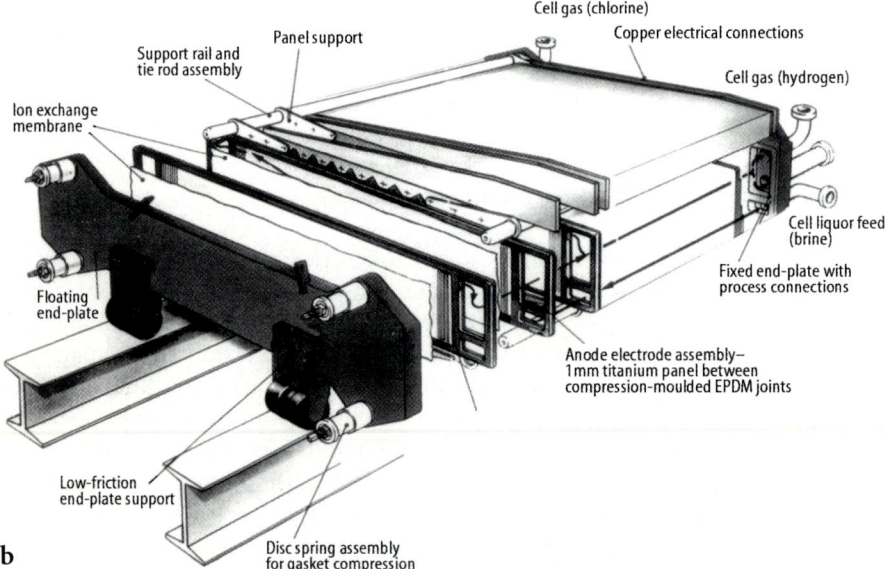

Fig. 11.22 a,b. Multipurpose electrolyzers: **a** SU Electrosyncell system; **b** ICI cell FM 21 Courtesy of SW-AB and ICI.

11.11.3
Process and Reaction Techniques of Some Examples of Industrial Organic Electrosyntheses

11.11.3.1
Adipodinitrile Production by the Monsanto–Baizer Process

Worldwide there are produced 180,000 tons per year of adipodinitrile by cathodic hydrodimerisation of acrylonitrile, according to Eq. (11.48) in two plants of approximately equal capacity in Decatur, Alabama, USA and in Seal Sands, United Kingdom.

$$2CH_2=CH-CN+2e^-+2H^+ \rightarrow NC-CH_2-CH_2-CH_2-CH_2-CN \quad (11.48)$$
$$\text{acrylonitrile, AN} \quad\quad\quad\quad \text{adipodinitrile, ADN}$$

The adipodinitrile, being a precursor of nylon (6,6) is eventually catalytically hydrogenated to hexamethylene diamine. Initially the electroorganic synthesis was performed in divided cells. With separated sulfuric acid anolyte and an aqueous catholyte saturated with acrylonitrile and containing relatively high concentrations of tetraalkylammonium salts. The cathode was made of cadmium, the anode of Pb/PbO_2. The presence of tetra-alkylammonium cations is essential as they are adsorbed at the Cd-cathode rendering the cathode surface non-protic and hydrophobic and thus preventing the fast protonation of initially generated acrylonitrile radical anions. These react in a Michael addition with acrylonitrile very selectively to form the dimer. The cells were equipped with Nafion-cation exchange membranes and were stacked together in bipolar filter press arrangement. Today this very costly type of cell which is also expensive to maintain has been abandoned due to decisive changes in electrolyte composition which allow to use undivided cells and additionally facilitate the work-up procedure a lot. The most important improvement was to replace simple tetraalkylammonium salts by hexamethylene-bis-(ethyldibutylammonium)-phosphate as surface active agent. This surfactant, which is much stronger adsorbed on the cathode, can be used in considerably lower concentrations than the simpler tetraalkyl ammonium salts, is nonetheless much easier to recover from the product by a simple backwash procedure and is more stable against anodic oxidation and degradation than the formerly used tetra-alkylammonium salts.

The undivided cell in modern equipment for adipodinitrile production is a bipolar capillary gap cell with no more than 2 mm gap width and the bipolar plate electrodes are made of 1 mm stainless steel plates. This capillary gap stack is shown in Fig. 11.23. One side of the bipolar steel plates is covered galvanically by a cadmium coating of 50–100 μm thickness which represents the cathode and exhibits sufficiently high hydrogen overpotential. The anodic side is not protected by any additional coating since it is reliably passivated in the electrolyte which contains disodium hydrogen phosphate in several tenths mol dm^{-3} con-

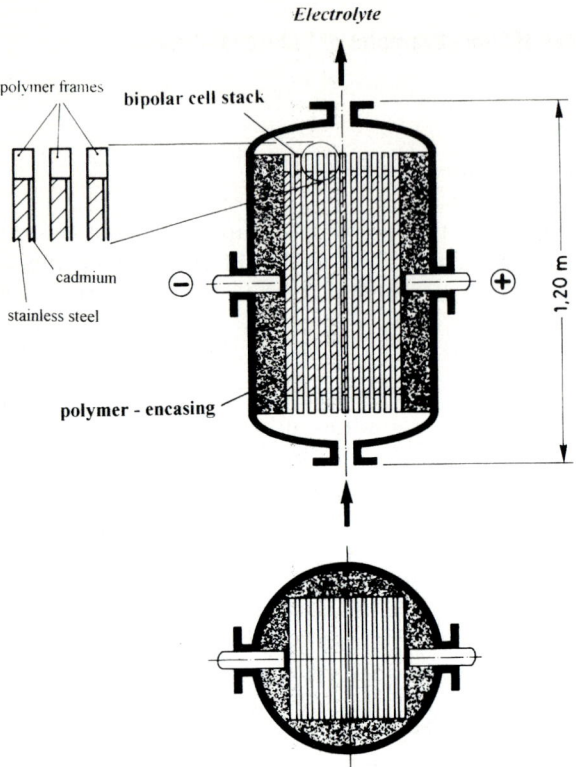

Fig. 11.23. Bipolar capillary electrode stack for electrosynthesis of adipodinitrile

centration and additionally boric acid. The electrolyte contains further a lower concentration of ethylendiamino-tetraacetic acid, (complexon), as complexing agent which has the purpose to complex and mask any traces of dissolved iron. By this means a cathodic deposition of iron upon the cadmium cathode is safely prevented. This is necessary since iron deposition would dramatically lower the hydrogen overpotential giving rise to excessive losses of current efficiency due to hydrogen evolution. The bipolar capillary gap cell is encased in a shell made of glass-fibre-reinforced polymer. Two cross sections together with a schematic presentation of three bipolar electrodes, the gap width being defined by polymeric spacers, demonstrate the main features of this very simple type of cell construction which reduces the investment costs for the cell by more than a factor of ten compared to the formerly used bipolar filter press cell equipped with frames, manifolds and cation exchange membranes. Table 11.17 compares the characteristic electrical data of the old and the advanced cell which demonstrate that due to the small interelectrodic gap width, and due to enhanced electrolyte conductivity and simultaneously decreased current densities the specific electrical energy consumption could be reduced by a factor of almost three. Lowering

11.11 Organic Electrosynthesis Processes

Table 11.17. Comparison of characteristic data of the initial version with divided cell and advanced version with undivided cell for adiponitrile electrosynthesis

	Current density	Cell voltage	Energy consumption
	A cm^{-2}	V	kWh kg^{-1}
divided cell	0.45	11.65	6.6
undivided cell	0.20	3.84	2.4

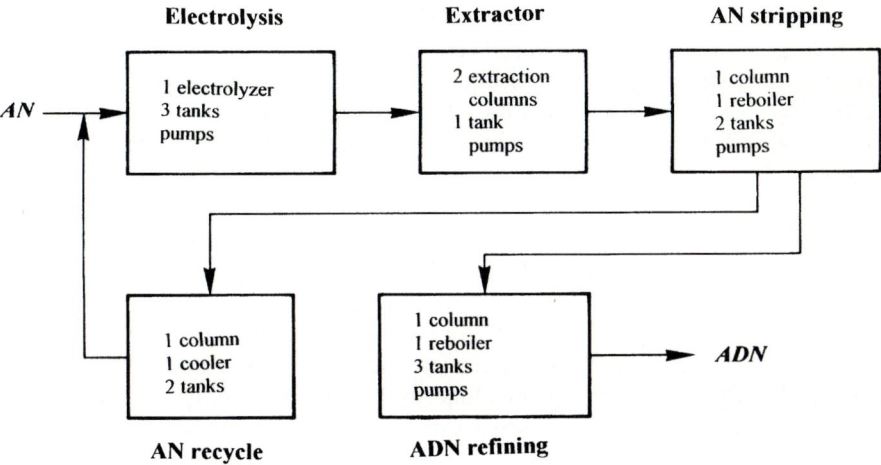

Fig. 11.24. Simplified flow sheet of the adipodinitrile electrosynthesis

the current density – which with the advanced cell is possible without enhancing the investment costs intolerably – is achieving several additional advantages:
- Current efficiencies are increased due to a decrease of hydrogen evolution rates.
- Cathodic formation of oligo- and polyacrylamides – which is due to further Michael attack of the adipodinitrile carbanions to acrylonitrile – Eq. (11.49) – is repressed.
- Acrylonitrile losses which are due to hydroxyl ion catalysed hydrolysis of acrylonitrile at the cathode – Eqs. (11.50) and (11.51) – are diminished.

$$NCCH_2CH_2\overline{C}HCN + CH_2 = CHCN + e^- \rightarrow NCCH_2CH_2CH_2CH_2(CN)CH_2 - \overline{C}HCN$$

$$\rightarrow \text{polymers} \qquad (11.49)$$

$$CH_2 = C - CN + H_2O \xrightarrow{OH^-} HOCH_2 - CH - CN \qquad (11.50)$$

$$2CH_2 = CH - CN + H_2O \rightarrow NCCH_2CH_2 - O - CH_2CH_2CN \qquad (11.51)$$

The high flow velocity established in the capillary gap provides high mass transport rates with consequently diminished hydroxyl ion concentrations at the cathode. It must be emphasised that due to the changed electrolyte composition and the improvements in yields and selectivities the work-up procedure was also facilitated and simplified with the advantage of a sizeable reduction in equipment and investment costs for the non-electrochemical part of the plant. None the less the "chemical" part of the plant is still very involved and contains several distillation, extraction and backwash columns, so that today nearly nine tenths of the investment costs of an adipodinitrile plant is due to the non-electrochemical part of the process. But the advanced electrochemical layout of the process has reduced the overall costs of the process to an extent which allows to successfully compete with the heterogeneous catalytic process for adipodinitrile production which is based on catalysed oxidative coupling of hydrogen cyanide to butadiene. Figure 11.24 shows the schematic of the improved process.

11.11.3.2
Electrosynthesis of Sebacic Diesters by Kolbe Synthesis

Sebacic acid is the second precursor for production of Nylon(6,6), a polyamide, which is produced by condensation of sebacic acid and 1.6 hexamethylendiamine. Already in 1960 in a patent filed by Russian researchers on the anodic electrosynthesis of sebacic diesters from succinic monoesters was reported by Fioshin. This process is based in its electrochemical part on the Kolbe synthesis, Eq. (11.52), which is performed with smooth platinum electrodes.

$$2ROOCCH_2CH_2COO^- \rightarrow ROOCCH_2CH_2CH_2CH_2COOR + 2e^- \qquad (11.52)$$

The Russian scientists did not report on the technical details of cells and electrochemical engineering of the process. In particular they did not report on the solution of the cost problem involved – since expensive platinum electrodes are necessary. Rather Pt-coated than massive Pt anodes would have been used. An improved electrochemical process technique which is similar to that used for the advanced adipodinitrile process has been elaborated in the sixtieth at BASF for production of sebacic acid diesters. It is, however, not performed commercially. Undivided circular capillary gap cells in bipolar cell stacks composed of carbon plates of several millimetres thickness had been used. (The circular bipolar capillary gap cell is depicted in Fig. 5.3 b.) Platinum foil, no thicker than 50 µm, was attached by a conductive glue to the anode side of bipolar carbon plates diminishing substantially the amount of platinum needed per unit area of the electrode to approximately 60 mg cm^{-2}. The cathode surfaces of the carbon plates where hydrogen is evolved did not carry any metal coating. Any platinum as cathodic electrocatalyst is avoided since its potential dissolution and deposition on the Pt-anode would spoil the smooth Pt surface by Pt-black formation and decrease the current yield for the anodic dimerization.

11.11.3.3
Benzaldehydes by Direct Anodic Oxidation of Toluenes

Benzaldehyde and substituted benzaldehydes are important synthons for the pharmaceutical industry and a number of commercially important small scaled organic syntheses. Benzaldehyde is obtained from toluene by heterogeneously catalysed gas phase or homogeneously catalysed liquid phase oxidation. By catalytic oxidation, however, only low yields with substituted toluenes are achieved. Therefore the anodic oxidation of substituted toluenes, like p-methoxy- or p-chlorotoluene to the respective benzaldehydes is of particular interest, because it is a more selective process than catalysed oxidation by oxygen. The anodic oxidation of toluene and substituted toluenes proceeds in two separate steps according to Eqs. (11.53) and (11.54). If the oxidation is performed in methanol solution benzyl-methylether and benzaldehyde-dimethylacetal are produced subsequently.

$$XC_6H_5CH_3 + MeOH \rightarrow XC_6H_5CH_2OMe + 2H^+ + 2e^- \tag{11.53}$$

substituted toluene   benzyl-methylether

$$XC_6H_5CH_2OMe + MeOH \rightarrow XC_6H_5CH(OMe)_2 + 2H^+ + 2e^- \tag{11.54}$$

benzalhyde-dimethylacetal

The selectivity-problem of this reaction – caused by the fact that alcohol and aldehyde are formed in subseqent steps and that a further 2e-oxidation of the acetal or the benzaldehyde to benzoic acid and anodic polymerisation occurring as a parallell reaction cannot be excluded – had been dealt with at the end of Chap. 6.5.7. It had been shown that the relative positions of half wave potentials of the toluene, the ether and the acetal are decisive for the selectivity. The performance of batch-synthesis allows for a higher selectivity than continuous electrolysis in a backmix reactor. At close to complete conversion 60–70% selectivities and materials yields can be obtained in this way. On these principle BASF developed the benzaldehyde synthesis to great perfection. The electrolyzer is the cylindrical capillary gap electrolyzer with bipolar carbon electrode according to Fig. 5.3 b through which the electrolyte is recycled till almost complete conversion is achieved. The electrolyte is proprietary – in the respective patents methanol, ethanol and *tert*-butanol are mentioned. The converted electrolyte contains some anodically generated polymer, the benzaldehyde diacetal and some unreacted toluene together with benzyl alcohol-ether and benzoic acid ester. After distillative separation the electrolyte, non-converted toluene and the ether are recycled, the aldehyde recovered from the diacetal by hydrolysis and the benzoic acid used for other, particular purposes.

11.11.3.4
The Selective Anodic Oxidation of L-Sorbose in Commercial Vitamin C Synthesis

Vitamin C, ascorbic acid, is commercially produced from L-glucose in appreciable quantities – from 15,000 to perhaps 20,000 tons per year world-wide. The

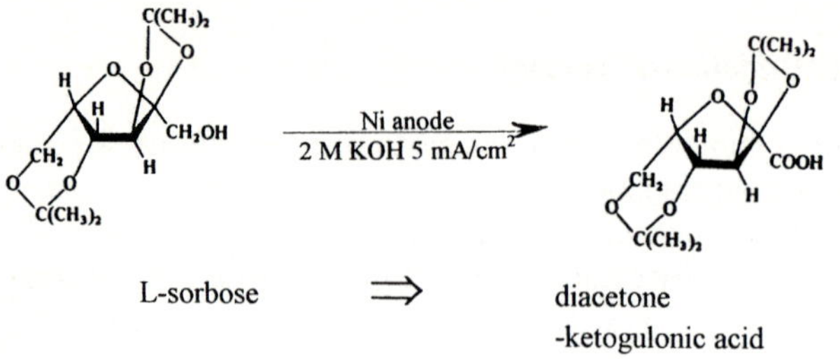

Fig. 11.25. Raction scheme of the anodic production of diaceton-ketogulonic acid by anodic oxidation of sorbose-diketaldiether at Ni/NiOOH anodes

synthesis comprises seven steps, beginning with reduction of glucose to sorbitol, oxidation to L-sorbose, which is converted with acetone to the di-ketal-diether. The fourth step, Eq. (11.55) and Fig. 11.25, consists of oxidising the terminal alcohol group to a carboxylic acid, thereby converting 4 F per molecule of ascorbic acid and producing diaceton-ketogulonic acid.

$$R-CH_2OH + H_2O \rightarrow RCOOH + 4H^+ + 4e^- \qquad (11.55)$$

The further steps are hydrolytically generating ascorbic acid as the end-product in two separate hydrolysis reactions. The 4e-oxidation, Eq. (11.55), which used to be performed by oxidation with hypochlorite, is an environmentally unbenign procedure as it loads the effluent with sodium chloride and also induces the generation of traces of chlorinated organic compounds. It was therefore a remarkable advancement to substitute two thirds of the expended oxidation equivalents introduced in the form of active chlorine by direct anodic conversion, or electric charge respectively. The reaction is an anodic oxidation mediated by nickel oxidehydrate, NiOOH, Eq. (11.56), which is continuously recuperated at the nickel anode, Eq. (11.57).

$$4NiOOH + RCH_2OH \rightarrow 4NiO + RCOOH + 3H_2O \qquad (11.56)$$

$$4NiO + 4H_2O \rightarrow 4NiOOH + 4H^+ + 4e^- \qquad (11.57)$$

A pilot plant with a Swiss-roll cell (Fig. 5.9 b) equipped with a nickel sheet anode had been reported to be operated on behalf a Hoffmann–La Roche, Basel, Switzerland [2] at ETH Zürich.

Merck AG, Darmstadt, Germany, performed this anodic electrosynthesis for many years on commercial scale. Interesting enough in this case not a specialised electrolyzer like the Swiss-roll cell was used, but a commercial water electrolyzer of not reported, unknown origin, had been adapted to the particular task in exchanging the mild steel anodes by sintered nickel anodes. In order to obtain commercially viable space velocities, the sorbose conversion did not exceed 70%, the rest still being oxidised by hypochlorite. Today this anodic electrosynthesis is no longer performed, because biochemical, enzymatic, oxidation of glucose turned out to be at a clear cost advantage over anodic oxidation.

11.11.3.5
Anodic Formation of Perfluoro-Propylene Oxide

Catalytic gas-phase oxidation is used for the production of ethylene-epoxide from ethylene according to Eq. (11.58)

$$C_2H_4 + 1/2 O_2 \xrightarrow{cat.} C_2H_4O \tag{11.58}$$

This method fails for propylene and all other higher olefins. Propylene epoxide, one of the most important epoxides, monomer for PO-polymerisation, is formed according to the Bayer process by a mediated anodic process, beginning with anodic chlorine and hypochlorite formation, the addition of HClO to propylene ending with epichlorohydrine formation, Eq. (11.59), and its eventual saponification to the epoxide, Eq. (11.60).

$$CH_2 - CH - CH_3 + HClO \rightarrow CH_2OH - CHCl - CH_3 \tag{11.59}$$

propylene   propylene-epichlorhydrine

$$CH_2OH - CHCl - CH_3 \xrightarrow{OH^-} \begin{array}{c} CH_2 - CH - CH_3 \\ \backslash \ / \\ O \end{array} \tag{11.60}$$

propylene-epoxyde

It would be desirable to substitute this multi-step synthesis by a one-step process. But anodic oxidation of olefins in aqueous or mixed water-alcohol solution usually yields in diol rather than epoxide formation. Only perfluoro-propylene can be oxidised in one singular anodic step to the respective propylene-epoxide, Eq. (11.61), which is an important monomer, as it forms the building-block of polymeric perfluorinated propanediol-ethers.

$$CF_2 = CF - CF_3 + H_2O \rightarrow \begin{array}{c} CF_2 - CF + 2H^+ + 2e^- \\ \backslash \ / \\ O \end{array} \tag{11.61}$$

This anodic electrosynthesis, which is performed in divided cells, had been developed at Hoechst AG up to the pilot-plant state. The anodic oxidation of the perfluorinated olefin is performed in sulfuric acid solution at lead dioxide anodes, which are deposited as relatively thick coatings on the inner wall of a titanium tube. The cathode, which is separated by a ionomer membrane from the anode is a massive cylindrical carbon block. This rather unconventional and little convenient to produce cell-design is dictated by the necessity to coat the an-

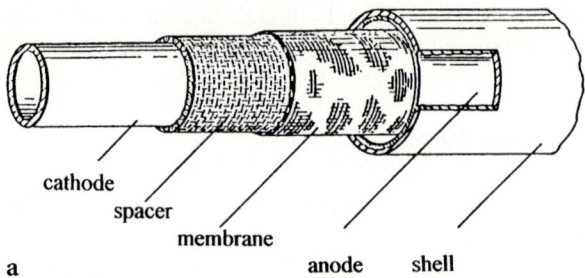

Fig. 11.26 a,b. Cylindrical divided tube cell for anodic perfluoro-epoxide synthesis from perfluoropropylene: **a** schematic of cylindrical cell design; **b** the appearance of the cell from outside. Courtesy of Hoechst AG

ode completely and evenly by lead dioxide without forming edges with little controlled coverage and in particular because in the process of electroforming this approximately 100 μm thick coating of PbO_2 internal stress develops in the coating which leads to bending, cracking and detachment. This is only reliably prevented if the coating is formed on the inner wall of a cylinder rather than on a flat electrode. Figure 11.26 a shows the schematic and cross section of the cell and Fig. 11.26 b shows the whole cell with a height of approximately 2 m.

11.12 Selected Electrochemical Procedures Outside the Chemical and Metallurgical Industries

11.12.1 Electrochemical Wastewater Treatment by Electrodeposition and by Electroosmosis

11.12.1.1
General Considerations

The increasing number and strictness of the legal limitations of effluent pollution call for reliable and cost-efficient processes for the purification of effluents containing salts of heavy metals. The minimum metal concentrations obtainable by conventional hydroxide precipitation at pH 8 and the maximal allowed metal concentrations for effluents in three countries are compiled in Table 11.18. As an alternative to hydroxide precipitation, ion exchangers are gaining in importance. However, ion exchangers are too costly for many types of effluents, and regeneratable exchange resins are not available for all metals. There is an incentive to develop electrochemical deposition and extraction processes for pollution control. In the polarisation curves of cathodes with cathodic deposition of most heavy metals there exists a potential range within which the metal deposits at the diffusion-limited current density. The extension of this potential range depends on the equilibrium potential of the metal and the hydrogen overpotential of the electrode material. Because of the low metal concentrations in effluents, often only approx. 100 ppm, electricity costs for electrochemical wastewater purification are insignificant. Space time yield, however, determines the specific investment and the process costs. The space-time yield is proportional to the product of volume specific electrode area, a, and mass transfer limited current density, i_l.

Table 11.18. Legal effluent limits and metal concentrations obtained by hydroxide precipitation at pH 8 (in ppm)

Metal	Effluent limits			Concentrations after precipitation
	F.R. of Germany	United States	Switzerland	
Pb	2	0.5	0.5	21
Cd	0.5	0.3	0.1	1500
Cu	2	0.5	0.5	1
Ni	3	0.5	2	340
Hg	0.05	–	0.01	–
Ag	2	–	0.1	–
Zn	5	0.5	2	2.6
Sn	5	–	2	–

With low concentrations of metal ions i_l is low. It is therefore essential to use cheap cells with a high value of electrode area per volume and/or to increase the effective current densities.

11.12.1.2
Particular Cells for Removal of Metal Ions from Effluents

1) Mass transfer and current density can be intensified by setting the electrodes in motion or by using turbulence promoters. Examples of moving electrodes are the pump cell, the ECO cell and the beat-rod cell. In the Chemelec cell, a fluidized bed of glass beads enhances the turbulence and increases the mass-transfer rate (see Fig. 11.27 a–c).
2) Large electrode areas can be accommodated in a small space, resulting in developments such as the multicathode cell, the Swiss-roll cell or in capillary gap cells.
3) High mass-transfer coefficients and large specific electrode areas are realised in three-dimensional electrodes. Examples are the porous flow-through cell and the packed bed cell (Enviro cell), the fluidized bed cell and the rolling tube cell.

The Chemelec cell, depicted schematically in Fig. 11.27 a makes use of the mass-transfer enhancement to a plate electrode or an expanded metal sheet electrode by means of a fluidized bed of inert particles. Since the electrolyte flow velocity must exceed a minimum value to keep the bed fluidized, the residence time and the degree of conversion per pass are limited. Therefore, this cell is suitable for pretreatment or recycling operation. The cell can be, for instance, used to remove or limit metal from electroplating rinsing baths.

The ECO cell, Fig. 11.27 b, is equipped with a rotating-cylinder electrode and has a segmented electrolyte chamber to improve the residence-time distribution by cascading. This cell has been used to recover silver and copper from wastewaters. The copper concentration of an 8 m^3 h^{-1} effluent stream, passed over a 1.7 m^2 electrode, is reduced from 100 to 2 ppm with a current of 1000 A at a current efficiency of 65%. In this case, the power consumption per m^3 of wastewater is 4.5 kWh, including the electrical energy necessary for electrode rotation. Another process applicable for electroplating wastewater treatment uses the beat

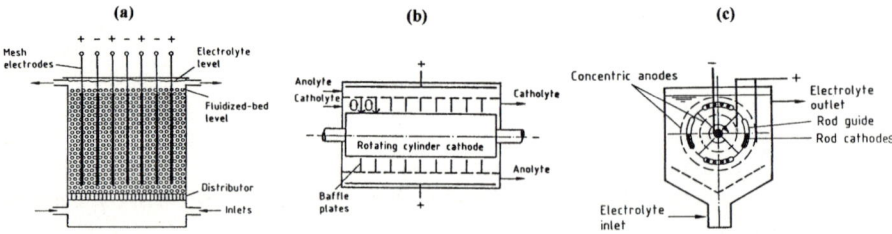

Fig. 11.27 a–c. Schematic of: **a** the Chemelec-cell; **b** the ECO-cell; **c** the beat rod cell

11.12 Selected Electrochemical Procedures Outside...

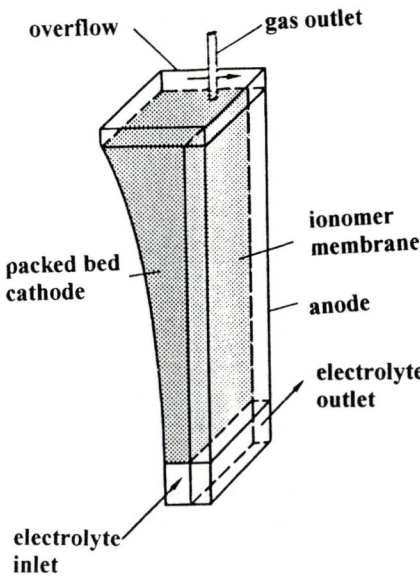

Fig. 11.28. Schematic of the enviro cell

rod cell which is shown in Fig. 11.27 c. Metal rods mounted into a rotating cylindrical carrier act as cathodes. Because of the rotation of the carrier, these rods beat against one another, causing mechanical release of the deposited metal, which is collected as a powder at the bottom of the cell. Such a cell can reduce the silver content of a cyanide solution from 7.05 to 3.5 mg dm^{-3} in a batch operation mode.

In a three-dimensional electrode the penetration depth of the current is restricted in the direction parallel to the current flow (compare Chap. 8.) The penetration depth of the limiting current density increases as the concentration, which is proportional to i_l, decreases. The application of this principle led to the design of the Enviro cell. Usually the packed bed contains graphite spheres. This cell, shown in Figs. 7.12 and 11.28, allows inlet concentrations to be reduced by a factor of 1000 with residence times of only a few minutes. Table 11.19 lists some applications. The packed bed can be loaded with 200 g dm^{-3} of metal of bed volume. Bed regeneration is by chemical or anodic leaching.

For all of these cells there are various operational modes. Not all the cells produce a high degree of metal extraction per pass of electrolyte; for example, the Chemelec cell does not. If for a particular application, the conversion per pass is not adequate, several alternatives are available:
1) Closed recycle loops with wastewater restriction.
2) Batch operation of a discontinuous recycle circuit until the desired concentration is reached.
3) Cascading of several cells.

Table 11.19. Applications of the Enviro cell for effluent treatment

Application	Metal	Throughput m³/h	Concentration, ppm inlet	Concentration, ppm outlet	Energy consumption kWh/m³	Anode area m²
Production of measuring instruments	Hg	0.3	300	0.05	1.2	1
Film processing	Ag	0.2	15	1.0	0.15	1
Salt production	Pb	0.5	2	0.1	0.07	1
Electroplating	Cd	0.2	20	1.0	0.18	1
Battery production	Hg/Cd	0.08	500	0.01	1.7	3
Cellulose acetate production	Cu	20	20	1.9	0.08	40
Pickling (recycling of solution)	Cu	3	150	50	0.19	5
Dye production, I	Cu	6	400	2.0	4.0	90
Dye production, II	Hg	2	4	0.5	2.5	15

11.12.1.3
Electrodialysis

For desalination or removal of ionic pollutants from aqueous wastewaters, electrodialysis is also useful. The scheme for recovery of nickel salts from electroplating rinsing waters is shown in Fig. 11.29. An electrodialysis cell is constructed like a bipolar filter-press electrolyzer, but only the two end chambers of the stack are fitted with electrodes. The compartments are separated from each other alternately by cation and anion exchange membranes. The process is based on the migration of ions in the electric field established by the cell voltage between the end electrodes. The molar material flux density of a species $\dot{n}$ is $it_i/v_e F$. The alternating cation and anion-exchange membranes cause the contaminated electrolyte to be depleted, while the concentration, in this case of nickel and counter ions, is enhanced in the concentrate. This separation is due to transport numbers, t_i, close to unity for anions in the anion-exchange membrane and for cations in the cation exchange membrane. Nickel-sulfate recovery from spent galvanic baths and rinsing waters leaving galvanic plants is performed by electrodialysis. Electrodialysis does not result in the recovery of solid metals, but

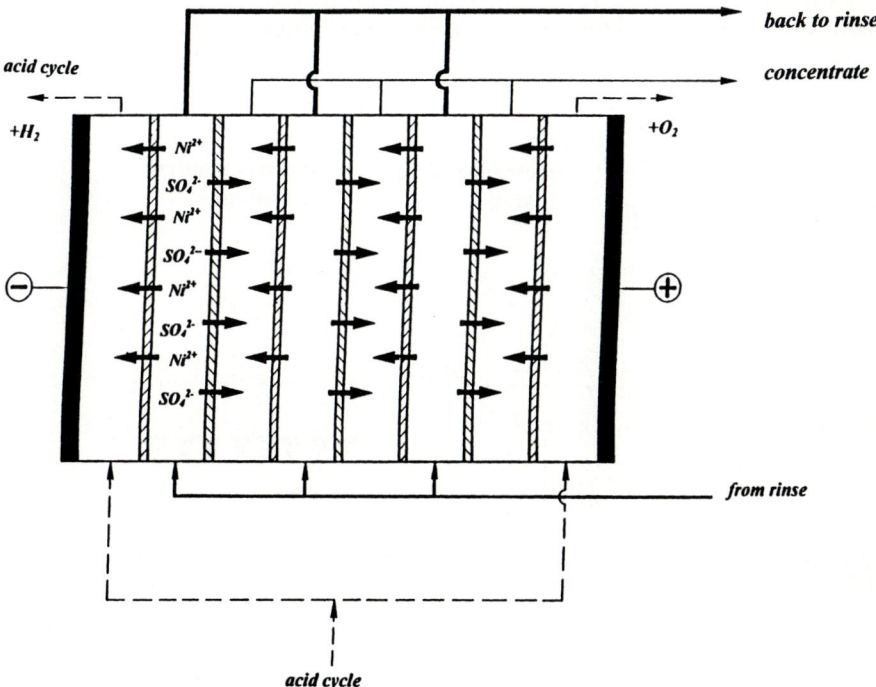

Fig. 11.29. Schematic of removal of nickel ions from galvanic rinsing waters by electrodialysis

concentrated salt solutions and depleted effluents. Although the concentration requirements for effluent treatment cannot be met by this method, the concentration of the treated solution can be maintained on a level, low enough for recirculation of the rinsing water.

11.12.2
Electrochemical Surface Treatment and Shaping of Metals[1]

11.12.2.1
Electrochemical Shaping

Shaping a piece of metal by controlled anodic dissolution – controlled with respect to dissolution depth in three dimensions – is the essential aspect of electrochemical metal shaping. This technique includes electropolishing, electrolytic or electrochemical machining and electrochemical grinding. Electrochemical machining is the fastest, most vigorous form of spatially controlled anodic metal dissolution. Anodic corrosion is the fundamental phenomenon behind these three techniques. Anodic dissolution, which proceeds at a sizeable rate in the subpassive potential region, increases with anodic polarisation. Its rate drops sharply, by many orders of magnitude, at the passivation potential, at which a dense insulating or semiconducting or insulating metal oxide hydrate film forms and prevents further dissolution. At much higher potentials, however, this pas-

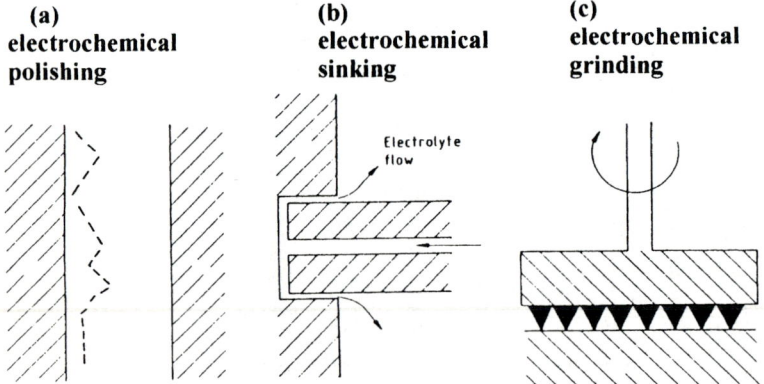

Fig. 11.30 a–c. Electrochemical shaping procedures; **a** electropolishing; **b** electrochemical drilling; **c** electrochemical grinding

---

1 The technologies used to electroplate pieces of any shape and material with metal coatings for corrosion protection, surface finish or decoration are not descibed in this context as the problems, methods and technologies are too variable on one hand and too specific on the other hand for different galvanic metal deposition process, so that there is little chance of a general process description.

sivating film breaks down and vigorous transpassive dissolution of the metal is observed. If a compound of low solubility precipitates on the surface as a more or less porous deposit, instead of a passivating film, then active anodic dissolution is inhibited but the current density remains several orders of magnitudes higher than in the case of passivation. In such cases anodic dissolution usually is limited by mass transfer of the corrosion products through the pores of the deposited layer. This mass transfer-controlled dissolution through layers of precipitates is the principle governing electropolishing (Fig. 11.30 a), whereas transpassive, very high current density corrosion is used for electrochemical sinking or machining and electrochemical grinding (Fig. 11.30 b,c). Electrochemical grinding, which combines anodic dissolution with limited mechanical grinding, is advantageous whenever a component of an alloy is not dissolved electrochemically but sticks to the surface and must be removed continuously along with the anodically dissolved base metal and anodically generated debris. This is the case for cemented carbides as well as nitrides and borides of transition metals or refractory metals in steels. For processing superalloys and refractory metals, which came into use in the aerospace industry and for the construction of steam and gas turbines in the 1960s and early 1970s, the electrochemical methods – electrochemical machining, electrochemical grinding and electrochemical polishing – are especially suitable. However, these methods, with the exception of electropolishing, are not likely to replace mechanical processing of conventional metals and alloys because the electrochemical techniques are expensive and too sophisticated to be used for softer alloys.

11.12.2.2
Electropolishing

Anodic metal dissolution, which is hindered by the formation of porous layers of hydroxides and other little soluble matter, is the principle of electropolishing. This dissolution is mass-transfer controlled and is caused by formation of a relatively thin layer of porous oxide hydrate. The metal cations that are formed by anodic dissolution can leave the metal surface only by diffusion through the pores. Since there is good reason to believe that the pores are randomly distributed and formed and closed at random, dissolution should also be random with no preference for particular crystal faces. Consequently, etch patterns are not observed. The polishing effect is attributed to higher local dissolution rates of the elevated parts of the initially rough surface because of the lesser thickness of the porous oxide hydrates there. Table 11.20 lists the conditions and electrolyte compositions for electropolishing of different metals and alloys. All the electrolytes contain components or establish a pH that cause the formation of only moderately soluble oxide hydrates or salt deposits. Such deposits form on the metal surface under steady-state conditions at current densities of approx. 300–3000 mA cm^{-2}. Electropolishing titanium is particularly difficult and impossible in/ aqueous electrolytes because of passivating TiO_2 formation. Only in organic solvents like formamide with electrolytes like $NaClO_4$ – a hazardous combination – Ti can be electropolished today.

Table 11.20. Electrolytes for electropolishing metals

Metal	Electrolyte	Current density mA cm^{-2}	Temperature, °C	Polishing time, min
Aluminium	20 wt% HClO$_4$+80 wt% acetic anhydride[a]	200–400	<=38	15
Copper	63 wt% H$_3$PO$_4$+37 wt% H$_2$O	20–50	22	5
	65 wt% H$_3$PO$_4$+15 wt% H$_2$SO$_4$+6 wt% CrO$_3$+14 wt% H$_2$O	200–250	22–25	2–5
Lead	30 vol% HClO$_4$+70 vol% glacial acetic acid	1300–1600	22	5
Nickel	73 wt% H$_2$SO$_4$+27 wt% H$_2$O	1000–1300	22	2
	65 wt% H$_3$PO$_4$+15 wt% H$_2$SO$_4$+6 wt% CrO$_3$+14 wt% H$_2$O	200–250	22–25	2–5
Tin	20 vol% HClO$_4$+80%vol acetic anhydride	600–1000	32	8–10
Tungsten	10 wt%NaNO$_3$+90 wt% H$_2$O	200–400	22	20–30
	Na$_3$PO$_4$ (160 g/L)	600	38–49	10
Zinc	25 wt% KOH+75 wt% H$_2$O	1000	22	15
	17 wt% CrO$_3$+83 wt% H$_2$O	2000	–	–
Brass	16 wt% CrO$_3$+84 wt% H$_2$O	2000	26	5
	70 wt% H$_3$PO$_4$+30 wt% H$_2$O	500	–	–
Carbon Steel	48 wt% H$_3$PO$_4$+40 wt% H$_2$SO$_4$+12 wt% H$_2$O	3000	35–49	10
	10 wt% HClO$_4$+90 wt% glacial acetic acid*	1500–2500	26	0.5–2
Stainless Steel	30 wt% H$_3$PO$_4$+60 wt% H$_2$SO$_4$+10 wt% H$_2$O	1800	49	2
	42 wt% H$_3$PO$_4$+45 wt% glycerol+13 wt% H$_2$O	100–600	93–149	8–15

[a] Especially hazardous

11.12.2.3
Electrochemical Machining (ECM)

For a practical electrochemical machining process, the metal must be removed at the cutting rate of 0.25 cm min^{-1} or more, corresponding to current densities >160 A cm^{-2}. At these current densities, the electrolyte gap between the cathode and the workpiece must be adjusted from 0.005 to 0.13 cm to conserve electrical energy and in order to obtain precise cuts. In addition, the anodically generated metal ions, hydroxides and debris, the hydrogen evolved cathodically at the tool cathode and the dissipated Joule heat must be removed efficiently by vigorously pumping the electrolyte through the narrow gap between the cathode and the workpiece. In order to keep the ohmic voltage drop across the interelectrodic gap as low as possible, the electrolytes must have high conductivity. Commercial ECM machines apply a maximum voltage. This is distributed between the sum of the anodic and the cathodic overpotentials, which determines the current density and hence the dissolution rate and the ohmic voltage drop across the interelectrodic gap. The ohmic voltage drop increases with gap width and current density. Therefore, for every cutting rate and voltage there is one gap width that is self-adjusting and that can be kept steady during the process, the so-called equilibrium gap. Short circuits, which can severely damage the tool, must be avoided and most ECM machines are protected against such damage by special sensing devices and ultrafast electronic switches.

Figure 11.30b shows how the electrochemical machining works in the case of electrochemical drilling or sinking. The tool is a cylindrical cathode, the wall of which is covered with an electrically insulating coating. Current densities around 300 A cm^{-2} flow between the bottom of the hole and the cathode and a vigorous stream of electrolyte flushes the dissolved metal ions and solids out of the hole. Immediately behind the cathode face there is a region where electropolishing takes place at the walls of the drilled hole because of the reduced current densities. Consequently, electrochemical machining in most cases produces an excellent surface finish. Most important for successful electrochemical machining is the choice of the electrolyte. Its composition should allow metal dissolution with current densities up to several hundred amperes per square centimetre while preventing the precipitation of insoluble salts and oxides. Alkali metal chlorates and perchlorates, which can be hazardous, have proved to be suitable ECM electrolytes. Table 11.21 lists the electrolyte components and their concentration in typical ECM electrolytes. Electrochemical machining became popular for processing high-strength metals and alloys and refractory metals. Some of these procedures for which ECM can be used are shown schematically in Fig. 11.31. Die sinking is also possible. Sinking the cathode tool into the workpiece produces the cavity used for casting.

Current density distributions in the different electrochemical shaping procedures due to high current densities and subsequent overwhelming ohmic potential drops is essentially primary distribution (compare Chap. 5). In all these operations overcuts due to the action of stray currents cannot be avoided. Comput-

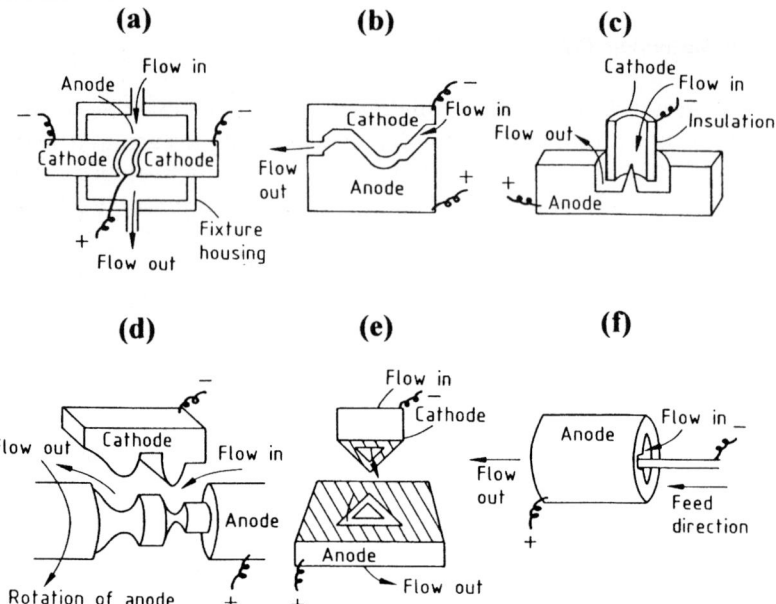

Fig. 11.31. Some applications of electrochemical machining

ed current-density distributions have always turned out to be insufficiently accurate to prevent overcuts. Therefore exact shaping is carried out by empirically changing the tool dimensions, current density, electrolyte concentration, electrolyte flow velocity, etc.

11.12.2.4
Electrochemical Grinding

Diamonds, α-alumina (corundum), or another hard material, the grinding powder, is embedded into the surface of a grinding wheel of copper or copper bronze, which serves as the cathode. The grinding powder is pressed into the metal surface and held in place by plating the studded surface with a layer of nickel. With an optimum load of 3.0 carats cm^{-2}, the grinding powder is exposed to a height of only 10–60 μm by etching the nickel away in an electropolishing bath. The diamond (or corundum) grains do not serve themselves for grinding, but rather define and maintain the width of the electrolyte gap between the grinding wheel and the workpiece. But they serve to scratch away loose debris and solid residues (see Fig. 11.30 c). Therefore, only light pressure is used to press the grinding wheel and the workpiece together. Around 90% of the metal is removed by anodic dissolution. Fully hardened steels can be processed without difficulty and without detectable wear of the tools because nearly 100% of the metal is removed by anodic dissolution. Current densities of 12–40 A cm^{-2}

11.12 Selected Electrochemical Procedures Outside...

Table 11.21. Electrolytes for electrochemical machining

Metal	Electrolyte	Remarks
Aluminium and aluminium alloys	NaNO$_3$ (100–400 g dm^{-3})	excellent surface finish
Cobalt and cobalt alloys	NaClO$_3$ (100–600 g dm^{-3})	excellent dimensional control, excellent surface finish
Molybdenum	NaOH (40–100 g dm^{-3})	NaOH consumed and must be added continuously
Nickel and nickel alloys	NaNO$_3$ (100–400 g dm^{-3}) NaClO$_3$ (100–600 g dm^{-3})	good surface finish good surface finish, good dimensional control, low metal removal rate
Titanium and titanium alloys	NaCl (180 g dm^{-3}) + NaBr (60 g dm^{-3}) + NaF (2.5 g dm^{-3}) NaClO$_3$ (100–600 g dm^{-3})	good surface finish, good dimensional control, good machining rate bright surface finish, good machining rate above 24 V
Tungsten	NaOH (40–100 g dm^{-3})	NaOH consumed and must be added continuously
Steel and iron alloys	NaClO$_3$ (100–600 g dm^{-3}) NaClO$_3$ (100–400 g dm^{-3}) NaNO$_3$ (100–400 g dm^{-3})	excellent dimensional control, brilliant surface finish, high metal removal rate, fire hazard when dry good dimensional control, lower fire hazard, good surface finish, good machining rate good dimensional control, fire hazard when dry, low metal removal rates, rough surface finish

approach those used for ECM. Mixtures of chlorides and nitrates serve as electrolytes. Vigorous pumping and copious electrolyte exchange is vital for electrochemical grinding. Unlike mechanical grinding, electrochemical grinding leaves no scratches, grooves or tool registry marks. Current densities must not be too high if increased surface roughness is to be avoided.

11.12.3
Electroreforming of Microdies and Microtools by the LIGA-Process

Electroreforming of microprofiles and other delicate structures by cathodic metal deposition is impossible as the enhanced current density concentration at sharp edges and lines impairs the formation of precisely defined surfaces, edges and corners of small scale. If, however, galvanic metal deposition confines to filling hollow precisely shaped microvoids, whose dimensions are measured in millimetres or even micrometers, the deposited metal copies the inner surface of the hollow template with high precision. The so called LIGA-process (LIGA in German means Röntgen*l*ithographie, *ga*lvanische *A*bformung; translation: galvanic copying of X-ray photography-generated hollow templates) is based on this principle. The different steps of the process, which had been at first developed

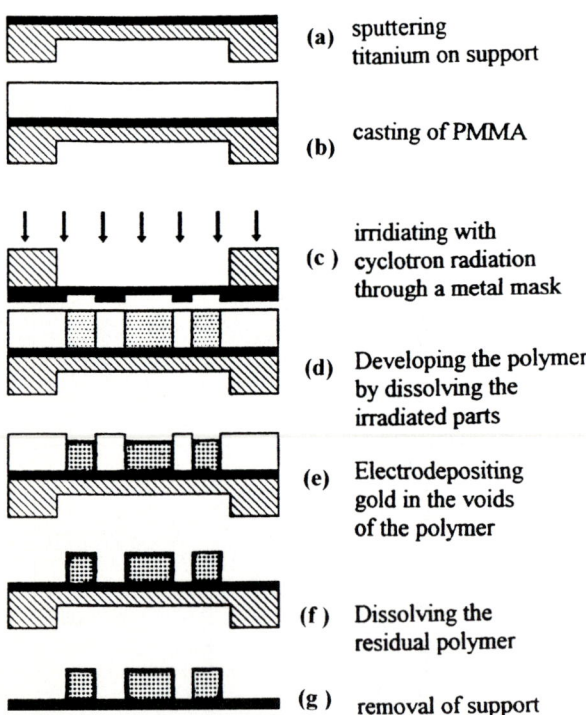

Fig. 11.32. The separate steps of the LIGA process

for producing micro-orifices for gas-diffusion separation of uranium isotopes at Forschungzentrum Karlsruhe are schematically depicted in Fig. 11.32 a–g.

One begins with casting a thin photoresist layer of from micrometer to millimetres thickness on a supporting metal plate. The next step is irradiating the photoresist, which usually is polymethylmethacrylate, PMMA, through the open surface of an X-ray impenetrable mask on top of the photoresist by synchrocyclotron radiation. This destroys the irradiated photoresist polymer. The depolymerized photoresist is dissolved or etched away, so that a positive polymer template is left. The third step consists of depositing galvanically a metal (silver, nickel or copper) around this electrically insulating template and eventually etching away or dissolving the polymer, which formed the positive template. The metal is now left with the void, which precisely reproduces the volume of initially non-irradiated photoresist. This template may serve, for instance, as a die for casting or extruding a polymer part of miniature dimension with high reproductive precision. Micrometer structures can be reproduced by the Liga process with precisions of better than 10^{-4} mm.

References

1. Aluminium Electrolysis, 2nd. Ed., K. Grjotheim, C. Krohn, M. Malinovsky, K. Matiasorvsky, J. Thonstad, Aluminium Verlag, Düsseldorf, 1982
2. P.M. Robertson, P. Berg, H. Riemann, K. Schleich, P. Seiler, J. Electrochem. Soc. *130*, 591, (1983)

Further Reading

Ullmann's Encyclopedia of Industrial Chemistry, F.T. Campbell, R. Pfefferkorn, F. Rounsaville Eds., VCH Weinheim 1985

D. Pletcher, Industrial Electrochemistry, Chapmann and Hall, London, New York 1982

F. Hine, Electrode Processes and Electrochemical Engineering, Plenum Press 1985

J.P. Hoare, M.A. La Boda, Electrochemical Machining in E. Yeager, A. Salkind eds, Techniques of Electrochemistry, J. Wiley and Sons, New York, Chichester, Brisbane Toronto, 1978, p 48–141

CHAPTER 12

Fuel Cells

12.1
Fuel Cells as Gas Supplied Batteries

According to the simplified schematic of Fig. 12.1 fuel cells can be described as gas-fed batteries, which convert the formation enthalpy of water, Eq. (12.1), to a substantial fraction into electrical energy, comparable to, for instance, the lead acid battery converting the stored chemical energy of the reaction $Pb+PbO_2+2H_2SO_4 \rightarrow 2PbSO_4+2H_2O$ into electricity.

$$H_2 + 1/2 O_2 \rightarrow H_2O_g; \quad \begin{aligned} \Delta H^0_{400K} &= -242.6 \text{kJmol}^{-1} \\ \Delta G^0_{400K} &= -223.7 \text{kJmol}^{-1} \end{aligned} \quad (12.1)$$

As the entropy of the reaction at Eq. (12.1) is negative, the Gibbs free energy is less negative than the enthalpy – compare Chap. 3 – and therefore definitely less than 100% of ΔH may be converted into electricity.

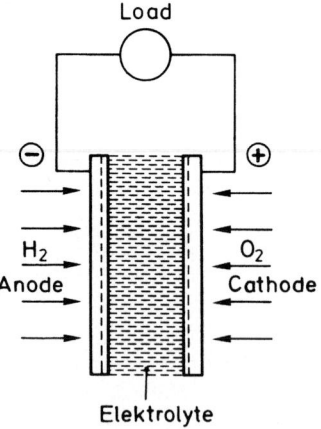

Fig. 12.1. Schematic of a fuel cell presenting the cell as a gas-supplied battery

Reaction Eq. (12.1), can be executed with relatively little catalytic efforts even at temperatures close to ambient temperature. Due to the reactivity of hydrogen, its anodic oxidation as in Eq. (12.2) is easy:

$$H_2 \rightarrow 2H^+ + 2e^- \tag{12.2}$$

and even oxygen, although demanding stronger catalytic activation than hydrogen, can also be reduced relatively easily at the cathode according to Eq. (12.3) with appropriate electrocatalysts.

$$1/2 O_2 + 2e^- + 2H^+ \rightarrow H_2O \tag{12.3}$$

Pt is the catalyst of choice for anodic hydrogen oxidation and cathodic oxygen reduction. Today this expensive noble metal can be utilized to an extent, and Pt loadings of the electrodes are so low, that its costs are no longer prohibitive for the application of low temperature fuel cells for stationary electricity generation where specific system costs are in the range of 1000 US $ kW^{-1}. But fuel cells are also considered realistic options as power sources even for automobiles, where specific costs of power sources must not exceed 100–200 US $ kW^{-1}, that means that their costs be not appreciably higher than those of internal combustion engines. Most important for the application of fuel cells is, that the oxygen can be extracted directly from air and needs not to be pure oxygen or oxygen-enriched air and that hydrogen can be extracted by the cell from a type of synthesis gas, which is produced insitu from natural gas and contains hydrogen and carbon dioxide in a 4:1 ratio, so that pure hydrogen, which is too expensive, is not necessary as fuel.

High temperature fuel cells can completely dispense with noble metals as electrocatalysts, because of the thermal activation of the electrode processes and can do with nickel as anodic and with oxidic p-type semiconductors like lithiated nickel oxide or lanthanum–strontium–manganese oxide as cathodic electrode materials and electrocatalysts.

12.2
Theoretical Efficiency of Hydrogen/Oxygen Fuel Cells

Referring to Chap. 3 and the equation $\Delta G = v_e FE_0$, with E_0 equalling the theoretical or equilibrium cell voltage, the theoretical efficiency of the fuel cell process, η_{th}, is given by the ratio of the Gibbs free enthalpy and the enthalpy of reaction Eq. (12.4):

$$\eta_{th} = \Delta G / \Delta H = v_e FE_0 / \Delta H \tag{12.4}$$

Referring to Fig. 3.4, Fig. 12.2 depicts the temperature dependence of η_{th} for the formation of water under standard conditions in the temperature range 373–1300 K. Plotted is also the theoretical efficiency of a Carnot process, defined by Eq. (12.5),

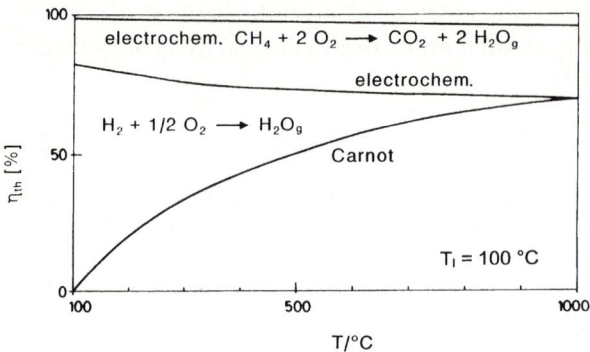

Fig. 12.2. Temperature dependence of the theoretical electrical efficiency of hydrogen/oxygen fuel cells and a methane-converting fuel cell in comparison with the Carnot efficiency of a process with T_u the upper and T_l the lower temperature of the cycle. T_l is always assumed to be 373 K

$$\eta_{th}(\text{Carnot}) = \frac{T_u - T_l}{T_u} \qquad (12.5)$$

with T_u and T_l the upper and lower process temperatures in dependence on T_u and with T_l always assumed to be 373 °C. Clearly the fuel cell process is at an advantage at lower temperatures and the theoretical efficiency of the Carnot cycle surmounts that of electrochemical energy conversion only above approximately 1,000 K.

Figure 12.2 shows with the uppermost line additionally the temperature dependence of the theoretical energy efficiency of the electrochemical methane combustion, which is almost temperature-independent and approaches almost a value of unity. Obviously methane, today – in the form of natural gas – a quite common gaseous fuel, would be, from the thermodynamic point of view, the most advantageous fuel for fuel cells. It can, however, not be converted directly at fuel cell anodes since this gas, kinetically spoken, is too inert and its anodic oxidation is too sluggish. Nonetheless, for stationary fuel cells it is today the primary fuel of choice, but it needs chemical conversion by steam reforming and shift conversion into a type of synthesis gas with CO_2 and H_2 as the main constituents, from which fuel cell anodes can extract the hydrogen. With respect to the economical chances of fuel cell technology today, it was the most important decision – taken at International Technology Corporation, ITC, in the 1970s – to develop fuel cell systems for natural gas in order to get rid of the necessity to prepare highly purified hydrogen.

12.3
Fuel Cell Types

The different fuel cell technologies are nominated according to the respective electrolyte of the different cell types. According to Table 12.1 one distinguishes the low temperature technologies
(i) the alkaline fuel cell (AFC), (working temperature 80–90 °C),
(ii) proton-exchange membrane cell (PEMFC), (80–90 °C), and
(iii) the phosphoric acid cell (PAFC), (200 °C)
   and the high-temperature applications
(iv) the molten-carbonate cell (MCFC), (650 °C) and
(v) the solid-oxide fuel cell (SOFC), (800–950 °C).

Figure 12.3 a–d shows the process schemes of PAFCs, PEMFCs, MCFCs and SOFCs as the technologies relevant for terrestrial commercial applications. To-

Cells with proton conducting elctrolytes

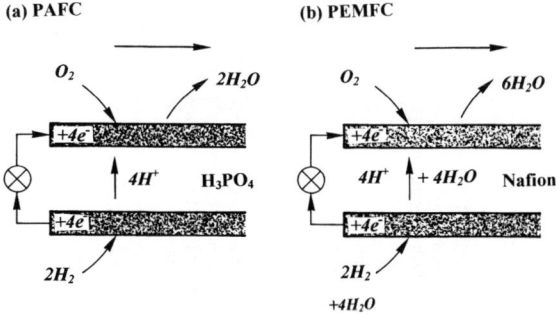

Cells with oxygen anion conducting elctrolytes

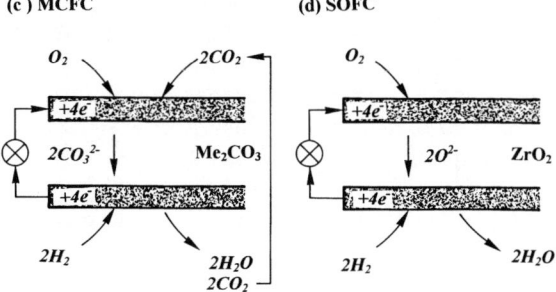

Fig. 12.3a–d. Process schemes of the four fuel cell technologies relevant for terrestrial applications: **a** phosphoric acid fuel cells, PAFC; **b** proton exchange membrane fuel cells, PEMFC; **c** molten carbonate fuel cells, MCFC; **d** solid oxide fuel cells, SOFC

Table 12.1. Basic data and applications of different fuel cell types

Type	Temperature °C	Primary fuel gas	Oxidant	System components	Electr. efficiency Cell theor. %	Electr. efficiency Cell pract. %	Electr. efficiency System[1] %	Remarks	Application
Alkaline (AFC)	60 to 90	pure H_2	pure O_2	cell, water removal, inverter	83	60		sensitive to CO_2	space and military
Membrane (PEMFC)	ca. 80–90	H_2	O_2, air	cell, inverter	83	60		sensitive to CO (<100 ppm)	electrotraction, small scale stationary power generation
Phosphoric acid (PAFC)	160–220	natural gas, H_2	air	reformer, shift converter, steam generator, cell, inverter, heat exchange system	80	55	40	sensitive to CO (<2%)	cogeneration (several 100 kW to MW)
Molten carbonate (MCFC)	650–700	natural gas, coal gas	air	steam generator; cell, inverter balance of plant for heat and rest enthalpy usage for instance boilers, steam turbines	78	55–65	48 to 55[2] ca. 60	CO_2 recycling is accomplished by mixing anode off-gas with air	cogeneration with high temp. heat several 100 kW
Solid oxide (SOFC)	800–1000	natural gas, coal gas	air	preformer, steam generator, cell, inverter, balance of plant for high temperature heat usage	73	60–65	55 to 60[3]	preforming necessary	cogeneration with high temp. heat 10 kW to several 100 kW

[1] Basis: methane, 80% utilisation of fuel gas in the cell; oxidant: air (with coal gas, the efficiency decreases by 8–10%).
[2] With internal reforming.
[3] Internal reforming, 70% efficiency on improvement of cell.

day alkaline cells seem to be definitely reserved for space applications. Even for this special purpose their use in future space missions might be questionable, because they are too expensive and their operation and particular their water management seems to be too complicated and not reliable enough on long term. In the future they might therefore be substituted in space vehicles by membrane fuel cells. We will therefore not deal here with the alkaline technology in greater detail, but will state that without having developed this type of fuel cell as a precursor one would never have tried to embark into the difficult and very expensive business of developing the other four types of fuel cells, which are now believed to offer realistic options and opportunities for terrestrial electrochemical electricity generation from fossil fuels.

12.3.1
Low-Temperature Fuel Cells – Their Technological State

Low temperature cells differ significantly from high temperature cells with respect to applicable production technologies and materials, because the relatively low working temperature allows to construct them from polymer-materials, which are easy to shape and can be processed with established, relatively inexpensive technologies. These technologies like film- and foil-casting or for instance die-casting or extruding are available for mass production of commodities fabricated from polymers for more than 40 years.

12.3.1.1
Phosphoric-Acid Cells

The phosphoric-acid fuel cell today is singular insofar as it is fully developed to a commercially available product in the form of fully automated cogenerating systems, whose performance is of significant reliability and which are commercial since more than ten years. Natural-gas-fired cogeneration plants of 200 kW nominal electric power, developed by United Technology and produced by ONSI Corporation are offered and run by private business proprietors and utilities. More than 100 units of 200 kW_{el} power – several now of the most advanced version PC 25 C –are in operation and by and large fulfil technically the expectations of the users and the promises of the producers. Figure 12.3 a shows the process scheme of PAFCs. The cells are composed of prefabricated PTFE-bounded electrodes – compare Chap. 10 - and a PTFE-bounded silicon carbide electrolyte matrix, which is composed of ≈0.1 µm SiC particles and contains the electrolyte, 103% phosphoric acid ($H_3PO_4+P_2O_5$). The cell stacks whose effective cell cross sections measure approximately 70×70 cm^2 are composed of substacks comprising seven cells, which are divided from each other by cooling plates. These plates are fabricated of dense carbon. The cooling liquid is an appropriate organic liquid of high boiling point. The total cell stack contains 260 cells, each having a nominal cell voltage of 0.65 V resulting in a stack voltage

Fig. 12.4. Three phosphoric acid fuel cell stacks from left to right: 40 kW, 200 kW (PC 25 A) and 750 kW of ONSI Corp.-Courtesy of ONSI Corp

of 170 V. Figure 12.4 shows three different PAFC stacks of 40 kW, 200 kW and 750 kW.

12.3.1.2
Membrane Cells

The membrane cell, based on the use of Nafion cation-exchange membranes as electrolyte, had been invented in the 1970s at ITC and was developed by Siemens to a power source of high reliability and performance for submarines. The electrocatalysts and electrodes were unsupported Pt black or finely dispersed platinum metal alloys, which had been deposited directly on the membrane surface. The costs, typical for highly qualified military equipment (70,000 bis 100,000 US $ kW^{-1}), were simply prohibitive for usual commercial application. Ballard was the first company demonstrating that PEMFCs would, in principle, be cheaper. The most important step to bring this type of cell closer to commercial application was an invention of Raistrick, later further developed by Gottesfeld and Srinivasan to technical applicability. It was the idea to impregnate active carbon supported catalysts with solutions of Nafion in *tert*-butanol – compare Chap. 10. By this method, which yields in "wetting" of the internal catalyst surface by the polymer electrolyte, a significant increase in electrocatalyst utilisation was achieved, which lowered the necessary load of platinum or platinum alloys from several milligrams per square centimetre to now fractions of milligrams per square centimetre (0.3–0.5 mg cm^{-1}) for hydrogen/oxygen membrane cells see section 10.8. Since that innovation the costs of noble metals do no longer determine the cell costs. As membrane cells exhibit a higher power density and lower weight per volume than any other cell type, they are particularly suited as power sources for electrotraction of automobiles. Figure 12.3 b shows the process scheme of PEMFCs.

12.3.1.3
Direct and Indirect Methanol-Combusting Membrane Cells

Roughly since 1990 big car producing companies are expending great efforts to develop membrane cells operated on hydrogen for electrotraction. Pure hydrogen is four times more expensive than natural gas and it is not easy to store. Methanol is from the logistic point of view a much better fuel than hydrogen as it stores roughly twice as much energy per volume as liquid hydrogen, but under normal pressure and temperature. It is also on an energy basis cheaper than hydrogen. Methanol being an oxidised and activated, highly reactive hydrocarbon, is much more suitable for anodic and also for chemical conversion than any saturated or unsaturated hydrocarbon of general formula C_nH_{2n+x}. Unfortunately the anodic oxidation of methanol on platinum electrodes, although in principle easy because methanol is quite reactive, is severely hampered as the anodic degradation products of this fuel, carbon monoxide in particular, are poisoning the catalyst. Therefore direct anodic oxidation of methanol would only become a realistic option after development of a new type of anodic electrocatalyst which would not be poisoned. Producing a type of synthesis gas from methanol on board the vehicle by low-temperature steam reforming and anodically extracting hydrogen from this gas mixture – reformate, as it is called – is a second option, called indirect anodic methanol combustion. Reformate fuelled membrane cells have to cope with the problem, that the carbon monoxide contents of the reformate gas cannot easily be kept below several tens of ppm also with the consequence of severe catalyst poisoning. By preferential catalytic CO-oxidation with added air in the reformer gas it is possible to lower the CO-contents to between 20 and 30 ppm. Figure 12.5 a shows the flow sheet of fuel processing of the so called indirect methanol fuel cell, which can be housed together with the cell in passenger cars at tolerable costs and low space demand. The most advanced option would certainly be to operate the membrane cell with reformate-gas obtained by reforming gasoline or kerosene on board the vehicle followed by low temperature shift conversion and selective catalytic CO-oxidation. This technology is being developed since 1996 by Chrysler. It would, however, also be a very reasonable option to oxidise methanol directly at the anode as chemical processing of the fuel could be avoided. Figure 12.5 b shows the simple flow sheet of PEMFCs with direct anodic methanol oxidation. Still electrocatalysts which perform well enough for anodic methanol oxidation are not yet at hand. But the enhanced electrocatalytic activity of Pt/Ru (1:1) alloy catalysts point the way to mitigate self-poisoning on Platinum-metal catalysts by using binary or ternary catalysts.

12.3.1.4
Process Principles of the PAFCs and PEMFCs with Proton Conducting Electrolyte

The process scheme of the two low-temperature cells are depicted in Fig. 12.3 a,b. The electrolytes of the PAFC and the PEMFC are proton conductors. The migrating protons are generated at the anode by hydrogen oxidation ($H_2 \rightarrow 2H^+ + 2e^-$). They convey the positive electric charge injected at the anode into the electrolyte to the cathode and there they produce water with the reduced oxygen ($1/2 O_2 + 2e^- + 2H^+ \rightarrow H_2O$). Figure 12.5 b shows the process scheme for a membrane cell in which methanol is oxidised directly at the anode. Apart from the anodic release of carbon dioxide and the necessity to recycle the water/methanol mixture, which might be in the form of vapours or – in pressurised cells – a liquid mixture there is no fundamental difference between H_2/O_2 and methanol/O_2 PEMFCs. Common to both and also to PAFCs is the necessity to supply

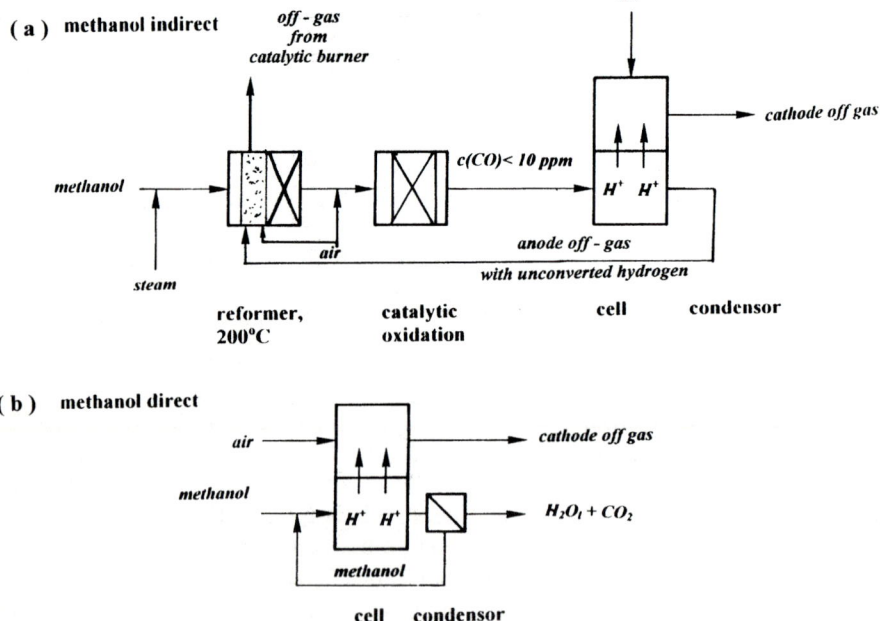

Fig. 12.5. a Flow sheet for producing CO-depleted reformate from methanol. The process comprises low-temperature methanol reforming (350–400 °C), low temperature shift conversion (200 °C) and catalytic preferential CO-combustion. **b** Flow sheet of direct methanol combusting fuel cells

sufficient water to the anode as the transport of protons from the anode to the cathode is unavoidably associated with water transport which would lead to drying out and subsequent failure of the anode and the membrane without supply of water or water vapour.

12.3.2
High-Temperature Fuel Cells

12.3.2.1
Molten-Carbonate and Solid-Oxide Fuel Cells

There exist two different types of high-temperature fuel cells. The molten-carbonate fuel cells (MCFC) working with the molten Li_2CO_3/K_2CO_3 (0.62/0.38 mol/mol) eutectic as liquid electrolyte at a cell temperature of 650 °C and the solid-oxide fuel cell, SOFC, which uses yttria stabilised zirconia (YSZ), which is an oxygen-anion-conducting solid-state compound as electrolyte. The yttria added in amounts from 4 to 6% stabilises the tetragonal modification vs. the monoclinic and allows to take advantage of the higher ionic conductivity of the tetragonal form of zirconia. The working temperature ranges – depending on the technology, in particular depending on the thickness of the zirconia diaphragm – from 800 °C for membranes thinner than 50 µm to 1000 °C for membranes of several hundred micrometers. At such high temperatures highly disperse metals, with the dispersity in the nanometer range, are no longer stable as anode catalysts. Instead sintered porous nickel composed of nickel granules with micrometer diameters, which are additionally dispersion hardened in order to prevent their slow compaction and sinter growth are used for MCFC anodes and in the form of a nickel/zirconia cermet in anodes of solid oxide fuel cells. In both high temperature cell types p-type semiconducting porous oxide ceramics are used as cathode materials – lithiated nickel oxide $Li_xNi_{1-x}O$ in the case of the MCFC; lanthanum manganite, $LaMnO_3$, in case of the SOFC.

12.3.2.2
Process Schemes of MCFCs and SOFCs

The process schemes of Fig. 12.3 c,d are typical for fuel cells with oxygen-ion conducting electrolyte. The oxygen anion is produced at the cathode and injected into the electrolyte in MCFCs and SOFCs. There is, however, an important difference as in MCFCs O^{2-} ions are binding to carbon dioxide to form carbonate

$$1/2 O_2 + 2e^- + CO_2 \rightarrow CO_3^{2-} \tag{12.6}$$

which is essential for preserving the integrity of the electrolyte at the cathode because without addition of carbon dioxide one would experience precipitation of lithium oxide in the cathode accompanied by irreversible deterioration of cathode structure and performance, as Li_2O is only little soluble in the carbonate melt. Therefore it is indispensable to supply sufficient carbon dioxide together with the oxidant, which is usually air, to MCFC cathodes. At the anode the carbon dioxide is released together with water as the anodically produced protons combine with oxygen anions which have been shuttled by migration of the carbonate anions from the cathode to the anode:

$$H_2 + CO_3^{2-} \rightarrow H_2O + CO_2 + 2e^- \tag{12.7}$$

In SOFCs the situation is simpler. At the cathode oxygen anions are injected into the electrolyte, zirconia, ($1/2 O_2 + 2e^- \rightarrow O^{2-}$), they migrate to the anode, where by oxidation of hydrogen, water vapour is released ($H_2 + O^{2-} \rightarrow H_2O + 2e^-$).

12.3.2.3
Internal Reforming in High-Temperature Fuel Cells

The working temperature of MCFCs (650 °C) and even more of SOFCs (800–1000 °C) is sufficiently high to make heterogeneously catalysed steam reforming of natural gas so fast, that this reaction can be performed in the cell. In SOFCs with at least 800 °C the reforming reaction, Eq. (12.8) is even so fast, that the nickel anode may serve as reforming catalyst, so that no dedicated reforming catalyst in SOFCs is needed.

$$CH_4 + H_2O_g \rightarrow CO + 3H_2; \quad \Delta H^0_{1000K} = +204.64 \, kJ/mol \tag{12.8}$$

$$CO + H_2O_g \rightarrow CO_2 + H_2; \quad \Delta H^0_{700K} = -37.8 \, kJ/mol \tag{12.9}$$

The exothermic shift-conversion reaction, Eq. (12.9), which converts carbon monoxide to hydrogen is much faster than steam reforming and is always operative and the shift-equilibrium is fully established if the reforming reaction is fast enough on the time scale of the residence time of the anode gas in the cell. At the lower temperature of MCFCs the surface of the anode is not sufficiently active for catalysing steam reforming and a particular low-temperature reforming catalyst is necessary for performing this reaction in the cell with sufficient reaction rate. Steam reforming within the cell by so called internal steam reforming has two big advantages: first the heat generated in the cell, with cell voltage U_{cell} and cell current, I,

$$\text{Heat per cell} = \dot{Q} = \dot{N}_{CH_4} \cdot \Delta H_{OX,CH_4} - I \cdot U_{cell} \tag{12.10}$$

with $\dot{N}_{CH_4}$ the molar flux of anodically converted methane and $\Delta H_{OX,CH4}$, the molar heat of combustion of methane, can be used in-situ for steam reforming avoiding an external reformer, which usually must be fuelled by the anode off-gas; second, additional heat exchangers for heat exchange between cool and heat adsorbing vapour/fuel mixtures or fresh air and hot anode and cathode off-gases are not necessary as the internal structure of the thin cell, which by itself is the heat source, and the anode lumen through which the fuel together with water vapour is circulated and in which the reforming catalyst is stored is optimal for heat exchange. Furthermore, the anodic oxidation of hydrogen, which is generated in the cell according to Eq. (12.7) generates additional water vapour enhancing the rate and the degree of methane conversion by the reforming reaction Eq. (12.8). Considering the cell with integrated catalyst as a black box, the application of the internal reforming principle would mean to combust methane anodically which, according to Fig. 12.2, offers the great advantage of an almost 100% theoretical efficiency. It must be stressed, however, that in practice the electrical efficiencies of methane-fired fuel cell power plants of any technology do not yet exceed 50% significantly. The data collected in Table 12.1 read 42% for phosphoric-acid cell power plants and 48–55% for MCFCs.

12.3.3
Cell Technologies of MCFCs and SOFCs

12.3.3.1
Molten-Carbonate Fuel Cells

Molten-Carbonate Fuel Cells are far more advanced than SOFCs. Since 1996 there had been established at least three different demonstrations – one of a 2 MW-power plant and two multi-100 kW-power plants – so that these cells exhibit today already a remarkable degree of technical maturity. The reason for this difference in technical status is the longer history of MCFCs, which dates back to the early 1950s, when Ketelaar and Broers built the first molten carbonate cells. Also the lower operating temperature, which simplifies in a sense the solution of certain materials problems helped to develop MCFCs faster than SOFCs. The cell, shown schematically together with the stack in Fig. 12.6, is approximately 1.5 mm thick with cathode (NiO), electrolyte matrix ($LiAlO_2$) and anode (sintered, dispersion hardened nickel) each being approximately 0.5 mm thick. Current collectors are made of perforated stainless steel at the cathode and hardened nickel sheet at the anode. The bipolar plate is made of stainless steel, which is clad by a thin nickel layer at the anode side, because under the reducing atmosphere of the anode gas and in contact with the aggressive melt, stainless steels are not reliably passivated. In contrast under the oxidising atmosphere of the cathode, stainless steel becomes passivated by a dense $LiFeO_2$ layer and can therefore be exposed to the melt. Gas tight sealing of the anode chamber against the cathode chamber is accomplished by the electrolyte-flooded matrix and at the rim by the so called wet seal, where the capillary forces of the flooded matrix prevent the release of working gases to the outside as shown schematically in Fig. 12.6 b. Figure 12.6 a shows the schematic of the cell stack design for cross-flow operation by arranging the anodic and cathodic current collectors perpendicularly to each other and blocking the side of each anode and cathode section respectively by so called rails, so that anode and cathode gases can be fed into the cell on vicinal edges of the cell through gas collecting hoods and at

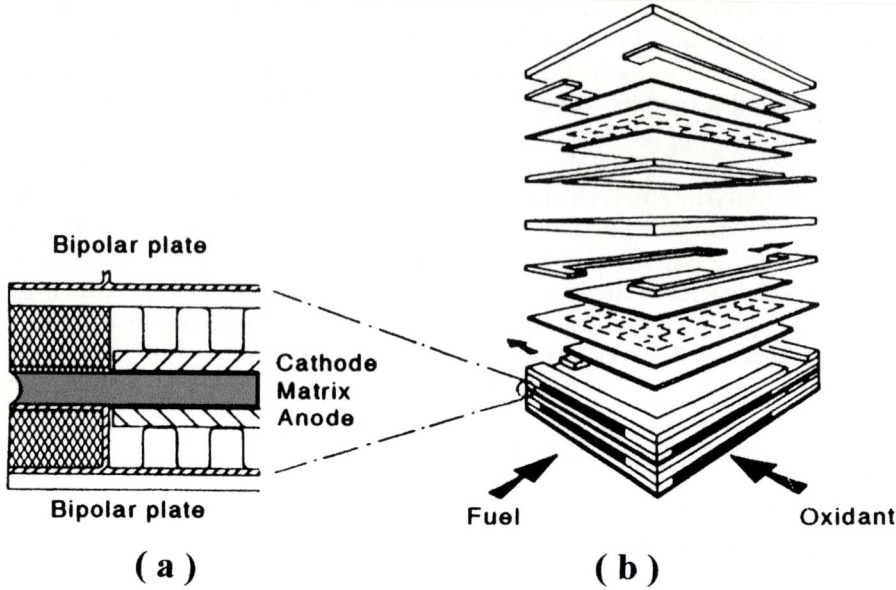

Fig. 12.6a,b. a Wet seal at the rim of the cell. **b** Schematic of a MCFC stack with gas-tight electrolyte-flooded matrix: arranging cells and rails for cross-flow operation

the respective opposite sides hoods are gathering the anode and cathode off-gases.

12.3.3.2
Solid Oxide Fuel Cells

Figure 10.15 show the cross section of a solid oxide fuel cell consisting of a nickel/zirconia cermet anode, a dense zirconia solid electrolyte layer of 50–150 μm thickness and a relatively thick, porous $LaMnO_3$ anode. Figure 12.3 d demonstrates the process scheme of an SOFC.

12.3.3.3
The Westinghouse Technology

The initial technical development of SOFCs had been performed at Westinghouse. The Westinghouse technology is based on a one-side closed tubular cell concept, which is schematically depicted in Fig. 12.7. The zirconia cylinders closed at one side, which today are fabricated with a length of 1 m and would eventually be 2 m long, have an inner diameter of approximately 3 cm and are self-supported on their inner surface by the porous $LaMnO_3$ cathode covered by the dense zirconia electrolyte membrane. The zirconia is coated at its outer side

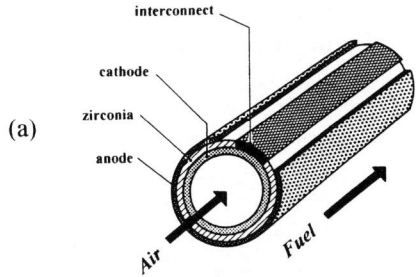

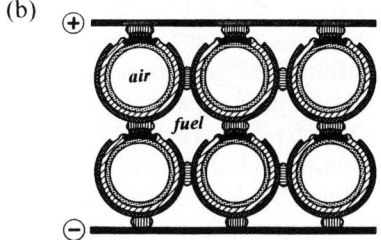

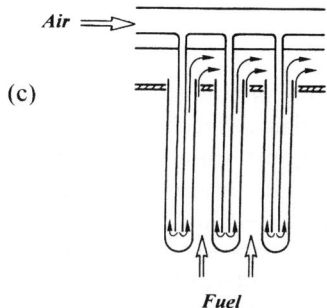

Fig. 12.7. Schematic presentation of the tubular SOFC developed by Westinghouse: **a** schematic of concentric arrangement of the cell components and coflow of fuel and air; **b** cross section through a tube register; **c** fuel flow and final combustion of unconverted fuel by mixing with the cathode off-gas

by the porous cermet anode. As also shown in Fig. 12.7 the $LaCrO_3$ interconnect – a material which under reducing as well as under oxidising atmosphere is chemically stable and in contact with both atmospheres has a sufficiently high electronic conductivity – connects the internal air electrode with an outer metallic contact composed of porous sintered nickel. This outer metallic contact is more or less ductile at 800–1000 °C forming a plastically deformable connection between the inner air cathode of one tube via the interconnect and the outer fuel anode of the next tube, establishing the means for in series and in parallel connection of any desired number of tubes. Most significant for the Westinghouse concept is the avoidance of any gaskets and tightenings, which in plate cells would be necessary to prohibit intermixing and cross-over of the two working gases from respective opposite electrodes. As shown in Fig. 12.7 c the fuel gas enters into a register of cells from the bottom side of the cylindrical cells, flows along the anodes of the densely packed tubes in axial direction and leaves the anode chamber through a diffusion barrier of porous alumina felt. The cathode air enters the interior of the cylinder cell through a central alumina tube, which extends down almost to the cell bottom. It flows in parallel flow respective to the fuel along the cell axis and leaves the cell entering a combustion chamber, in which the rest of the anodically non-converted fuel is mixed with the air and combusted. In this way the heat is generated, which is necessary for preheating the fuel and steam and the air supplied for the cathode. Westinghouse had been able to demonstrate successfully several 100 kW SOFC power plants. Internal reforming is applied but an external pre-reformer, which reforms at least one third of the fuel, is necessary as otherwise the cells would be cooled by the heat absorbing reforming reaction at the fuel inlet to intolerably low temperatures because steam reforming is too fast and consumes more heat than is released by the cell reaction at the entrance of the cell register.

12.3.4
Flat-Plate Solid-Oxide Cells

During the last decade world wide several institutions and industries developed SOFC flat-plate cells. One of the most advanced flat plate cell technologies is that of Siemens, depicted schematically in Fig. 12.8. A large pre-sintered flat cell is inserted into and fixed by a glass forming oxide to a metal frame and is covered by so-called window plates with approximate window dimensions (20×20 cm^2) of stainless steel, which support four cells. The cells are contacted from either side by a grooved, bipolar stainless steel plate by which cathode and anode are electronically contacted to the bipolar plate whose grooves supply the working gases evenly to the electrode surfaces. The bipolar plate is made of a particular high chromium steel, which contains disperse rare-earth metal oxides in order to match the thermal expansion of the bipolar plate with that of the zirconia electrolyte. Frame, window-plate and bipolar plate carry the gas channels with appropriate openings for the release of fresh gases into and collection of spent gases from the cells. Sealing is accomplished by a type of oxidic glass of proprietary composition. A 10 kW unit of this cell type had been operated successfully in 1996 and scaling up will eventually lead to power plants of several tens to hundreds of kilowatts.

A radically simplified version of the SOFC with internally integrated heat exchange is the circular-disk stacked device of Sulzer's. The incoming air flows through the mantle of the circular cell stack and then enters the stack flowing ra-

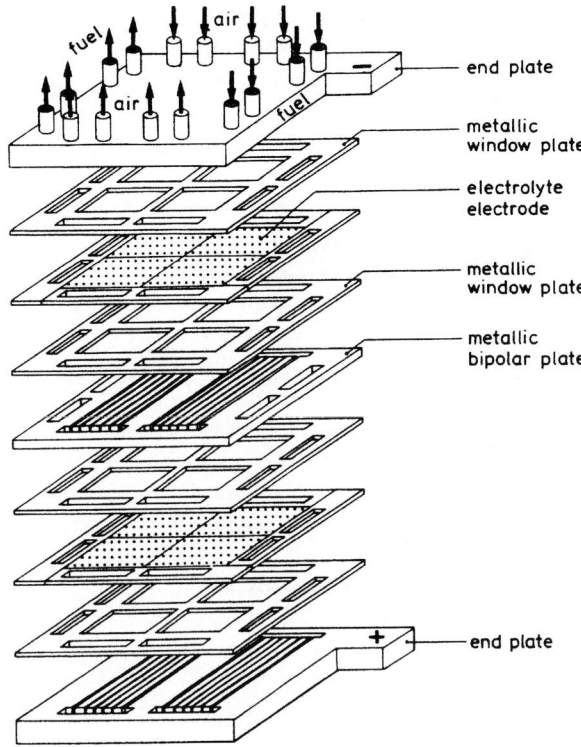

Fig. 12.8. Schematic of the Siemens flat plate SOFC

dially inwards being heated up by heat exchange with the surrounding cathode and anode gases which flow outwards. It then enters the cathode chamber at the central shaft and flows radially outwards. Also the anode fuel, which enters the stack and the cells through a central bore flows radially outward from the central tube as does the preheated air. At the outer perimeter partially spent air and anode gas leaving the cell are mixed and the non-converted fuel is combusted. The generated heat is used to preheat the fresh air. Sulzer develops small units of from tens to several hundreds of kilowatts electrical power for onsite cogeneration and domestic heating accompanied by electricity generation.

12.4 Current Voltage Curves of Different Fuel Cells

All electrochemical power sources loose voltage as the current load increases. As seen in Fig. 12.9 three different parts of the current voltage curves of fuel cells can be distinguished. The logarithmic potential decay at low current densities is mainly determined by charge transfer kinetics and the related overpotential according to the summed Butler–Volmer equations of the anode and cathode.

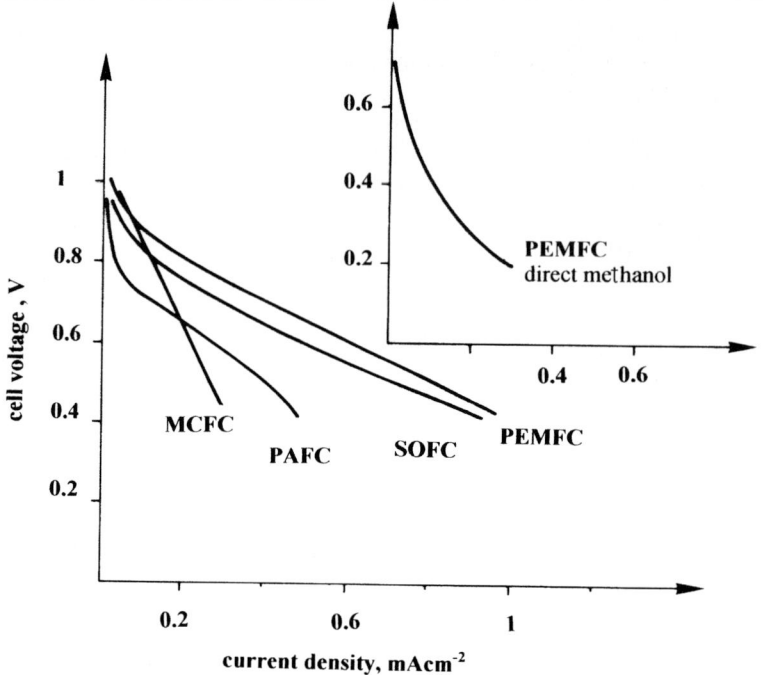

Fig. 12.9. Comparison of current voltage correlations of different types of fuel cells

Then there follows a close to linear cell voltage decay with increasing current, which is mainly determined by the internal resistance of the fuel cell. Eventually at high current densities there follows a more than proportional potential decay, which is indicative of significant mass transfer hindrance of fuel and/or oxidant which may end in a definitely mass transfer controlled current density determined by limited access of fuel or oxidant to the respective electrode.

Low temperature fuel cells as PAFCs and PEMFCs do exhibit in their current voltage curve always the charge transfer controlled logarithmic section as oxygen reduction is usually so strongly hindered, that its current voltage behaviour commands the current voltage correlation of the cell at low current densities. In PAFCs and PEMFCs the respective potential loss up to a current density of 100 mA cm^{-2} amounts to 200–300 mV. Both cells, however, differ significantly in their internal resistances. Commercial PAFCs have an internal resistance of 0.7–0.8 Ω cm^2. Typical in PAFCs is a current density of 0.3 A cm^{-2}, which would not be exceeded because one aims at cell voltages of 0.65–0.7 V, which would be the minimum practical value. This allows in PAFCs for power densities of only 0.2 W cm^{-2}. PEMFCs with one third of this internal resistance indeed allow for current densities around 1 A cm^{-2} and power densities of 0.6–0.7 W cm^{-2}. It is this low internal resistance and high power densities, which makes the PEMFC the cell of choice for electotraction. Figure 12.9 compares typical current–voltage curves of the two low-temperature and high-temperature cells working on O_2/H_2 together with the current voltage correlation of a direct methanol PEMFC. MCFC current voltage curves are typically linear even at low current densities, which is indicative of ohmic potential drops of the cell defining these correlation because anode and cathode kinetics are sufficiently fast but the electrolyte matrix and the cathode have a relatively high ohmic resistance resulting in an internal resistance of 1.8–2 Ω cm^2. SOFCs have typically internal resistances around 0.5 Ω cm^2. A particular feature of current voltage curves of methanol combusting membrane cells is their low open-cell potential – see insert in Fig. 12.9. With 0.6–0.7 V the open-cell potential is 300–400 mV lower than the equilibrium potential of the cell which almost matches that of hydrogen/oxygen cells – a clear indication that the anodic electrocatalyst (Pt/Ru) of today is not yet really a satisfactory solution to the problem of catalyst poisoning.

12.5
Fuel Cell Systems

Fuel-cell stack, gas processor, inverter for electric power conditioning and process command, are the integrative parts of a fuel-cell system. To make fuel cells a useful commercial commodity they have to be integrated into a whole system, which is able to convert the combustion energy of natural gas or other fuel into electricity of the usual quality and usable heat, and which is able to operate on its own with little maintenance and automatic control of its functions and automatic response in case of malfunction. The fuel-cell stack alone is almost useless, unless it is coupled to a gas processor, which produces hydrogen containing synthesis gas of relatively low CO-content from the primary fuel, which is usually natural gas. As the stack produces direct current of a given voltage, the inverter – a silicon based solid state device – has to transform this direct current into alternating current of a given frequency, voltage (AC of 60 or 50 Hz and 110 or 220 V according to US or European standards) and quality, i.e. very low percentage of overtones. Last but not least an automatic command unit, which adjusts the production rate of the chemical plant which produces the hydrogen containing feed gas to the power demand of the grid or other load, which detects reliably irregularities in the chemical process plant, the fuel cell stack, the cooling system or the converter, and which allows without human interference to operate, start and shut down the whole unit is indispensable for establishing the properties of a commercial energy converter.

12.5.1
Phosphoric-Acid Fuel Cell / PC 25

Figure 12.10 a shows the containerised 200 kW$_{el}$ unit, PC 25 C delivered by ONSI corporation (a) and the internal assembly of fuel processor, stack and inverter, (Fig. 12.10 b). The fuel processor consists of a steam reformer, which according to Eq. (12.8) converts at a maximal temperature of 800 °C in the endothermic re-

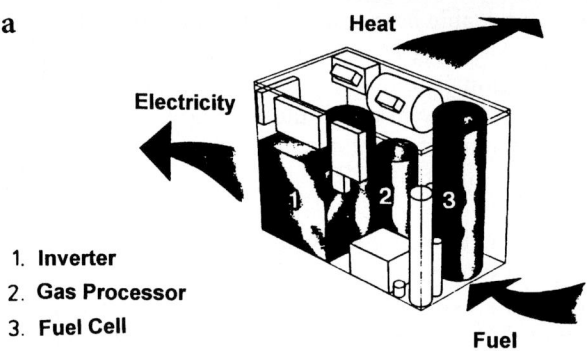

Fig. 12.10. 200 kW PAFC, type PC 25, fabricated by ONSI Corp.: **a** schematic of arrangement of system components; **b** the containerised system; Courtesy ONSI Corp.

forming reaction methane into a mixture of carbon monoxide, hydrogen and some carbon dioxide with approx. 1% unconverted methane in the endothermic reaction.

The heat of steam reforming is supplied by a burner in which the unspent anode gas (approx. 15% non-converted H_2 and 3–4% of the initial methane) are combusted with part of the oxygen contents of the cathode off gas. After heat exchange and cooling of the reformer gas the shift converter produces at approx. 200 °C according to Eq. (12.9) in an exothermic reaction a synthesis gas with a

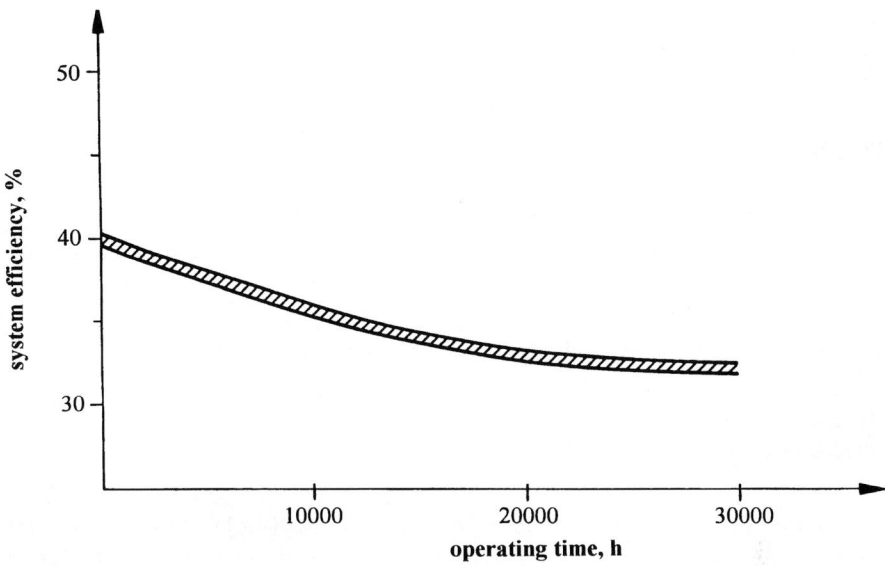

Fig. 12.11. Efficiency decay of a PC 25 A-unit over more than 30,000 h

CO_2/H_2 ratio of 1/4 and a CO-contents of roughly 1–1.5%. It is essential for obtaining an overall electrical system efficiency of 42% that the heat generated by the cell is recovered at a temperature level of approx. 160 °C securing the heat of evaporation of the steam, needed for establishing a steam to methane (S/C)-ratio of approx. 2/1 at the reformer inlet. The PAFC, model PC 25 C is the third version of cogenerating fuel-cell power plants brought to the market place by ONSI corporation and it represents 1998 still the only one fuel-cell technology, which is commercially available[1]. The containerised unit delivers 200 kW electricity with initially 42% electrical efficiency and delivers 200 kW thermal power at 80 °C for instance for domestic heating. Several units of the older type PC 25 A have finished the anticipated life time of 40,000 h.

Figure 12.11 shows the progressive efficiency loss of a PC 25 A unit beginning with approximately 40% and ending after more than 33,000 h at 34–35%. The greater part of the losses of the system efficiency is due to deterioration of the electrocatalyst at the cathode. From the experience gathered till the end of 1997 one can conclude, that provided it is possible to decrease system costs from now 3000 US $ kW^{-1} to half of that value, phosphoric-acid fuel cells will have good chances to compete in the market of small cogeneration plants successfully with power plants, which are equipped with Diesel engines or gas motors. Phosphoric-acid cells will certainly pave the way also for stationary application of the other fuel-cell types.

---

1 Japanese developers are expected to offer their own PAFC power plants in 1999.

12.5.2
Molten Carbonate Cells

12.5.2.1
ERC-2 MW Plant

In the year 1996 a 2 MW cogenerating MCFC power plant with internal reforming developed and fabricated by ERC (Energy Research Corporation), Danbury, Con., had been operated for almost one year. In spite of some mishaps which forced several of the eight stacks – each of 125 kW – out of service, this demonstration of the MCFC technology can be judged to be a success.

12.5.2.2
Hot Module of MTU

The cost contribution of the balance of plants to the whole costs of a fuel cell power plant is crucial and can approach 70% or more in particular for small power plants of the 100 kW class. As capital costs of such power plants contribute significantly to electricity cost, it is decisive to reduce the balance of plant costs drastically.

The design of the Hot Module of MTU (Maschinen-Turbinen-Union, Friedrichshafen) depicted schematically in Fig. 12.12, whose fuel cell stack is based on the design of the ERC technology, aims at a decisive reduction of balance of plant costs. As shown in Fig. 12.12a the fuel cell stack equipped with internal reforming catalyst is positioned not upright but horizontally in a heat-insulated cylindrical stainless steel container and the stack's entrance face for the fuel is sitting on a rigid frame on the bottom side of the cylindrical container wall, so that the fuel duct is self-tightening. The anode off gas is mixed with fresh air. In the mixed gas unspent fuel combusts immediately and the mixture containing excess oxygen and carbon dioxide is then used as cathode gas. This is the easiest way to recycle the carbon dioxide from the anode to the cathode. Only one hood is needed for the stack. It collects the cathode gas at the cathode outlet, which is recycled in a recycling ratio of one to the cathode inlet, which can be recognised by its entrance grill in Fig. 12.12a. Recycling is performed by two blowers at the ceiling of the container. Fig. 12.12b depicts the process scheme. Fresh air of ambient temperature enters at first a simple plate heat exchanger where it is heated by the anode off-gas and mixes then with the recycled cathode gas adjusting the temperature of the whole Hot Module to approximately 650 °C with almost negligible temperature gradients across the cell. Oxygen and carbon-dioxide contents of the recycled lean cathode gas match that of the 650 °C hot off-gas which leaves the cathode gas collecting hood by an additional exit and whose sensible heat might be used eventually for any desired purpose. Certainly MCFC technology is still relatively far from marketing maturity. Nevertheless one would expect that MCFCs may appear on the market place during the next ten years.

12.5 Fuel Cell Systems

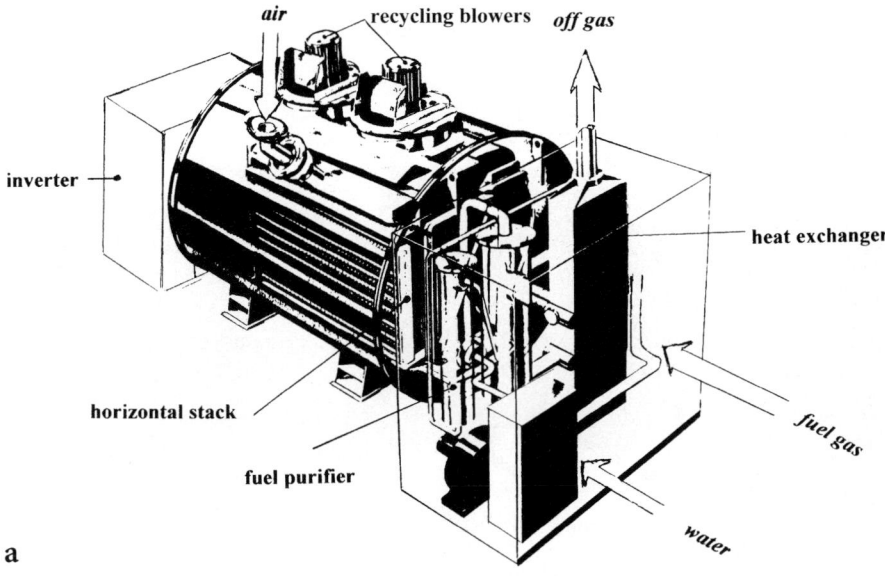

Internal Reforming MCFC - Process

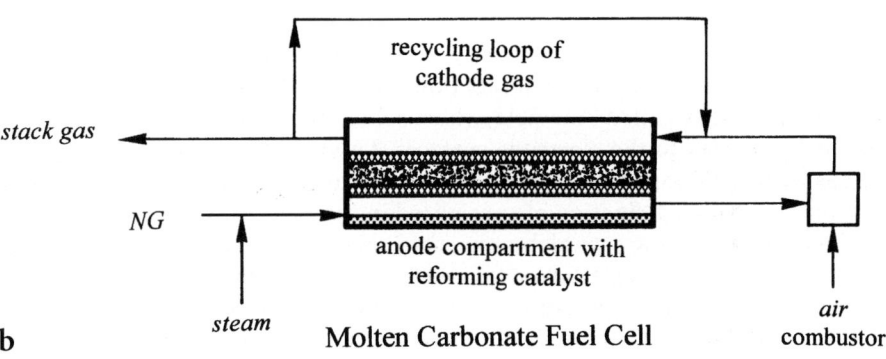

Molten Carbonate Fuel Cell

Fig. 12.12a–b, a Schematic of the MCFC Hot Module of MTU, courtesy of MTU Friedrichshafen AG. **b** Schematic of the process flow sheet in the Hot Module

12.5.3
Proton Exchange Membrane Cells

Development of PEMFCs for electrotraction of cars and also for stationary electricity generation is fast, the competition between the different developers is tough and therefore details of the technology are proprietary and little is known about crucial details. One of the most important points of interest is the cheapest

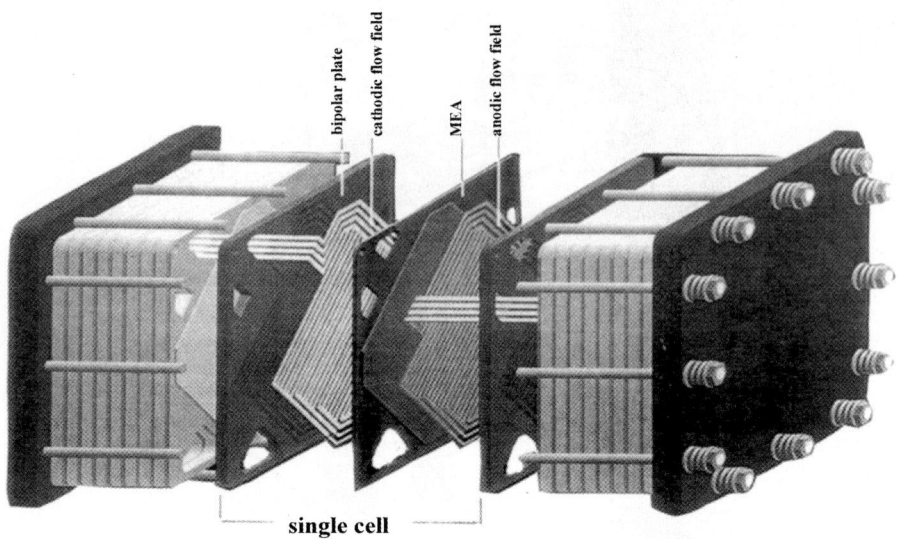

Fig. 12.13. Exploded view of the Ballard cell with flow fields for anodes and cathodes. One single cell comprises the membrane electrode assembly-MEA and the anodic and cathodic flow fields which are grooved in the bipolar plates. Courtesy Ballard Corp.

and most reliable method for water or water vapour supply to the anode and water removal from the cathode. Another point is the detailed structure of the cells, in particular the so called flow-field, which provides evenly distributed access of the working gases to the electrode surfaces. An example is given in Fig. 12.13. Details of production technologies are not disclosed, but it seems to be clear, that an essential participation of specialised industries, from which an influx of production know-how would be expected, has not yet been fully initiated. Therefore the technological advances necessary for mass production of PEMFCs are still to be waited for but are soon expected to become effective.

12.5.3.1
The Ballard Cell

Ballard Company is the most prominent and most experienced developer in the field. Figure 12.14 shows a prototype of Ballard's water cooled 205 kW fuel cell engine, which powers transit buses transporting 75 passengers. In buses hydrogen is stored under 200 bar pressure.

Another PEMFC-powered car is the Necar III passenger car of Daimler-Benz, whose hull is that of the A-class Daimler–Benz car and which is powered by a 40 kW Ballard cell, being operated on syngas from methanol and on slightly compressed air. Figure 12.15 shows the assembly of gas processors, vaporisers, water and methanol tanks and the PEM-cells in the NECAR III. Gas processing

12.5 Fuel Cell Systems

Fig. 12.14. Ballard's prototype 205 kW fuel cell engine for a 75-passenger commuter bus. The 205 kW PEMFC is situated in the rear of the vehicle above the electronic power drive controls and air blowers. Courtesy Ballard Corp.

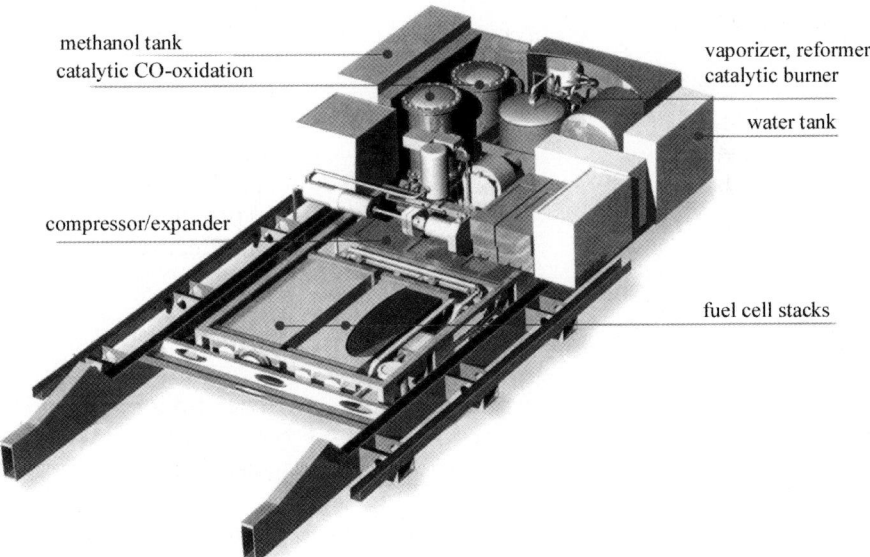

Fig. 12.15. View and schematic of cell, gas process units and power train of the passenger A-class car, NECAR III, of Daimler–Benz. The anode is supplied with synthesis gas obtained from methanol reforming with catalytic CO-combustion. Courtesy Daimler–Benz

consists of methanol reformer and low-temperature shift converter followed by catalytic preferential CO-oxidation. The cells require much less space than gas processors and the tanks and are situated below the floor of the car.

12.5.3.2
De Nora's Cell

A competitor and PEM developer is DeNora SpA, Milano, Italy, who develops PEMFCs as well for stationary as for mobile power generation. In contrast to Ballard their bipolar plate is not made of carbon or polymer-bonded carbon but of aluminium. Most big car-producing companies are now actively engaged in development of membrane cells and PEMFC-systems for passenger cars, buses and trucks. But it might last 10 years till we can expect fuel cell technology to become competitive with internal combustion engines. The promise is there, that a new solution of the mobility problem of modern mankind can be solved by emission-free PEMFC engines which are operated on fuels like pressurised or liquefied hydrogen or methanol. This will be a continuing challenge also for electrochemical engineers. It should be emphasised that fuel cell systems are only developed by close cooperation of chemical, electrochemical and materials engineers. This is a very convincing example, that electrochemical engineers must be truly interdisciplinary oriented if they are trying to cope with the challenging task to establish new key technologies for electrochemical energy conversion.

Further reading

K. Kordesch, G. Simader, Fuel Cells, VCH, Weinheim, 1996

Subject Index

ABB Membrel cell 274
absorption columns 220
absorption rate 184
Acheson graphite 72, 260
Acheson-process 193
activation energy 49
activity 19
activity coefficient 19, 31
adimensional correlations 92, 236
adimensional numbers 85
adipodinitrile 78, 346, 349
adipodinitrile electrosynthesis flow sheet of 351
adsorption 72
advanced alkaline water electrolysis 265
ageing of Pt catalysts 280
ageing of Raney-Nickel coatings 272
Alcoa electrolyzer 341
Alcoa-process 206
alcohols 68, 71
aldehydes 68
alkali hydroxides 291
Alkali metals from chloride melts 341
alkaline fuel cell (AFC) 274, 373
alkaline water electrolysis, 270, 317, 325
– improved 323
alkaline water electrolyzer 209, 318
alkyl radicals 77
alloy catalysts 281
alloyed steels 191
aluminium 129
aluminium electrowinning 209
aluminium production 290, 336
amalgam process 292
amalgam technology 133
ammonium peroxidisulfate 314

amorphous carbon 196
amorphous coke 193
anion exchange membranes 202, 361
anode 9
anodic chlorine evolution 223, 258
anodic evolution of oxygen 66
anodic fluorine generation 196
anodic hydrogen oxidation 274, 285
anodic hydrogen oxidation, catalysts 276
anodic hypochlorite oxidation 307, 308
anodic mediators 71
anodic metal dissolution 187
anodic methanol oxidation 281
anodic organoelectro syntheses 268
anodic oxidation 8, 71
– of hydrogen 61
– of l-sorbose in commercial vitamin C synthesis 353
– of toluenes 353
anodic oxygen evolution 45, 64
– electrolysis 267
arenes 72
arylonitrile 78
asbestos diaphragm 203, 297

Baizer-Monsanto process 216, 346
balance of plant 310
Ballard Cell 392
batch operations 129
batteries 9
battery electrode 171, 172
bed electrodes 5
benzaldehydes 68, 353
benzene 67
benzene oxidation heterogeneously catalysed 74

benzoquinone   67
bipolar capillary gap cell   349
bipolar cells   209
bipolar electrode stacks   257
bipolar electrodes   257
bipolar gap electrolyser   98
bipolar plate   381
bipolar technology   320
blowers   220
brine recycling   302
Bunsen   341
Butler Volmer   46, 161
Butler Volmer equation   41, 42, 49, 55, 56

cadmium-clad steel   194
C-anodes   77
capillary gap cells   216
carbenium cations   77
carbon   192, 196
carbon anodes   192, 260
carbon electrodes   72
carbon materials   193
carbonaceous anode material   339
carbon-supported Pt   277
carbonyl compound   77
carboxylate radicals   76
carboxylic acids   68
Carnot efficiency   372
catalyst   74
– loadings   253
– morphology   68
– particles   69
– utilisation   69, 282
catalytic coating   252, 255
catalytic hydrogenation reactions   68
cathode   9
cathodic copper deposition   11
cathodic H_2 evolution   45
cathodic hydrogen evolution   44, 188, 268
cathodic hydrogenation   77
cathodic hypochlorite reduction   308
cathodic immunity of iron   194
cathodic mediators   71
cathodic O_2 reduction   187
cathodic oxygen reduction   45, 64, 274, 285
cathodic reduction   8, 71
cation   361

cation-exchange membrane   202, 262, 376
caustic soda   291
cell and electrode design   208
cell cascades   136
cell voltage   117
ceramics   187
channel flow   97
charge exchange   17
charge transfer   45, 81
charge transfer coefficient   42, 49
charge transfer control   55
charge transfer, multielectron   45
charge-transfer overpotentials   123
chemical consecutive reactions   79
chemical engineering   6
chemical equilibria   20
chemical heterogeneous catalysis   69
chemical potential   19
chemical process industry   5
chemical reactions   45
Chilton–Colburn analogy   107
chloralkali   192
chloralkali electrolysis   3, 5, 129, 133, 195, 206, 209, 220, 261, 262, 268, 270
– membrane process   209
– plants   244
chloralkali electrolyzers   132
chlorate   306
chlorate electrosynthesis   310
chlorate industries   192
chlorinated polymers   205
chlorine   66, 291, 303
chlorine dioxide   306, 311
chlorine electrode   19
chlorine evolution   260
chlorine evolving anodes   263, 293
chlorine production   290
chlorine-alkali electrolyzers   187
chlorine-derived bleaches   306
chloro-oxoacids   291
chromate/bichromate   68
chromium   189
chromium-alloyed steels   192
circular-capillary gap cell   89
coatings   252
– of flame sprayed, doped nickel oxide   273
– of platinum-metal oxides   273

Subject Index

collection efficiency 227, 228
commuter bus 205 kW PEMFC 393
competitive reactions 76
compressors 220
concentration profile 52, 84
conditioning of the product gases 301
consecutive reaction, selectivity problem of 142
consecutive reactions 60, 225
continuous 129
continuous stirred tank reactor (STR) 132, 309
continuously-stirred tank 135
convective diffusion 1, 81, 92
– by free convection 328
convective heat conduction 108
convective mass transfer 90, 102
conventional alkaline water electrolysis 320
copper 330
copper electrorefining 331
copper refining 1, 129
copper refining cells 132
corrosion, general 190
corrosion of metals 187, 188
corrosion, intercrystalline 190
corrosion-protection coatings 191
cost analysis 243
cost optimisation 245
C-radicals 76
crevice corrosion 190
cryolite 336, 337
cryoscopy 33
current concentration 211
current densities 11, 41, 42, 51, 81, 91
– distribution 122, 255, 257
– potential correlations 42, 51, 222
current voltage correlation 55, 129, 386
– voltage correlations 51
current efficiency 11, 132
current transients 234
current voltage curves 42, 54, 128, 129, 131
current-density distribution 1, 117, 118, 124, 126, 131
cyclic voltammetry 231, 232

decarboxylation 76

deposit corrosion 190
diacetone-2-keto-l-sorbic acid 74
diaphragm 199, 203, 295
diaphragm cell 294
diaphragm materials 200
diaphragm process 292, 297
diffusion coefficient 82
diffusion length 58
diffusion potential 18
diffusional mass transfer 90
diffusive mass transport 81
dimensionally stable anode, DSA 261, 293
dimensionally stable electrodes 260
direct electrochemical conversion 72
distillition 220
divided cells 199, 209
Dow process 343
DSAs 262, 264

ebulloscopy 33
eddies 87
electrical control of cells 131
electrical efficiency, theoretical 372
electrocatalysis 4, 5, 6, 39, 61, 64, 71, 252, 254, 285
– of oxygen 265
– of the second kind 78
electrocatalyst 252
– deterioration 222
electrocatalyst coatings, longevity of 253
electrocatalysts, structural features of 260
electrocatalytic hydrogenation 74
electrocatalytic oxidations 72
electrocatalytic RuO_2 coatings 260
electrocatalyzed mediated reduction 74
electrochemical adsorption equilibrium 45
electrochemical cell 187
electrochemical double layer 2, 39
electrochemical drilling 362, 365
electrochemical gas evolution 179, 254
electrochemical grinding 362, 363, 366
electrochemical kinetics 50
electrochemical machining (ECM) 365
electrochemical methane combustion 372
electrochemical polishing 363
electrochemical reaction orders 49

electrochemical sinking 363
electrochemical surface treatment 5
electrochemical technologies 3
electrochemical thermodynamics 17
electrochemical Thiele modulus 70, 166, 170
electrochemical wastewater treatment 357
electrocoating 122
electrode design 196
electrode kinetic measurements 102
electrode kinetics 2, 4, 39, 45, 81
electrode materials 193
electrode potential 21
electrode processes 2
electrode reaction 9
electrodeposition 122
electrodialysis 361
electrolyte conductivities 12
– temperature dependence of 111
electrolyte ions 12
electrolyte matrix 275
electrolyte matrix ($LiAlO_2$) 381
electrolyte recovery 251
electrolyte seperation 251
electrolytes for electrochemical machining 367
electrolytic conductivities 12
electrolytic corrosion 2
electrolytic hydrogen production 265
electrolyzer 8, 108
– modelling of 138
electrolyzer cascades 137
electromotive force 18
electroorganic synthesis 71, 76
– processes 250
– reactions 222
electropolishing 362
electrorefining 326, 330
electroreforming of microprofiles 368
electrosorption 35, 75
electrosynthesis 2
Electrosynthesis of Sebacic diesters 352
electrowinning 129, 330
– of metals 97, 265, 326
– processes 329
electrowinning cells, design of 329
emission-free PEMFC engines 394

enviro cell 360
environmental protection 5
equilibrium cell potential 17, 18, 20, 21
equivalent conductivities 13
ERC-2 MW Plant, MCFC Technology 390
erosion 254
ethylendiamino-tetraacetic acid 350
Euler number 85
exchange current density 42, 44, 49, 50
expanded metal electrodes 207
expanded metal gauze 196

factorial design 222, 239
– of experiments 240, 241
Falconbridge process 334
Faraday's Law 10, 41
fast preceding reaction 58
Fick's law 90
finned 196
flat-plate solid-oxide fuel cells 384
flow pattern 131
flow-through electrodes 133
fluid dynamic 81, 84
fluid flow 81, 221
fluidized bed electrodes 173, 178
fluorine 315
fluorine electrolyzers 315
forced convection 92, 97, 98
fore-electrodes 258
fractional conversion, X 150
fuel cell 6, 9, 108, 114, 274, 370
– electrodes 274
– systems 387
– types 373
fuel cell engine, 205 kW 393
fugacities 19
fugacity coefficient 19

galvanic bathes 122
galvanic coating 122
gas bubbles in 179
gas conditioning of Cl_2 and H_2 303
gas consuming electrodes 68, 197
gas evolving 68
– electrodes 103, 196, 258
gas processor 387
gas purification 303
gas-diffusion electrodes 157, 275

Subject Index

gaseous reactants   183
gaskets   205
Gibbs energy   17, 19, 20
Gibbs–Duhem equation   33
glass-fiber reinforced polyester   205
glassy carbon   193
graphite   193, 293
Grashoff   5
Grashoff number Gr   93
Grotthus mechanism of H^+ and OH^- migration   111

H_2 evolution   44
hafnium   191
Hagen–Poiseulle   86
Hall–Heroult cell   339
Hall–Heroult Process   206, 336
Haring-Blum cell   122, 124
Hatta number   58, 60, 61
HCl electrolysis   19
heat balance   28, 29
– of the cell   81
heat exchangers   219
heat generation   108
heat transfer   108
Heat transport   81, 107
Henry Beer   261
heterogeneous catalysis   6, 61, 71
heterogeneous mediators   72
heterogeneous reactions   72
heterogeneous redox catalysis   66, 67
hexafluoropropylene   67
hexafluoropropylene oxide   67
Heyrovsky   48
Heyrovsky reaction   63
high density polystyrene   204
high-nickel alloys   191
high-temperature fuel cells   284, 379
high-temperature steam electrolysis   320
Hofer–Moest   77
homogeneous chemical reaction   54, 57
homogeneous chromate oxidation   71
homogeneous redox catalysis   67
horizontal electrodes   206
hot module of MTU, MCFC Technology   390
Hunt cell   122
hydrochloric acid   18, 19

hydrodimerisation   78
hydrodynamic boundary layer   93
hydrogen   291, 304
hydrogen electrode   19
hydrogen embrittlement   195
hydrogen evolution from alkaline solution   268
hydrogen evolving cathodes   268
hydrogen/oxygen fuel cells   371
hydrogenation rate   74
hydrogen-evolution overpotentials   194
Hydrometallurgical electrorefining   97
hydrometallurgy   5
hydrophilic electrodes   72
hydrophobic carbon electrodes   72
hypochlorous acid   306

ICI chlorine electrolyser   257
industrial electrodes   252, 255
industrial organic electrosyntheses   349
initial polarisation curves   233
inner Helmholtz layer   39
interelectrodic gap   97
interfacial charge transfer   41
interfacial potential   21, 47
intermediate chemical reaction   61
internal reforming   380
internal reforming MCFC-Process   391
internal resistance   386
intrinsic mobilities   116
intrinsic transfer number   116
inverter   387
ion exchange membranes   200, 201
ionic charge   110
ionic conductivity   110
ionic migration   110
ionomers   201
IR compensation   236
IR Drop   235
IrO_2-coated anodes   267
iron   188, 189
iron poisoning of electrocatalyst   273
isolated planar electrodes
free convection at   97

JANAF Tables   31
jarosite zinc electrowinning process   328

Kalrez   206
ketones   68
ketyl radicals   77
Kirchhoff's rules   14
Kolbe electrosynthesis   268
Kolbe reaction   76
Kolbe synthesis   76, 77, 195, 345, 352
Kolbe-dimer   76

laboratory methods   221, 222
laminar flow   86, 87, 92
laminar flow along a plate   87, 94
laminar free convection   98
$LaMnO_3$ 284
Langmuir isotherm   50
Laplace equation   118, 119
lead   195
LIGA-Process   368
linear potential sweep method   231
lithiated nickel oxide   284, 285, 287
lithium   342
louvered electrode   196
low-temperature fuel cells   375
Luggin probe   235
Luggin-capillary   234

macrokinetic models   131
macrokinetics   51, 81
magnesium   129, 342
Magnesium electrolysis   342
magnesium electrolysis cells   344
magnetite anodes   260
magneto-hydrodynamics   6
mass   135
mass balances   81
mass transfer   1, 51, 81, 97, 98, 103, 106, 131, 221
mass transfer coefficient   184
mass transfer control   52
mass transfer hindrance   69
mass transfer limited current density   53, 81, 100, 135
mass transfer measurements   237
mass transport   82, 92, 95
mass transport control   56
materials choice   187
mathematical modelling   224, 239

MCFCs, molten carbonate fuel cells   379
mean activity coefficients of ions   33
mediated electrochemical conversion   71
melt electrolysis   335
membrane (PEM) Fuel Cells   282
membrane cells   300, 376
membrane chloralkali electrolysis   248
membrane electrode assemblies (MEAs)   283
membrane electrolyzer   207, 274, 294
membrane process   292, 295, 298
membrane water electrolysis   324
membranes   199, 201
mercury cells   262
mercury cells, operational data of   296
mercury process   294
metal electrowinning   129
metal winning   291
methanol oxidation   6
methanol oxidation, direct anodic   282
methanol reformer   394
methanol-combusting membrane cells   377
Michael addition   78
microkinetic investigations   128
microkinetic models   128
microkinetics   51, 81
microporous electrodes   156
migration   82
mild steel   265
minimum cost   240
MnO_2 71
molar heat of evaporation   30
mole fraction   32
molten salt electrolysis   193
molten-carbonate cell (MCFC)   114, 274, 285, 373, 379, 381, 390
molten-carbonates   136
momentum   86
momentum exchange   87
monopolar cells   209
monopolar diaphragm cells   297
monopolar electrode   255, 256
monopolar technology   320
Monsanto process   78, 245
Monsanto-Baizer process   149, 349
multieffect evaporation   311
multiphase electrolyte systems   153

Subject Index

Nafion   201, 282, 376
Nafion membranes   4, 324
nanopores   74, 197
nanoporous catalyst coatings   74
nanoporous electrocatalyst   69
nanoporous electrode particles   156
nanoporous Raney nickel catalyst coatings   165
Navier/Stokes equation   84, 85, 95
NCE, normal calomel electrode   23
Nernst diffusion layer   91, 93, 100
– thickness   91, 94
Nernst equation   25, 51, 53, 61
Nernst's law   54
NHE   23
$Ni(OH)_2$ 68
nickel   188, 189, 191, 194, 265, 330
nickel anodes   68, 74
nickel cathodes   295
nickel chloride leach process   333
nickel electrodes   194
nickel electrowinning   331
nickel matte   331
nickel refining   334
nickel-base alloys   192
nickel-coated steel cathodes   295
NiOOH   68
NiOOH-mediated oxidation   74
noble metals   195
non-steady state methods   230
normal potential   24
normal calomel electrode see NCE
Norsk-Hydro process   343
Nusselt number   107

Ohm's Law   12
ohmic potential drop   129
ohmic resistors   12
ohmic-voltage drop   110
olefins   72
one-electron oxidation   76
open cell potential   18
optimisation   239
optimum finding by experiment   239
organic compounds, fluorination of   196
organic electrosynthesis processes   345
organic electrosynthesis reactions   239
organic oxidants   71

organic polymers   187
organic substrates   71, 72
organic substrates, electrochemical conversion of   71
organic syntheses   72
organo-electrochemical processes   192
organo-electrosynthesis   5, 6, 67, 194, 211
organo-electrosynthesis processes   5
osmometry   33
outer Helmholtz layer   39
overall current efficiency   12
overpotential   42, 49, 53, 64, 129
overpotential η   41
oxide ceramics   284
oxo compound   77
oxygen electrode   190
oxygen evolution from acid solutions   266
oxygen evolving anodes   265, 329
oxygen reduction   188
– catalysts   276
oxygen-evolution catalyst   254
oxygen-evolution overpotential   267

packed beds   141, 142, 173
packed grids   133
packed-bed electrodes   133
parabolic velocity profile   87
parallel plate cells   97
parallel plates   86
parallel-plate electrodes   84
passivating oxide layers   187, 189
passivation   190
passivity breakdown   190
PbO_2 71
$PbSO_4$ solubility   192
Peclet-number Pe   141
PEM fuel cells   202
PEMFCs   391
penetration depth   175
perchlorate cells   313
perchlorates   312
perchloric acid   312
perfluorinated ion exchange membranes   298
perfluoroethylene   192
perfluoro-propylene oxide   355
perforated plate electrodes   258
performance criteria   149

permselectivities 202
peroxidisulfates 313
phosphoric acid fuel cell (PAFC) 274, 279, 373, 375
phosphoric acid fuel cell / PC 25 387
physical passivation 190
physisorption 77
pilot plant measurements 221
pilot plant methods 236
pinacol 77
pitting corrosion 190, 192
planar electrodes 88, 93, 95, 196
planar plate cells 98
plant engineering 187
plate electrode 87
plug-flow reactor (PFR) 133
point of zero charge 77
polarisation curve 14
polarisation resistance 14, 121
polyarylethersulfone 204
polyesters 187
polyether ketones 203
Polyetherether ketone 204
polyethers 187
polyethylene (PE) 203
polyethylene high density 204
polyethylene low density 204
poly-fluoroethylene- propylene 204
polymer 187
polymeric materials 203, 204
polymides (PI) 203
polyperfluoroalkylvinylether 204
polyphenylensulfide (PPS, polysulfide, Ryton) 203
polyphenylensulfidec 204
polyphenylensulfon (polysulfon) 203
polypropylene (PP) 203
polypropylene oxide (PO) 203
polystyrene 204
polysulfone, UDEL 204
polytetrafluoroethylene 204, 205
polyvinyl chloride (PVC) 203, 204, 291
polyvinylidene fluoride 205
polyvinylidene fluoride ethers 205
porous catalysts 69
porous electrocatalyst particles 69
porous electrocatalytic coating 69
porous electrodes 153, 154

porous flow through electrodes 174
potential transients 233
potentiodynamic polarisation curves 230
potentiostatic step 234
Pourbaix 188
Pourbaix diagrams 189
power densities 386
power supply for electrochemical plants 218
Prandtl number Pr 107
prebaked carbon anodes 339
pre-electrodes 196
primary current-density distribution 119
process design 3
process development 221
process engineering 6
product separation 251
production rate 11
productions costs 243
propionitrile 78
proton exchange membranes 282
proton-exchange membrane cell (PEMFC) 198, 274, 373, 391
Pt 77, 278
Pt-activated active carbon 277
Pt-doped active carbon 198
PTFE 197
PTFE-bonded active-carbon electrodes 279
Pt-wire anodes 195
pump cell 217
PUREX process for reprocessing nuclear fuels 186
purification of effluents 357
purification steps for electrowinning 328
PVC 205
pyrolytic carbon 193

quaternary electrocatalyst 282
quaternary ammonium cations

radical anions 76
radical cations 76
radical dimerization 77
radicals 72, 76
Raney nickel 194, 269, 274, 295
Raney nickel coated cathodes 74

Subject Index

Raney nickel coatings   69, 266, 269, 270, 271
rational potential   35, 36
reaction control   54
reaction controlled current voltage curves   57
reaction engineering   128, 129
reaction enthalpy   28
reaction entropy   28
reaction layer thickness   60
reactive intermediates   72
reactive organic intermediates   75
rectangular ducts   87
rectifiers   218
redox catalysis   66
redox systems   71
redox-couples   17
reductants   71
reduction   76
reference electrode   21
refractory metals   284
residence time distribution   139, 141, 149, 221, 238
reversible hydrogen electrode (RHE)   22
Reynolds   5
Reynolds number Re   85, 92, 103
RHE, reversible hydrogen electrode   23
ring disc electrode   226, 228
rock salt   301
rotating circular capillary gap cell   217
rotating cylinder   102
rotating disc   103
– electrode   102, 126
– measurements   223
rubber-coated steels   187
RuO_2 267
RuO_2 coatings   195
RuO_2 on titanium   252
RuO_2-activated Ti anodes   207
RuO_2-activated titanium electrode   253
RuO_2-coated anodes   263
ruthenium metal   273
ruthenium-dioxide   293

SCE, saturated calomel electrode   21, 23
Schmidt   5
Schmidt number Sc   92, 94
sea salt   301

secondary current distribution   121
segregation in stagnant electrolytes   114
selectivity   76, 79, 143
separators   199
Sherwood   5
Sherwood number Sh   92, 94, 107
shift converter   388
shift-conversion reaction   380
Siemens flat plate SOFC   385
sintered nickel anodes   284
sintering   284
slotted electrodes   196
smooth platinum anodes   313
Söderberg anodes   339
sodium   342
sodium chlorate   307
sodium chlorate cells   308
sodium hypochlorite   306
sodium perchlorate   312
SOFC flat-plate cells   384
SOFC-cathode   288
SOFCs   379
solid oxide fuel cell (SOFC)   274, 287, 373, 382
solid state ion   2
solid polymer electrolyte (SPE) water electrolysis   318
solubility of $PbSO_4$ 193
solute activities   33
solvent molecules   72
solvent recovery   251
space time yield ρ   74, 133, 173, 211
specific conductivity   12, 13
specific resistivity   12
square wave pulses   233
stabilised alloy catalysts   279
stainless steel   191, 194, 265
stainless steel cathodes   194
standard concentration   19
standard hydrogen electrode, SHE   21
standard potential   23
standard pressure   19
statistical models   243
steam electrolysis   324
steam reformer   387
steam reforming   380
steel   187, 190
stirred-batch tank reactor   131

stoichiometric coefficient   10
stoichiometry   8
stress corrosion cracking   190
styrene oxidation   55
surface concentration   37, 50, 51, 72, 77
surface corrosion   190
sweep voltammograms   231
Swiss-roll cell   216
symmetry factors   42

Tafel equation   42
Tafel reaction   48, 63
Tafel slope   43
tantalum   191
Teflon   197
ternary   282
ternary current-density distribution   125
tetraalkyl ammonium cations   78
thermal activation   49
thermodynamic data   31
thermoneutral cell voltage   28, 29
Thiele modulus   69, 168, 169
three-dimensional electrodes   106, 154, 217
throwing power   122
tissues   133
titanium   189, 191, 192, 195
titanium anodes   260, 267
titanium anodes, coated   293
titanium electrodes   195
titanium electrodes, catalyst-coated   266
toluenes   68
transformer wiring   218
transition time   83
transport   110

tubes   87
turbulent flow   86, 87, 92, 98

unalloyed steels   191
undivided cells   209
unsaturated hydrocarbons   72, 76
utilisation   68, 69, 251

vacuum dechlorination   302
velocity field   84
vertical/horizontal electrodes   209
viscous flow   84
Viton   206
volcano curve   62, 65
Volmer   48
Volmer reaction   62
Volmer–Heyrovsky mechanism   46
voltage series   21, 23

Wagner Number, Wa   124
waste water treatment   138
water electrolysis   209, 291, 316
water electrolyzers   129
Westinghouse SOFC technology   382

yield   76
Young–Laplace   160
Young–Laplace equation   166

zero distance cell   258
zero gap electrolysis cells   208
zero distance arrangement   259
zinc   129, 330
zinc electrowinning   334
zirconium   191, 192

Devoted to all aspects of solid-state chemistry and solid-state physics in electrochemistry

Journal of Solid State Electrochemistry

Current Research and Development in Science and Technology

Editor-in-Chief:
F. Scholz, Greifswald

Regional Editor Australia, New Zealand and South East Asia:
A.M. Bond, Melbourne

Regional Editor Japan:
K. Itaya, Sendai

Regional Editors North America:
H.B. Mark, Jr., Cincinnati, OH
D.E. Tallman, Fargo, ND

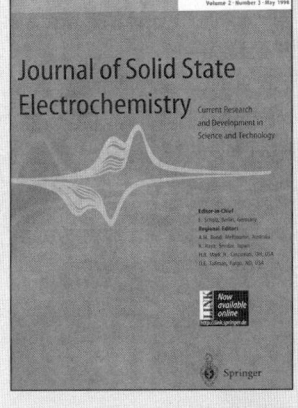

Focuses on the following fields:
mechanisms of solid-state electrochemical reactions, semiconductor electrochemistry, electrochemical batteries, accumulators and fuel cells, electromechanical mineral leaching, galvanic metal plating, electrochemical potential memory devices, solid-state electrochemical sensors, ion and electron transport in solid materials and polymers, electrocatalysis, photo-electrochemistry, corrosion of solid materials, solid-state electro-analysis, electrochromism and electrochromic devices, new electrochemical solid-state synthesis.

Subscription information 1999:
Vol. 3, 8 issues
DM 528,00

Plus carriage charges.
Errors and omissions excepted.
Prices subject to change without notice.
In EU countries the local VAT is effective.

ISSN 1432 8488 (print)   Title No. 10008
ISSN 1433-0768 (electronic)

■ ■ ■ ■ ■ ■ ■ ■ ■ ■

Please order from
Springer-Verlag Berlin
Fax: + 49 / 30 / 8 27 87- 448
e-mail: subscriptions@springer.de
or through your bookseller

Springer-Verlag, P. O. Box 14 02 01, D-14302 Berlin, Germany.   Gha.

Springer and the environment

At Springer we firmly believe that an international science publisher has a special obligation to the environment, and our corporate policies consistently reflect this conviction.

We also expect our business partners – paper mills, printers, packaging manufacturers, etc. – to commit themselves to using materials and production processes that do not harm the environment. The paper in this book is made from low- or no-chlorine pulp and is acid free, in conformance with international standards for paper permanency.

Computer to plate: Mercedes Druck, Berlin
Binding: Buchbinderei Lüderitz & Bauer, Berlin